METHODS in MICROBIOLOGY

METHODS in MICROBIOLOGY

Edited by

J. R. NORRIS
Borden Microbiological Laboratory,
Shell Research Limited,
Sittingbourne, Kent, England

D. W. RIBBONS
Department of Biochemistry,
University of Miami School of Medicine,
and Howard Hughes Medical Institute,
Miami, Florida, U.S.A.

Volume 7B

 1972

ACADEMIC PRESS
London and New York

ACADEMIC PRESS INC. (LONDON) LTD
24–28 Oval Road
London NW1

U.S. Edition published by
ACADEMIC PRESS INC.
111 Fifth Avenue
New York, New York 10003

Library of Congress Catalog Card Number: 68–57745
ISBN: 0–12–521547–9

PRINTED IN GREAT BRITAIN BY
ADLARD AND SON LIMITED
DORKING, SURREY

LIST OF CONTRIBUTORS

D. A. HOPWOOD, *The John Innes Institute, Colney Lane, Norwich, England*

J. MEYRATH, *Department of Applied Microbiology, Hochschule für Bodenkultur, Vienna, Austria*

M. T. PARKER, *Central Public Health Laboratory, Colindale Avenue, London, England*

L. B. QUESNEL, *Department of Bacteriology and Virology, University of Manchester, England*

D. F. SPOONER, *The Boots Company Limited, Nottingham, England*

GERDA SUCHANEK, *Department of Applied Microbiology, Hochschule für Bodenkultur, Vienna, Austria*

G. SYKES, *The Boots Company Limited, Nottingham, England*

v

ACKNOWLEDGMENTS

For permission to reproduce, in whole or in part, certain figures and diagrams we are grateful to the following—

Gillett & Sibert, London; Ilford Limited, London; E. Leitz, Wetzlar; Nippon Kogaku, Tokyo; Olympus Optical Company Limited, Tokyo; C. Reichert, Vienna; Vickers Instruments Limited, York; Wild Heerbrug, Switzerland; Carl Zeiss, Jena and Oberkochen.

Detailed acknowledgments are given in the legends to figures.

PREFACE

It was inevitable with a Series as large and wide ranging as "Methods in Microbiology" that some topics would fall outside the themes on which the majority of the Volumes were based, and that some contributions intended for inclusion in earlier Volumes would be unavailable at the time they were required. To a certain extent therefore, Volume 7 is a miscellany of disconnected topics.

As with other Volumes, the material has been divided into two parts; a step necessitated primarily by the amount of material presented, but which enabled us to group together related contributions. Thus, Volume 7A contains Chapters dealing with the use of computers in microbiology and a treatment of the mathematical bases of assay methods. Two Chapters concern bacteriophage and one bacteriocins. The rest of the material comprises topics which are of considerable interest and importance but whose themes are unrelated to one another.

Volume 7 completes the initial Series of "Methods in Microbiology" and this preface affords us a welcome opportunity to express our thanks and appreciation to our many contributors whose ready help, co-operation and patience has done so much to make our task of editing an enjoyable one. We are also grateful to numerous of our colleagues who, although not themselves contributing to the Series, have provided valuable advice and comment concerning the subject matter. Our thanks are due to Shell Research Limited, The University of Miami and The Howard Hughes Medical Institute, without whose material assistance in many ways the production of the Series would have been far more difficult, if not impossible. Finally, we would like to acknowledge the assistance of the publishers and that faithful army of typists, secretaries, technicians, research students and sympathetic wives and husbands whose painstaking work and attention to detail earns them little recognition but provides the essential basis for a work of this kind. We would particularly like to mention our appreciation of the co-operation of Dr. C. Booth who edited Volume 4 and enabled the Series to cover techniques in Mycology; an area which was outside our own experience.

The question of continuing the Series comes to the fore at this time. Several contributions have been offered for a further Volume and we have decided to produce one more, a single Volume 8 which should appear early in 1973 following a manuscript date of May 1972. After that we have no plans but we will reconsider the situation from time to time as and when advances in techniques and methodology suggest that the production of a further Volume will be useful. Of course, not all possible topics have been covered and needless to say we would welcome comments and suggestions for future articles from our colleagues in the field of microbiology.

J. R. Norris

D. W. Ribbons

October 1972

CONTENTS

CONTENTS OF PUBLISHED VOLUMES

xi

CHAPTER I

Phage-Typing of *Staphylococcus aureus*

M. T. PARKER

Central Public Health Laboratory, London, England

I. INTRODUCTION

Between a quarter and a half of the human population are carriers of *Staphylococcus aureus*. Little progress could therefore be made in studying the sources of infection with this organism, and the routes by which it is spread, until a way had been found of distinguishing between different strains. The absence of a suitable method for the serological typing of

2

S. aureus led medical bacteriologists to develop phage-typing as a means of investigating staphylococcal disease in man. Phage-typing has now been used for this purpose for over 25 years, and the method has been continuously modified in response to changes in the epidemiological situation.

At first phage-typing was used mainly in short-term investigations of limited outbreaks of sepsis or enterotoxic food-poisoning. Fortunately, the workers concerned with developing the method realized the importance of standardizing the typing procedure, and within a short time it became obvious that strains with similar phage-typing patterns were becoming prevalent in hospitals in widely separated parts of the world. Phage-typing thus became an indispensable means of investigating the evolution of endemic strains of *S. aureus* in hospital populations.

Widespread interest in staphylococcus phage-typing developed later among veterinary bacteriologists, and it became apparent that the current method of typing is unsuitable for studying staphylococcal disease in several domestic animals. This, in its turn, has stimulated interest in taxonomic differences between "human" and "animal" staphylococci.

II. DEVELOPMENT OF THE METHOD

A. Principle

Cultures of *S. aureus* are classified according to their susceptibility to a set of phages chosen to make as many epidemiologically valid distinctions as possible between strains. It is therefore a method of bacterial classification based on a single class of characters. The one thing it can never do, therefore, is to show that two organisms are "the same". What it can establish, with varying degrees of certainty, is that they are "different". Its use in field investigations is to narrow down the field of enquiry by the exclusion of alternative sources of infection. In this it is most effective when the number of cultures to be considered is limited.

B. Early typing systems

Fisk (1942a, b) found that temperate phages from *S. aureus* cultures often lysed other members of the same species and that many of them had a relatively narrow host-range. He showed that susceptibility to such phages could be used as the basis of a typing system.

Wilson and Atkinson (1945) obtained a different series of temperate phages from *S. aureus* cultures, made high-titre preparations of them by propagation in broth, and separated them from the bacteria by filtration. To keep the host-range of a phage constant, it was always propagated on the same strain of staphylococcus (the propagating strain). The strength of each phage preparation was measured in terms of the routine test dilution

(RTD)—the highest dilution of a filtrate that produced confluent lysis when a standard loopful was placed on a lawn of the propagating strain. Attempts were also made to adapt phages to cultures on which no plaques were seen at RTD by applying undiluted phage suspensions to them. Some true adaptations resulted, but subsequent serological examination suggest that other phages obtained in this way were unrelated to the original phage.

With 18 phages at RTD, over 80% of cultures from human sources showed some lysis, and 60% could be allotted to one of 21 types or subtypes; but to do this all reactions of less than confluent lysis had to be ignored.

C. Phage-typing patterns

Williams and Rippon (1952) used Wilson and Atkinson's 18 phages with six others. They confirmed the value of the method, but their conclusions led to substantial changes in the way in which the typing system was to be used. They defined the RTD as the highest dilution of a phage that gave just less than confluent lysis when a 0·02 ml drop was placed on a lawn of the propagating strain, and a strong (+ +) reaction as one of 50 plaques or more. Three-quarters of the cultures examined gave a + + reaction with one or more phages, but when groups of strains from one source were compared there was often a variation in the degree of lysis produced by individual phages. The strict application of a type designation based only upon strong reactions (whether confluent lysis or a + + reaction) would therefore have resulted in related strains being placed in different types. Weak reactions would therefore have to be taken into account, and a limited number of clear cut types could not be defined. Strains could however be characterized by the pattern of their lysis by the phages, if an estimate could be made of the frequency with which epidemiologically related strains showed a variation in typing pattern of a given magnitude. This proved possible (Section IV.A), and it was then found that the comparison of phage-typing patterns was a good guide to the relationships between strains. The number of possible patterns was large.

D. Use of strong phage preparations

To increase the proportion of typable cultures beyond three-quarters, Williams and Rippon (1952) re-tested with undiluted phages those cultures which gave no strong reactions at RTD. It later became the practice to use a strength of 1000 times the RTD (RTD × 1000) to type strains untypable at RTD. Typable cultures then generally exceed 90%, but the reading of reactions obtained with phages at RTD × 1000 presents difficulties (Sections III.D, 2 and IV.C). For this reason the strength of phage for secondary typing was later reduced to RTD × 100 (Report, 1971).

E. The basic set of typing phages

By 1953, interest in the method had developed in many countries outside Britain, and the International Subcommittee on Phage-Typing of Staphylococci held its first meeting. The principle was accepted that a common basic set of phages should be used by all. The set agreed upon included 13 of Williams and Rippon's phages (nine of them from the original collection of Wilson and Atkinson), together with six more (see Table IA). A set of 22 additional phages for optional use, either in primary typing or for the examination of untypable cultures, was also defined.

TABLE I

Constitution of the international basic set of phages for typing
Staphylococcus aureus

A 1st Meeting of International Subcommittee, 1953
B 5th Meeting of International Subcommittee, 1970

Lytic group	A 1953 (Phage numbers)					B 1970 (Phage numbers)				
I	29	52	52A	79		29	52	52A	79	80
II	3A	[3B]	3C	55		3A	3C	55	71	
III	6	[7]	42E	47	53	6	42E	47	53	54
	54	[70]	[73]	75	77	75	77	83A	84	85
IV	42D					42D				
Not allotted						187	81			

[ ] Phages removed from the basic set since 1953.
☐ Phages added to the basic set after 1953.
Phages 83A, 84 and 85 are used only at RTD.

F. The phage groups

Although typing with the basic set of phages revealed numerous patterns of lysis, there were indications of a small number of broad subdivisions. Certain combinations frequently occurred in patterns, and other combinations were rare. For example, lysis by phage 52 was often associated with lysis by phage 29 or phage 52A, but less often with lysis by phages 6 or 7, and very rarely indeed with lysis by phages 3A or 3B. It appeared that the staphylococci belonged to a series of **phage groups** which included strains lysed only by one or more of a restricted set of phages, and that the phages might be classified into corresponding **lytic groups**. Williams, Rippon and Dowsett (1953) recognized three phage groups (I, II and III).

As early as 1946, Macdonald had observed that over half of *S. aureus* cultures isolated from cows' milk in Britain were lysed by phage 42D. Smith (1948a, b) confirmed this, and also showed that "bovine" staphylococci carried a number of other phages that could be used to subdivide strains lysed by phage 42D, and to type other hitherto untypable "bovine" staphylococci. "Human" staphylococci were seldom lysed by phage 42D, and rarely lysed only by it. The only occasion on which such strains are at all commonly found in human material is when they are isolated from victims of staphylococcal food-poisoning due to milk products. A further phage group (IV) was later established to include strains lysed only by phage 42D.

In Table IA the basic-set phages of 1953 are allotted to lytic groups I to IV. Later it became necessary to introduce into the set some phages that did not fall clearly into these groups (Table IB). Phage 81 lyses many strains which otherwise have patterns in group I, but also often forms part of group III patterns. Phage 187, on the other hand, lyses strains that are sensitive only to this phage.

In addition to the strains falling into the four phage groups (and those lysed only by phage 187), there are strains—not numerous among "human" staphylococci—having complex patterns of lysis by phages of more than one group. It has been agreed, however, that lysis by phage 81 in addition to phages of lytic group I or III does not exclude a strain from the corresponding phage group. Complex patterns including phages of lytic groups I and III are more common than I–II or II–III patterns.

G. International organization

The International Subcommittee for Staphylococcus Phage-Typing was formed in 1953, and has concerned itself with the standardization of the method and its development to meet changing needs. The Staphylococcus Reference Laboratory of the British Public Health Laboratory Service (Central Public Health Laboratory, Colindale, London, NW9 5HT, England) became the international reference centre, and in 1961 was recognized as the World Health Organization Centre for Staphylococcus Phage-Typing.

The Subcommittee consists ordinarily of one representative from each country who takes responsibility for the distribution of materials and information to workers in his own country. The starting materials for propagation used in each country (phages and propagating strains), are drawn from large freeze-dried batches prepared at Colindale. All other laboratories are expected to obtain their supplies from the respective national laboratory.

Acceptable methods for the propagation and testing of phages, and for the typing test, have been agreed upon (Blair and Williams, 1961; see Section III), and regular comparative tests of phage typing in national laboratories have been carried out every 3–4 years since 1955.

Criteria have been laid down for the usefulness of phages. A new phage might be considered for introduction into the typing system if it lysed a significant percentage of otherwise untypable strains, or if it was of value in subdividing a common phage-typing pattern, and if it could be readily propagated to at least RTD × 1000 and was stable in its characteristics. The Colindale laboratory examines such new phages submitted by national laboratories.

The nomenclature of the phages has been standardized. The serial numbers used by the earlier British workers were accepted, and further numbers have been given to phages that have appeared to be useful enough to warrant distribution to other laboratories. This system of numbering is not entirely consistent. At first, phages thought to be adaptations of other phages retained their original number followed by a letter (e.g. phage 29A was obtained by growth of phage 29 on a fresh propagating strain, which is now known as propagating strain (PS) 29A), but it is not always possible to tell from this designation whether or not the new strain resulted from a single adaptation (e.g. phage 42B was an adaptation of phage 42, but phage 42F was an adaptation of phage 42E). More recently, entirely new numbers have been given to apparent adaptations (e.g. phage 80 was adapted from phage 52A, phage 84 from phage 77, and phage 87 from phage 42D).

H. Later changes in the basic-set phages

Table I shows the present basic set of typing phages in comparison with the set agreed upon in 1953 (see Report, 1959, 1963, 1967, 1971). Seven phages found to be useful over wide geographical areas have been added to the set, and four of the original phages have been removed from it. The objective has been to keep the total number of phages below 25, because this is the largest number of drops that can be conveniently accommodated on a single plate.

In 1953, a total of 22 additional phages were recommended for optional use, but experience showed that this resulted in little increase in the percentage of typable strains. All but 2 of the additional phages have now been discarded, and these (phages 71 and 187) have been upgraded to the basic set. Lysis by phage 71 is characteristic of certain otherwise untypable phage-group II strains that cause vesicular skin lesions in man (Parker *et al.*,

1955), and phage 187 is of value for the recognition of a small but distinct group of strains lysed only by it.

These two phages were added for the purpose of typing strains found in the general population, but the other five additions (phages 80, 81, 83A, 84 and 85) were made in response to the appearance in hospitals of apparently "new" strains of *S. aureus* that were untypable with the basic-set phages. Had these phages not been introduced, the percentage of typable staphylococci in hospital populations would have fallen precipitately.

The first of these additions was phage 80, which was produced by adapting phage 52A to an untypable staphylococcus that had caused outbreaks of sepsis in Australia in 1954 (Rountree and Freeman, 1955). Staphylococci lysed by phage 80 were recognized soon after in many other countries. Phage 81, adapted from phage 42B (Bynoe *et al.*, 1956), lysed most of the strains also lysed by phage 80, but also lysed a number of otherwise untypable but related strains. Later, staphylococci appeared in hospitals which had somewhat similar phage-typing patterns (e.g. 52/52A/80/81, 52/52A/80, 80 etc.) and resembled the 80/81 organisms in pathogenicity and ability to spread. Evidence was soon obtained that these members of what is now called the 52, 52A, 80, 81 complex were really all "the same" staphylococci that had undergone various changes in phage-typing pattern as a result of the loss or gain of phages (Rountree, 1959; Asheshov and Rippon, 1959; Rountree and Asheshov, 1961; Asheshov and Winkler, 1966).

Three further phages have since been introduced into lytic group III to characterize otherwise untypable strains which have appeared in hospitals. The first of these (phage 83A) had been isolated some years earlier (Blair and Carr, 1953) and called VA4. At this time it was used in the U.S.A. for the subdivision of typable group III strains. When untypable strains again became common in Europe after 1958, it was found that they were lysed by this phage (Williams and Jevons, 1961), which was given the number 83. Unfortunately the phage had undergone a good deal of unofficial distribution between 1954 and 1960, and it appeared that two entirely different phages were circulating under the name VA4. The number 83A was therefore given to the original phage VA4, and the mutant was designated 83B (Report, 1963).

In 1961, further untypable strains appeared, and it was shown that they had arisen from 83A strains by lysogenization with a variety of phages (Jevons and Parker, 1964). Phage 84 (adapted from phage 77), and phage 85 were introduced for the recognition of these strains (Report, 1967).

In order to avoid unnecessary complication of the phage-typing patterns given by members of phage-group III, the International Subcommittee agreed that phages 83A, 84 and 85 should be used only in typing at RTD.

Phages 70, 73, 3B and 7 were removed from the basic set between 1958 and 1966, after calculations had shown that this would lead to little reduction in the percentage of typable strains. No change in the basic set of phages has been made since 1966, but certain additional phages for local use have been officially recognized in recent years (see Section VII.B).

III. TECHNICAL METHODS

A. Media and conditions of growth

The International Subcommittee does not specify that a single standard medium should be used for staphylococcus phage typing. Broth and agar made from various dehydrated products, such as Difco "Bacto" Nutrient Broth and Oxoid Nutrient Broth No. 2, and from infusion broth made in the laboratory from fresh meat have been used with success. This is not to say that the results of typing are uninfluenced by the composition of the medium. Several variables do have a profound effect on the susceptibility of staphylococci to lysis by phage, but the more important of them appear not to be directly connected with the type of nutrient in the medium. For example, we have found no significant difference in the results obtained, either in propagation or in typing, on the Difco and the Oxoid products.

The medium chosen should not necessarily be the one that gives the most luxuriant growth of staphylococci; indeed, this may result in small plaques and difficulty in seeing the plaques produced by some of the phages. For this reason we do not favour the use of rich digest media. It is also important that the agar should be as soft as is practicable, and that the plates are not dried more than is necessary to remove the surface moisture.

Two other variables are of great importance. The first is that sufficient calcium is present in soluble form, because a number of the phages need calcium ions at a concentration of at least 300 μg per ml. for absorption and replication. Some solid media prepared with Japanese shred agar do not require the addition of calcium, but excess heating should be avoided because this appears to reduce the amount of free calcium. It is safer to add 400 μg per ml $CaCl_2$ to the melted and cooled agar before pouring the plates. With some media—including combinations of certain commercial dried products—the phosphate content is so high that the addition of calcium salts results in an immediate heavy precipitate, and a medium may result that is both cloudy and deficient in calcium ions. In such cases it is necessary to remove the excess phosphate by heating under alkaline conditions and then filtering before adding calcium.

The second is the type of agar used. For many years we recommended the use of Japanese shred agar, and there is no doubt that it gives reliable results. But its preparation for use is time-consuming. The use of New

Zealand powdered agar and Oxoid No. 3 Agar result in considerably narrower phage-typing patterns. Recent experiments have shown that Oxoid No. 1 Agar give results that are practically identical with those obtained with Japanese shred agar.

A medium with the following composition has given good results in our hands.

Oxoid Nutrient Broth No. 2 (CM67)	2·0%
Sodium chloride	0·5%
Oxoid Agar No. 1 (L11)	0·7%

Calcium chloride to a final concentration of 400 μg per ml is added just before the plates are poured.

A complete dehydrated phage-typing medium prepared by Messrs. Oxoid Limited is at present under investigation and is giving encouraging results.

We use a broth medium of the following composition for phage propagation; Oxoid Nutrient Broth No. 2, with 400 μg per ml $CaCl_2$ added at the last moment.

Soft agar for phage propagation is made by adding 0·5% of Japanese shred agar to nutrient broth; 400 μg per ml of $CaCl_2$ is added at the last moment.

Unless stated otherwise in the succeeding sections, incubation is carried out overnight in air at 30°C.

B. Propagation and testing of phages

Full details of recommended methods for propagating the phages and for testing their lytic spectra will not be given. The reader is referred to a paper by Blair and Williams (1961) which was prepared on behalf of the International Subcommittee. Information about phages and propagating strains introduced into the basic set since this date may be obtained from national typing laboratories (see Report, 1971).

1. *Propagation*

Phages may be propagated in broth or by the soft-agar method of Swanstrom and Adams (1951). The first method is simpler to perform, but the second is to be preferred because the yield of phage is generally greater.

The source of the phage and its propagating strain should be freeze-dried material, either from Colindale, or dried from a first propagation or subculture of material obtained from Colindale. When the ampoule of phage has been reconstituted, it should be stored at 4°C until propagation and testing have been completed. The first subculture of the propagating strain from the reconstituting broth should be similarly stored on a series of nutrient agar slopes, and in national laboratories should also be freeze-dried.

The aim should be to produce a phage suspension of at least RTD × 1000 (about 10^8 particles per ml). If this is not achieved, the suspension should be discarded and the propagation repeated from the original materials. Under no circumstances should the phage be propagated or the propagating strain subcultured in series.

TABLE II

Phage-typing patterns of the propagating strains for the basic-set phages

Strain No.	RTD	RTD × 100
29	29 + +	29 + + 80°
52	52 + + 52A ± 80 ±	52 + + 52A + + 80 + +
52A/79	52 ± 52A + + 79 + + 80 ±	52 + + 52A + + 79 + + 80 + +
80	80 + + 81 + +	80 + + 81 + +
3A	3A + + 55 ± 71 ±	3A + + 3C + + 55 + + 71 + +
3C	3C + + 55 + + 71 + +	3A + + 3C + + 55 + + 71 + +
55	3C + + 55 + + 71 + +	3A + + 3C + + 55 + + 71 + +
71	3C + + 55 + + 71 + +	3C + + 55 + + 71 + +
187	187 + +	187 + +
6	6 + + 42E ± 47 + + 53 + + 54 + + 75 + + 77 + + 83A + + 84 + + 85 + + 81 ±	6 + + 42E + + 47 + + 53 + + 54 + + 75 + + 77 + + 81 + +
42E	42E + + 81 ±	42E + + 53 + 81 + +
47	47 + + 53 + + 75 + + 77 + + 84 + + 85 + +	29 + 52 + 52A ± 79 + 80 + + 47 + + 53 + + 54 + + 75 + + 77 + +
53	53 + + 54 + + 75 + + 77 + + 84 + 85 + +	53 + + 54 + + 75 + + 77 + +
54	47 + + 53 + + 54 + + 75 + + 77 + + 84 + + 85 + + 81 ±	42E + + 47 + + 53 + + 54 + + 75 + + 77 + + 81 + +
75	53 + + 75 + + 77 + + 84 + + 85 + +	79 ± 47° 53 + + 75 + + 77 + +
77	77 + + 84 + 85 +	47 + + 53 + + 77 + +
81	80 + + 81 + +	80 + + 81 + +
42D	42D + +	42D + +
83A	6 + + 47 + + 53 + + 54 ± 75 + 77 + + 83A + + 84 + + 85 + + 81 + +	52 + 52A ± 79 ± 80 + 6 + + 42E + 47 + + 53 + + 54 + + 77 + + 81 +
84	84 + + 85 + +	77 ±
85	84 + + 85 + +	75 ± 77 ±

± = less than 20 plaques
+ = 20–49 plaques
+ + = 50 or more plaques
° = inhibition (used at RTD × 100 only)
Phages 83A, 84 and 85 are used only at RTD in routine typing.

Before propagation is begun, it is essential to check that the propagating strain is the correct one and that its host-range is unchanged, and to measure the titre of the phage. The propagating strain is phage-typed at RTD and at RTD × 100 and the result shown in Table II should be obtained. The appearance of + + reaction at RTD × 100 where it is not recorded in the standard pattern, or the absence of a + + reaction that should be present, is evidence that something is wrong with either the medium or the propagating strain. If the propagating strain has the correct typing pattern the phage is titrated on it (Section III.C,2), and a plaque count is made.

In propagation the aim is to obtain a cell : phage ratio of about 100 : 1, on the assumption that a plaque represents one phage particle and a colony represents a single bacterium. A standard suspension of the propagating strain is made by emulsifying with 2 ml broth the growth on an 18 h agar slope. Soft agar (III.A) is melted and cooled to 45°C–48°C; bacterial suspension to give a final dilution of 1 : 100, and phage to give a final concentration of 10^5 particles per ml are added. This agar is poured on the surface of a nutrient agar plate to give a layer 1–2 mm in thickness. After incubation there should be evidence of near-confluent lysis. Sterile broth is added to the plates (20 ml for a 15 cm plate), and the soft agar layer is scraped off and emulsified in the broth by rapid pipetting. The mixture is centrifuged to remove lumps of agar and most of the bacteria, and is stored at 4°C while the strength of the phage is measured. If this exceeds RTD × 1000, the suspension is filtered through a sintered glass (5/3) or membrane filter. Sterility tests should be carried out. Seitz filtration causes unacceptable loss of phage and chemical sterilization may result in a residual bactericidal effect in undiluted lysates.

2. Testing

After a plaque-count has been carried out on the filtered lysate, the lytic spectrum of the phage must be determined, to make sure that no mutations or other changes have occurred during propagation. The action of the phage on a set of 16 test strains (some but not all of which are propagating strains) is determined in a two-stage test.

(a) A 0·02 ml drop of phage, at a concentration of 1 to 5×10^9, is applied to a lawn of each of the test strains.

(b) A titration of the phage is carried out on each of the test strains that showed any lysis or inhibition, with 10-fold dilutions of the phage-suspension used in (a).

A record is made of the highest dilution of the phage that gives a + + reaction on each test strain, and this is compared with the dilution that gives the same strength of reaction on the propagating strain. The differences are scored on an arbitrary scale and compared with the scores given in a table of lytic spectra (see Blair and Williams, 1961) which has been compiled from the experience of many propagations made at Colindale. In general, gross discrepancy in the reaction with a single test strain is considered to be grounds for the rejection of a batch of phage.

The freeze-dried ampoules issued from Colindale contain phage in high titre, and it is possible to use this material at a concentration of about 10^9 particles per ml as a "model" in parallel tests of lytic spectrum, and so to assess whether variations are due to change in the phage or in the test strain, or to differences in the medium.

C. Typing

In Britain, liquid suspensions of phage are issued from Colindale to all typing laboratories, and with rare exceptions have a titre of between RTD $\times$ 1000 and RTD $\times$ 10,000. They are despatched by post in screw-capped bottles and show no loss of potency in transit. They should be stored at 4°C but not allowed to freeze. Some phages show little loss of titre in a year, but the effective life of others is only a few months.

Propagating strains are issued as freeze-dried ampoules at two-yearly intervals, or more frequently on request. After reconstitution, they should be subcultured on blood agar, and a series of 3–4 agar slopes should be prepared from a single colony. At this stage it is wise to check their identity by phage-typing them (Table II) .The slopes should be kept at room temperature in the dark; survival at room temperature is at least as good as in the refrigerator, and growth occurs more promptly on subculture. A rather poor meat-extract agar is best for storage. One set of slopes should be used for routine purposes until most of the growth has been removed. When growth on subculture begins to be delayed or fails, a fresh set of slopes is prepared from a set of unopened slopes.

1. *Control of reagents*

(a) Before a batch of phage is taken into use, it must be titrated (Section III.C, 2), and diluted to a strength of RTD and RTD $\times$ 100.

(b) The strength of the RTD suspensions must be checked on the day before each session if typing is done infrequently, or twice a week if it is done daily (Section III.C, 3). The RTD $\times$ 100 suspensions are not checked so frequently, because they are made up in smaller amounts and are quickly exhausted. If they are to be used more than one month after preparation they should be checked by making a $\times$ 100 dilution and spotting on the propagating strain. If a phage becomes contaminated with bacteria, it should be discarded.

(c) When a new batch of typing medium is prepared, and several days before it is to be used, an RDT suspension of each phage should be tested on its propagating strain on a test plate.

(d) Should any doubt arise about the typing results, confidence may be restored by including the propagating strains in the next batch of cultures to be typed. Their patterns should correspond to those shown in Table II.

2. *Titration of phage filtrates*

The RTD is the highest dilution of phage that gives just less than confluent lysis of the propagating strain when applied as a 0·02 ml drop. The relation of RTD to plaque count is influenced by the size of the plaques, but the RTD usually contains between 1 and 5×10^5 plaque-forming units per ml.

The propagating strain is inoculated into broth and incubated at 37°C for 4–6 h, and should then have produced some turbidity. A plate of nutrient agar is flooded with this broth and drained of excess moisture with a Pasteur pipette. The lid is left off until the surface is dry (about $\frac{1}{2}$ h). Ten-fold dilutions of the phage suspension are made, and one 0·02 ml drop of each is placed on the surface of the plate with a calibrated dropping-pipette. When the drops have been absorbed, the plates are incubated at 30°C and examined next morning for lysis. It often happens that no dilution corresponds exactly to the RTD, and this must be arrived at by interpolation.

3. *Checking the dilute phage suspension*

A small area (about 2 cm square) of the surface of the agar in a 15 cm Petri dish is inoculated with a 4 h to 6 h broth culture of each propagating strain. A 0·02 ml drop of the corresponding RTD phage preparation is placed on each area. Next morning, each should show just less than confluent lysis.

4. *The typing test*

Typing with the international basic set of phages is applicable only to coagulase-positive staphylococci; with rare exceptions, coagulase-negative staphylococci are not lysed by the phages. The practice of doing a slide-coagulase test on one colony and selecting an apparently similar one for phage-typing is to be condemned. Single colonies from primary plates occasionally consist of two distinct strains of *S. aureus*; all cultures should be purified by subculture and selection of a single colony before typing. The presence of two *S. aureus* strains in a culture may result in unusual typing patterns and may often be recognized because the areas of lysis are turbid. The occurrence of "spontaneous" lysis—indicated by plaques throughout the whole lawn—also raises a suspicion that two strains may be present.

All cultures are first tested with the basic-set phages at RTD. Those not lysed strongly (< 50 plaques) by one or more phage at this strength are re-examined next day with the phages at RTD $\times$ 100 (but phages 83A, 84 and 85 are omitted). Bacterial lawns for RTD typing are prepared from 4 h to 6 h cultures (Section III.C, 2), which are then left overnight at room temperature and used if necessary for the preparation of lawns for typing at RTD $\times$ 100. Phage is applied (Section III.C, 5) with a standard loop; it should be noted that the volume of phage suspension used in the typing test is smaller than that used in the measurement of the RTD. Plates are incubated overnight at 30°C.

5. *Application of phage*

Phages may be applied by hand with sterile loops, but this is very laborious. The use of pipettes leads to error; if the drops fall through the air they cannot be placed accurately, and if they are "touched off" staphylococci may be carried over from one plate to another; if the staphylococci on the first plate carry phages active on those on the second plate, "false" lysis will occur.

A number of mechanical devices for the simultaneous application of all the phages to the plate have been described. The multiple-loop applicator of Tarr (1958) has a set of spirally wound loops which slide freely in guides, and drops are deposited on the agar without cutting it. The loops are charged by being dipped into wells in a perspex block which contain phage suspension.

Lidwell (1959) designed a more elaborate model working on the same principle (Figs 1 and 2), and this is the most convenient apparatus for large-scale work. Information about manufacturers of this apparatus, and of the special loops, can be obtained from Colindale. There are two sets of loops, one of which is being sterilized and cooled while the other is picking up and depositing the drops of phage. In the original design, which is illustrated here, the loops were to be sterilized by burning-off over gas jets immediately after the drops of phage had been deposited, but this led to "coking" of the wires. Present practice is to pick up the phage in position D (Fig. 2), move the horizontal arm to position E and deposit the phage on the inoculated plate; then, with the loops in the same position, they are dipped in an open Petri dish of alcohol. When the arm is moved to position B and depressed, burning-off occurs without much heating of the loops, and it is not necessary to employ a special cooling device in position C. In current models there are no cooling tubes.

Another method is to use a set of metal pins in a template. Standard

FIG. 1. General view of the Lidwell phage-typing apparatus taken from the front and slightly to the left.

drops are satisfactorily picked up and deposited, but the pins cool slowly after sterilization by heat. The necessity to sterilize them after each application of phage is avoided in the "pre-stamping" method, i.e. the drops of phage are deposited on the plate and allowed to dry thoroughly before the plate is flooded with culture. In our experience, good results may be obtained in this way, but the plates must be flooded very gently. Higher concentrations of phage must be used to compensate for the loss of phage by absorption into the agar; the RTD must therefore also be measured by a "pre-stamping" method.

D. Reading and reporting of results

1. *Examination of plates*

The plates are examined by indirectly transmitted light, against a dark background, with the aid of a ×5 hand-lens. This is preferable to a colony counter, because the worker learns to manipulate the plate so as to obtain the best results, and examines the plates with the naked eye before using the lens. Little difficulty is experienced in reading the results of typing at RTD once experience has been gained in detecting very small plaques. It is not customary to record separately any degree of lysis greater than + + (50 plaques or more).

Reading the results of typing with strong phage presents some difficulties, and is subject to considerable "observer error". Discrete plaques

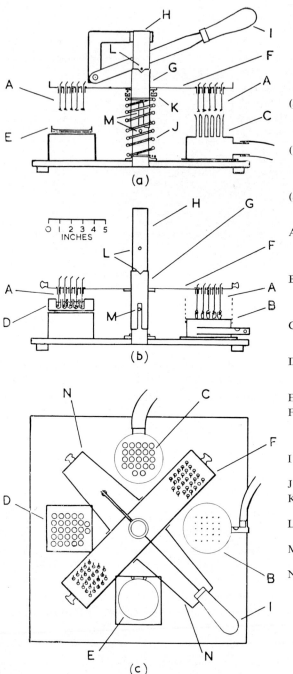

FIG. 2. Diagram of the Lidwell phage-typing apparatus.

or confluent lysis may be seen, but secondary growth on areas of confluent lysis is common. In addition, what are referred to as "inhibition reactions" are seen, and are often difficult to distinguish from lysis with secondary overgrowth.

2. *Inhibition reactions*

Inhibition reactions occur when strong phage preparations are used, and are apparently due to absorption of phage which leads to death of the staphylococcus without phage multiplication or lysis. The appearance is frequently of a circular area of thinning of growth coincident with the original phage drop, which may be confused with confluent lysis with secondary overgrowth. Indeed, reactions are occasionally so strong as to mimic confluent lysis. Plaques, often quite numerous, may be superimposed on areas of inhibition. Inhibition reactions at RTD × 1000 are frequent and often very strong. At RTD × 100 they are rarer and weaker, and can usually be recognized for what they are.

Whether a reaction is one of inhibition or lysis can be established with certainty in the following manner. A titration is carried out of the phage against the apparently inhibited staphylococcus at RTD × 100, RTD × 10 and RTD. With inhibition reactions, the area of thinning becomes less intense with the weaker phages and disappears without any plaques being seen; with confluent lysis, or lysis with secondary overgrowth, discrete plaques will be seen with weaker phage.

3. *Conventions for recording typing results*

The degree of lysis observed at RTD is recorded as follows.

+ + = more than 50 plaques; semi-confluent lysis; confluent lysis with or without secondary overgrowth.
+ = 20–50 plaques.
± = less than 20 plaques; when the number of plaques is 10 or less it is recorded as a superscript (e.g. ± [5]).

At RTD × 100 a more elaborate system is used.

CL = confluent lysis.
GR/CL = confluent lysis with secondary overgrowth.
+ + = more than 50 discrete plaques.
+ = 20–50 plaques.
± = less than 20 plaques.
○ = inhibition; the presence of discrete plaques is indicated by placing the appropriate + signs within the circle.

It is important to record the presence of secondary overgrowth, and of

3

inhibition reactions, because of the possibility of error in deciding whether
or not a reaction is one of true lysis must always be borne in mind.

4. *Reporting*

In reporting the results of typing it is customary to record the numbers
of the phages that produce strong lysis of the culture (> 50 plaques).
This is referred to as the typing pattern (Table III).

When two or more phages give strong lysis the numbers are reported
thus, 52/52A/80. The presence of additional weak reactions exceeding
10 plaques is recorded by the addition of a + sign thus, 52/52A/80/+ ;
but the phages giving the weak reactions are not specified. When weak
reactions are seen that are significant for the interpretation of the results
they are taken into account and may be mentioned in the notes that
accompany the report.

If the culture is not typable at RTD (i.e. does not show any + + reactions) it will have been typed at RTD × 100. Only lytic reactions with
the strong phages are reported, and discrete plaques occurring in areas of
inhibition are not mentioned (see Section IV.C).

Most laboratories record weak reactions with phages at RTD when there
are no strong reactions by quoting the numbers of the phages followed by
the letter w, thus 84/83A/w as in Table III. The result of typing at RTD
× 100 is also given. It must be remembered that phages 83A, 84 and 85
are used only at RTD. Thus, as in this case, a weak reaction with one of
these phages at RTD will appear in the report, but no reaction at RTD ×
100. Had a strong reaction with phage 83A been seen, secondary typing at
RTD × 100 would not have been performed.

All reports should be accompanied by a commentary. An alternative
method of reporting, to send photostatic copies of the records of the actual
results obtained, has certain advantages (Section IV.E).

IV. INTERPRETATION

A. Range of variability

The precision with which phage-typing distinguished between unrelated staphylococcal strains had first to be established by laboratory experiment and epidemiological enquiry. It was necessary to find out how big a
difference there must be between the phage-typing patterns of two cultures
before it could be concluded that they were unrelated (see Section II.C).

Williams and Rippon (1952) showed that when groups of cultures from a
common source were compared, variations occurred even if great care
was taken to standardize the technique. These variations increased with the
distance of the cultures from their presumed common source. Duplicate

TABLE III

Reporting and interpretation of phage-typing patterns

Culture no.	Lysis by phages								Report	
	6	42E	47	53	54	75	77	83A	RTD	RTD × 100
1	+ +	±⁴			+ +	+ +	+ +		6/54/75/77	
2	+ + +	±			+ +	+ +	+ +		6/54/75/77/+	
3	+ + +	+			+ +	+ +	+		6/54/75/+	
4	+ + +				+	+			6/+	
5	+ +	+		+ +	+ +	+ +	+ +	+ +	6/53/54/75/77/83A	
6		+	+ +		+ +	+ +	+ +	+ +	47/54/75/77/+	
7					±			+	54/83A/w	
	(+ +)	(+)			(+ +)	(+ +)	(+ +)			6/54/75/77/+

Reactions with all other phages negative; phage 83A not used at RTD × 100.
In parenthesis: lysis at RTD × 100.
Interpretation: Cultures 1–4 indistinguishable from each other: cultures 5, 6 and 7 distinguishable from cultures 1–4 and from each other.

tests on one culture showed less variation than tests on separate colonies from the same culture; and these in their turn showed less variation than successive cultures isolated from the same natural source, or from different but epidemiologically related sources. But when sets of cultures believed to be of common origin were all tested at RTD on one day, less than 5% differed by more than one strong reaction. If, with two or more phages, there was a + + reaction with one culture and no lysis of the other, there was a strong presumption that the two cultures were not related. This "two strong differences" rule has been generally accepted as a guide for interpreting the result of typing at RTD (Blair and Williams, 1961).

The use of this rule is illustrated in Table III. Cultures no. 1–4 would be considered indistinguishable, because in no case does one show two strong differences from another; despite differences in the strength of reactions, nos 1, 2 and 3 show no strong differences; no. 4 shows one strong difference from nos 1 and 2 (phage 77), but no strong difference from no. 3. Cultures no. 5 and 6 differ by two strong reactions from nos 1–4; no. 5 has two strong reactions (with phages 53 and 83A) not shown by nos 1–4; no. 6 gives a strong reaction with phage 47 but lacks the reaction with phage 6. Despite the fact that no. 7 has the same phage-typing pattern at RTD × 100 as no. 1 has at RTD, it differs from it by three strong reactions (with phages 6, 75 and 77) at RTD.

Interpretation is, however, not simply a matter of applying the "two strong differences" rule. Some strains are much more variable than others in their phage-typing patterns, and the experienced worker makes use of his knowledge of the behaviour of organisms with particular patterns. In long-term investigations he is also often able to form opinions about the stability of particular strains. Sometimes, therefore, he will consider as indistinguishable a series of cultures showing larger differences in pattern, and at other times he will subdivide strains on smaller differences.

B. Causes of variability

Variability in phage-typing may be due to one of many causes. More of it seems to be due to uncontrollable factors in the method than to "experimental error", such as that due to variation in the strength of the phage or to differences between batches of medium. Some is probably due to differences in the initial ratio of phage particles to actively multiplying cocci. This is almost impossible to control in any routine typing method. Some may be caused by small differences in the physiological state of cultures. Poorly growing cultures are less susceptible to lysis by phage than are actively growing ones. The size of plaques is also influenced by the thickness of the bacterial lawn, and a reaction may not be detected because the plaques are too small to be seen with a hand lens. There is considerable "observer error" in the detection of very small plaques, particularly when their presence is not expected. It is probably for this reason that somewhat more consistent results are obtained by typing related strains on the same rather than on different days. Certain strains are known to be unstable in respect of the phages they carry, and differences in the proportion of non-lysogenic individuals in sub-strains from the same culture may influence the phage-typing pattern considerably. Such differences in the physiological state and even in lysogenicity between individuals from the same clone account for the variability seen when different colonies from the same cultures are typed. When cultures of the same staphylococcal strain are recovered on separate occasions from the natural environment they may have been exposed to different physical conditions and selective influences, and may therefore differ even more from each other. Other studies of the range of variability in typing patterns are those of Williams (1957), Wentworth, Romig and Dixon (1964) and Pether (1968).

C. Variability in reactions with strong phage

Until recently, little has been recorded about differences in typing patterns when typing was performed at $RTD \times 1000$, but the general opinion is that they are considerably greater than in typing at RTD. An investigation by Pether (1968) suggested that two strong differences occur in the typing of about 20% of pairs of related cultures at $RTD \times 1000$.

A collaborative investigation organized by the International Sub-committee (to be published) showed that discrepancies in replicate typing are considerably less at $RTD \times 100$ than at $RTD \times 1000$, and that much of the improvement is due to a reduction of "observer error" due to confusion between inhibition and true lysis. It was concluded that this improvement in reproducibility compensated for the small reduction in the percentage of typable strains resulting from the use of the phages at a lower strength.

It might be argued that more consistent reports would be produced if no distinction was made between lysis and inhibition. This is not so, because the appearance of inhibition reactions is much influenced by the initial ratio of phage to bacteria, and by the growth-rate of the staphylo-coccus. It is therefore more inconstant than true lysis. The same may be said of plaques within area of inhibition.

D. Long-term changes in typing pattern

Gross changes in phage-typing pattern sometimes occur under natural conditions as a result of the loss or gain of a carried phage (Section II.H). The hospital environment, in which there are frequent changes in the carriage-state of patients, probably provides many opportunities for contact between staphylococcal strains, and thus for prophage exchange (Parker, 1966). It is unlikely, however, that this occurs often enough to interfere with the interpretation of the results of short-term investigations.

E. Commentaries on reports

It must be admitted that the form in which the results of phage-typing are reported (Section III.D, 4) often does not permit them to be interpreted by someone outside the laboratory. For this purpose it would be necessary to list all the weak reactions with their strengths. Ruys and Borst (1959) proposed that this should be done, but the system proved inconvenient in practice. As an alternative, we have recently adopted the practice of sending photostatic copies of the full laboratory records to those submitting cultures. This in our experience is a useful method, particularly in long-term investigations.

If this is not done reports must be accompanied by a commentary which indicates possible relationships between the cultures typed. Such a commentary can be written only if the sender of cultures provides the laboratory with information about the epidemiological questions being asked. These questions are usually in the form "Did organisms A and B arise recently from the same source?" or "Did A infect B?". They must always be answered indirectly, by showing which organisms can be excluded from consideration on the basis of their typing patterns. The commentary must therefore indicate as clearly as possible which of the relevant cultures are distinguishable from each other and which are not, but the temptation should be resisted to describe cultures as "the same".

The ability to make these distinctions becomes progressively more difficult as the number of cultures increases. It is for example, practically impossible to subdivide the hundreds of *S. aureus* cultures that could be isolated in an hospital ward in the course of one year into a definite number of separate strains, but this is in any case seldom profitable. Most relevant questions can be answered by considering only the cultures isolated from persons who could conceivably have been in contact with each other; in the average hospital ward this number will be of the order of 20 to 40, and the problems of comparison are not insuperable. It is also helpful to know which cultures come from suspected recipients and which from suspected donors of staphylococci, because this may reduce the number of comparisons to be made. In Table III, for example, if cultures no. 1–4 came from cases of post-operative sepsis, and nos 5, 6 and 7 from carriers in the operating theatre, it would be necessary only to consider the relation of nos 1–4 to nos 5, 6 and 7 individually and would not matter whether nos 5, 6 and 7 were distinguishable from each other.

V. STAPHYLOCOCCI FROM DOMESTIC ANIMALS

In recent years, many workers have investigated the susceptibility to phages of *S. aureus* cultures from other animal species, but few have yet used phage-typing to study the spread of infection among animals. The phage-typing system used by medical bacteriologists proved unsuitable for the classification of staphylococci from a number of other mammals for one or both of the following reasons; either (1) most of the animal strains were not lysed by the "human" phages at RTD, or (2) those that were typable gave wide patterns that were unstable and difficult to interpret.

The first generally appeared to be so with staphylococci from dogs (Coles, 1963, Blouse *et al.*, 1964), poultry (Smith, 1954, Harry, 1967) and sheep (Watson, 1965), but there is also evidence that these animals may at times be colonized by staphylococci with "human" phage-typing patterns, and it appears that when these animals are in close contact with man they indeed acquire "human" strains. Examples of this in veterinary hospitals have been reported for dogs (Pagano *et al.*, 1960, Live and Nichols, 1961) and goats (Poole and Baker, 1966).

With regard to staphylococci from cattle, the evidence is to some extent contradictory. Some workers have found mainly phage-group IV strains, lysed either by phage 42D or untypable with basic-set phages but lysed by other group IV phages (Smith, 1948a, b, Meyer, 1967) but others report a predominance of strains with wide patterns of lysis by phages of groups I and III, with or without phage 42D (Davidson, 1961, 1966). It is not certain whether this difference is geographical or is related to the sort of material examined.

It is clear that independent phage-typing systems would be necessary for the study of staphylococcal disease in several animal species. Whether these systems are developed will depend upon the economic importance of the diseases in particular domestic animals. As regards staphylococcal bovine mastitis there is no doubt about the need for such a method, and steps have been taken to establish an agreed basic set of phages.

A. Typing "bovine" staphylococci

At least six sets of phages, in addition to the basic set of "human" phages, have been used to type staphylococci from cattle (Smith, 1948a, b, Seto and Wilson, 1958, Coles and Eisenstark, 1959a, b, Nakagawa, 1960; Davidson, 1961; Coles, 1962).

In 1962, Davidson undertook on behalf of the International Subcommittee an investigation into the most suitable set of phages for use with "bovine" staphylococci (Report, 1963, Davidson, 1966). He assembled all the available phages from the six previous sets, and used them to examine collections of staphylococci from cattle in Britain, the U.S.A. and Italy. Later, he tested a larger set of British strains with the 43 "bovine" phages, and with the international basic set, in collaboration with the Staphylococcus Reference Laboratory. The phages were assessed both with regard to their ability to type otherwise untypable strains and give clearly distinguishable pattern reactions. A selection of phages was then chosen, and was evaluated collaboratively by veterinary bacteriologists in 17 centres. As a result, the 16 phages shown in Table IV have now been accepted as the international basic set for typing *S. aureus* strains from bovine sources (Report, 1971); and the original international basic set will in future be known as "the basic set of phages for typing *S. aureus* of human origin". The "bovine" basic set includes 9 members of the present

TABLE IV

International basic set of phages for typing *Staphylococcus aureus* **strains of bovine origin**

Lytic group	Phages nos†				
I	29	52A			
II	3A	116			
III	6	42E	53	75	84
IV	42D	102	107	117	
Not allotted	78	118	119		

† Phage 78: former "additional" phage in "human" typing set.
  Phage 102: Davidson (1961).
  Phage 107: Davidson (1961).
  Phage 116 = No. 883: Nakagawa (1960).
  Phage 117 = No. 1363/14: Smith (1948a, b).
  Phage 118 = No. S1: Seto and Wilson (1958).
  Phage 119 = No. S6: Seto and Wilson (1958).
  The remaining phages are members of the present "human" basic set of phages.

"human" basic set; the remaining phages include one former "additional" phage from the "human" set and 6 phages of bovine origin. It is recommended that the "bovine" phages be used at RTD in the first instance. Further studies are proposed, and will be co-ordinated by the Central Veterinary Laboratory, Weybridge, U.K., which will act as source for phages and propagating strains.

VI. TAXONOMIC SIGNIFICANCE

Phage-typing is used at a "higher" taxonomic level in the classification of staphylococci than of many other organisms. Among the salmonellas, for example, phage-typing systems are used to subdivide a single serotype.

Difficulties in the serological typing of *S. aureus*, however, resulted in the introduction of phage-typing at what we regard as the specific level.

The international set of staphylococcal typing phages appears to be almost entirely specific for coagulase-positive staphylococci. Within the group, however, the patterns of susceptibility to phages suggest the existence of a number of subdividions, some of which had previously been suspected. The three main phage-groups among the "human" staphylococci appear to have some taxonomic significance. There are indications of a limited correspondence between antigenic structure and phage-groups (Hobbs, 1948; Oeding and Williams, 1958; Hofstad, 1964). Phage group II appears to be relatively distinct in its biological properties, and to include strains with only a small ability to acquire resistance to antibiotics. It includes an interesting subdivision, characterized by lysis by phage 71 alone or in combination with phage 55, with a special ability to cause vesicular rather than pustular skin lesions (Parker *et al.*, 1955).

Phage-groups I and III appear to be rather more closely related, and will probably eventually be resolved into a number of complexes (such as the 52, 52A, 80, 81 complex) consisting of organisms with the same basic constitution but differing in lysogenic state.

Phage-group IV appears to be an ecological "escape" from the "bovine" staphylococci, and forms part of a larger group lysed by related phages. The fact that the staphylococci of cattle include two apparently distinct groups—one lysed by group IV phages and the other by a wide range of phages of groups I and III—has received striking support from the work of Meyer (1966, 1967). It appears that, in general, the group IV strains have the ability to coagulate bovine plasma, to absorb the dye when grown on crystal violet agar, and to form β-lysin but not staphylokinase in contrast to the group I–III strains which have characters rather more like those of "human" staphylococci.

In general, the susceptibility to phages of the staphylococci from several other animal species suggests that they may form a series of distinct bacterial populations.

VII. THE FUTURE

Despite the many disadvantages and illogicalities of the staphylococcus phage-typing system, it is still the only available method for large-scale use in the subdivision of *S. aureus*. Future developments may be along the following lines.

A. One system or several?

Several typing systems may be evolved for the study of staphylococcal disease in different animal hosts. The main problems will be of cross-reference, both of the phages in the separate systems, and of the staphylococci themselves. The first can be solved by characterizing the host-range of the phages in the different sets on a common set of test staphylococci, but the second presents difficulties. It will be necessary to decide which of the staphylococcal strains isolated from a particular animal host "really" belongs to it. This may not matter much in studies of the spread of infection within one species; but although inter-specific spread of staphylococci appears so far to have been rather limited, it might be very important in some circumstances. Susceptibility to the existing phages suggests broad divisions between the staphylococci from different animal hosts, but is generally not specific enough to identify an individual animal strain when isolated from an unusual host. The experience of Davidson (1966) with phage 117, which lysed 49% of bovine staphylococci but no human staphylococcus, suggests that a more extensive search might lead to the development of a set of "orientating" phages.

B. Future of the present basic set of phages

The "human" international basic set must be allowed to develop in the way best suited to the requirements of medical bacteriologists. The emergence of further new strains of *S. aureus* in hospitals may necessitate the introduction of more new phages. In the past the total number has always been kept below 25 by the removal of existing phages, but some limit must be set to this process if the system is not to become distorted. There is a potential conflict between the objective on the one hand of typing as great a proportion as possible of the "specialized" hospital staphylococci and on the other of distinguishing by means of pattern-reactions between the different strains found in the general population. Recently, therefore, the Subcommittee has shown reluctance to add further new phages, taking the view that this should be done only when a new hospital strain appears that is worldwide in distribution. In other cases, additional phages of local value may be used to characterize otherwise untypable strains.

It appears in retrospect that the decision to introduce phages 83A, 84 and 85 was justifiable because strains recognizable only by their use were widespread in Europe and later in Australia. They proved much less useful in the U.S.A., where strains with rather similar cultural characters and antibiotic resistance pattern were more often lysed only by phage 86 (originally called UC18). An unrelated hospital strain prevalent in Hungary, but rarely seen outside central Europe, could be characterized by lysis by phage 87. In 1966, the Subcommittee decided that these two phages should be recognized for optional use in addition to the basic-set phages (Report, 1967).

A more difficult situation arose after 1966, with the widespread appearance of methicillin-resistant strains in Europe. Some of these strains were completely untypable or gave weak cross-reactions with certain group I phages and inhibition reactions with group III phages. It proved comparatively easy to isolate phages which lysed methicillin-resistant strains from a particular hospital, or even from one country, but none of them merit inclusion in the basic set. Thus, phage 89 was useful in Denmark but less so in Britain, and phage 88 was useful in Britain but not elsewhere in Europe. In 1970, therefore, the Subcommittee recognized six new phages of this sort for optional use (Nos. 88–93: Report, 1971).

When additional phages are used, it is important that all cultures untypable with the basic-set phages at RTD should be tested with them at RTD × 100 even if they are lysed by additional phages. Failure to do this would make it impossible to compare the results obtained in laboratories in which different additional phages were in use.

The appearance in the future of further untypable strains of worldwide distribution will raise serious technical problems which may necessitate a more fundamental change in the typing method.

ACKNOWLEDGMENT

The author is grateful to Dr O. M. Lidwell, and to the Controller of Her Majesty's Stationery Office for permission to reproduce Figures 1 and 2 from the Monthly Bulletin of the Ministry of Health and the Public Health Laboratory Service (volume 18, p. 49).

REFERENCES

Asheshov, E. H., and Rippon, J. E. (1959). *J. gen. Microbiol.*, **20**, 634–643.
Asheshov, E. H. and Winkler, K. C. (1966). *Nature, Lond.*, **209**, 638–639.
Blair, J. E., and Carr, M. (1963). *J. infect. Dis.*, **93**, 1–13.
Blair, J. E., and Williams, R. E. O. (1961). *Bull. Wld Hlth Org.*, **24**, 771–784.
Blouse, L., Husted, P., McKee, A., and Gonzalez, J. (1964). *Am. J. vet. Res.*, **25**, 1195–1199.
Bynoe, E. T., Elder, R. H., and Comtois, R. D. (1956). *Can. J. Microbiol.*, **2**, 346–358.
Coles, E. H. (1962). Pers. comm. to Davidson (see Davidson, 1966).
Coles, E. H. (1963). *Am. J. vet. Res.*, **24**, 803–807.
Coles, E. H., and Eisenstark, A. (1959a). *Am. J. vet. Res.*, **20**, 835–837.
Coles, E. H., and Eisenstark, A. (1959b). *Am. J. vet. Res.*, **29**, 838–840.
Davidson, I. (1961). *Res. vet. Sci.*, **2**, 396–407.
Davidson, I. (1966). *In* "5 Coloquium über Fragen der Lysotypie" (H. Rische, W. Meyer and K. Ziesché, eds.), pp. 155–162. Wernigerode.
Fisk, R. T. (1942a). *J. infect. Dis.*, **71**, 153–160.
Fisk, R. T. (1942b). *J. infect. Dis.*, **71**, 161–165.
Harry, E. G. (1967). *Res. vet. Sci.*, **8**, 479–489.
Hobbs, B. C. (1948). *J. Hyg., Camb.*, **46**, 222–238.
Hofstad, T. (1964). *Acta. path. microbiol. scand.*, **61**, 558–570.

Jevons, M. P., and Parker, M. T. (1964). *J. clin. Path.*, **17**, 243–250.
Lidwell, O. M. (1959). *Mon. Bull. Minist. Hlth.*, **18**, 49–52.
Live, I., and Nichols, A. C. (1961). *J. infect. Dis.*, **108**, 195–204.
Macdonald, A. (1946). *Mon. Bull. Minist. Hlth.*, **5**, 230–233.
Meyer, W. (1966). *Zentbl. Bakt. ParasitKde.*, *I. Abt. Orig.*, **201**, 465–481.
Meyer, W. (1967). *J. Hyg.*, *Camb.*, **65**, 439–447.
Nakagawa, M. (1960). *Jap. J. vet. Res.*, **8**, 331–342.
Oeding, P., and Williams, R. E. O. (1958). *J. Hyg.*, *Camb.*, **56**, 445–454.
Pagano, J. S., Farrer, S. M., Plotkin, S. A., Brachman, P. S., Fekety, F. R., and Pidcoe, V. (1960). *Science, N.Y.*, **131**, 927–928.
Parker, M. T. (1966). *In* "The Scientific Basis of Medicine Annual Review" (J. P. Ross, ed.), pp. 157–173. University of London, The Athlone Press, London.
Parker, M. T., Tomlinson, A. J. H., and Williams, R. E. O. (1955). *J. Hyg.*, *Camb.*, **53**, 458–473.
Pether, J. V. S. (1968). *J. Hyg.*, *Camb.*, **66**, 605.
Poole, P. M., and Baker, J. R. (1966). *Mon. Bull. Minist. Hlth.*, **25**, 116–123.
Report (1959). *Int. Bull. bact. Nomencl.*, **9**, 115–118.
Report (1963). *Int. Bull. bact. Nomencl.*, **13**, 119–122.
Report (1967). *Int. Bull. bact. Nomencl.*, **17**, 113–125.
Report (1971). *Int. Bull. syst. Bact.*, **21**, 165, 167, 171.
Rountree, P. M. (1959). *J. gen. Microbiol.*, **20**, 620–633.
Rountree, P. M., and Asheshov, E. H. (1961). *J. gen. Microbiol.*, **26**, 111–122.
Rountree, P. M., and Freeman, B. M. (1955). *Med. J. Aust.*, **2**, 157–161.
Ruys, A. C., and Borst, J. (1959). *Antonie van Leeuwenhoek*, **25**, 237–240.
Seto, J. T., and Wilson, J. B. (1958). *Am. J. vet. Res.*, **19**, 241–246.
Smith, H. W. (1948a). *J. Hyg.*, *Camb.*, **46**, 74–81.
Smith, H. W. (1948b). *J. comp. Path. Ther.*, **58**, 179–188.
Smith, H. W. (1954). *J. Path. Bact.*, **67**, 73–80.
Swanstrom, M., and Adams, M. H. (1951). *Proc. Soc. exp. Biol. Med.*, **78**, 372–375.
Tarr, H. A. (1958). *Mon. Bull. Minist. Hlth.*, **17**, 64–72.
Watson, W. A. (1965). *Vet. Rec.*, **77**, 477–480.
Wentworth, B., Romig, W. R., and Dixon, W. J. (1964). *J. infect. Dis.*, **114**, 179–188.
Williams, R. E. O. (1957). *Zentbl. Bakt. ParasitKde.*, *I. Abt. Orig.*, **168**, 528–532.
Williams, R. E. O., and Jevons, M. P. (1961). *Zentbl. Bakt. ParasitKde.*, *I Abt. Orig.*, **181**, 349–358.
Williams, R. E. O., and Rippon, J. E. (1952). *J. Hyg.*, *Camb.*, **50**, 320–353.
Williams, R. E. O., Rippon, J. E., and Dowsett, L. M. (1953). *Lancet*, i, 510–514.
Wilson, G. S., and Atkinson, J. D. (1945). *Lancet*, i, 647–649.

CHAPTER II

Genetic Analysis in Micro-organisms

DAVID A. HOPWOOD

John Innes Institute, Norwich, England

I. INTRODUCTION

Until about 25 years ago genetic analysis of micro-organisms, as I shall use the term in this Chapter, had hardly begun. A wide variety of "higher" organisms had been studied, and very widespread similarities between the various organisms had emerged in regard to most aspects of their genetics, apart from the obvious differences concerning the structure of the organs of reproduction, the behaviour of the gametes, and the position of meiosis in the life cycle.

With the rise of microbial genetics, particularly of bacteria and viruses, a diversity of "genetic" phenomena came to light which appeared to contrast sharply with the comparative uniformity previously encountered in macro-organisms. Such bizarre happenings as infectious sex and *coitus interruptus* in *Escherichia coli*, multiple incestuous rounds of mating in T-even phages, and the transfer of bacterial genes between *Salmonella* cells by viruses appeared to be entirely novel phenomena, whose study in great detail has occupied the attention of microbial geneticists. In many cases the same scientists who found no facet of the sexual biology of *E. coli* unimportant might have scorned a similarly detailed study of the copulating organs of the fruit-fly, yet each is as much or as little relevant to genetic analysis of the organism. Perhaps the difference in emphasis is at least partly due to the fact that, in the case of very small organisms, knowledge of their sexual biology is barely attainable except by a study of its genetic consequences; thus the sexual biology is deduced from the results of genetic experimentation and the whole area appears part of genetics. In contrast, the sexual biology of a macro-organism is, to some extent at least, open to study by morphological and physiological means, and only experiments clearly to do with inheritance are set aside as genetics.

The foregoing remarks are not intended as a serious criticism; in any case it is at least arguable that knowledge of the anatomy and function of the penis of *E. coli*, being on a size scale at which structure and function are intimately related to the properties of molecules, may be more revealing at this stage in the development of biology than similar knowledge for the fruit-fly. My purpose was to introduce the distinction between the procedures and concepts of genetic analysis on the one hand, which are simple and uniform, and the sexual biologies of the various organisms on the other, which are complex and heterogeneous. (In this Chapter I am using the concept of sexual reproduction to include any process leading to the association of genetic material, originally present in separate individuals, in a single individual. This usage, which leads to the consideration of bacteria and viruses, among others, as having a "sexual biology", is broader than that commonly used. No justification is made for it here except one of convenience in emphasizing the similar genetic consequences of interactions between individuals in all these systems.) A genetic experiment must work within the sexual system of the organism concerned (unless it is an experiment of the kind now developing which is independent of genetic interactions between individuals: see page 148); thus two experiments performed with different organisms may appear to have little in common, yet the objective of each, in genetic terms, is identical. To give an example, suppose we wish to ask the same question of *E. coli* and *Drosophila*: do these two mutants of similar appearance differ from the wild-type by mutation in the same gene or in two different genes? In genetic terms, the question is simply: does a diploid cell or multicellular system containing both mutant genes have the phenotype of the mutants or of the wild-type, respectively? However the tricks that have to be used to get the two mutant genes together before the question can be answered are so different in the two organisms that the underlying unity and simplicity of the complementation test for allelism can become obscured.

In this Chapter I shall first describe briefly the basis of the few genetic tests which are the foundation of most experimental genetics, as largely practised in micro-organisms at present. (That is I shall exclude the particular methodology concerned with the genetics of quantitative characters and with population genetics which have hardly begun to be applied to micro-organisms.) These procedures consist essentially of the following:

obtaining a segregation as evidence of simple genetic control of a change in phenotype; mapping the site of the mutation to its general position on the linkage map of the organism ("gross mapping") and to its position relative to other nearby sites ("fine-structure mapping"); testing for dominance of a mutant allele relative to another allele of the same locus, usually the "wild-type" allele; and testing for allelism by complementation studies. Next I shall outline the practical procedures that have been used to apply these tests to some representative micro-organisms with differing sexual biology. In certain cases, such as *Aspergillus nidulans*, *E. coli*, or T4 phage, a wealth of detailed practical knowledge has been gradually built up in the laboratories concerned with the genetics of one or other of these organisms and the best way for a newcomer to acquire this expertise is to visit one of these laboratories for an apprentice period; if this is impossible, a careful study of methodological papers from these laboratories has to suffice. (A very detailed practical manual for key genetic experiments with phages, enteric bacteria and *Aspergillus* is by Clowes and Hayes, 1968.) Space in this Chapter does not permit a detailed account of the practical manipulations of the best-studied experimental systems, and in any case I am not competent to supply it. I am writing in more general terms, primarily for those interested in developing the experimental genetics of a hitherto little-studied organism. There is currently an upsurge of interest in the genetics of microbes outside the few best-studied examples, and I shall try to provide pointers, derived from the successes of the past, which might aid the search for the way-in to a new system. Thus the treatment will be unbalanced in perhaps giving undue emphasis to comparatively preliminary results with less well-known systems.

A prerequisite for genetic analysis is usually the availability of a stock of suitable mutants. The isolation of mutants in micro-organisms has already been discussed in an earlier Chapter in this series (Hopwood, this Series, Vol. 3A).

II. SEGREGATION AS EVIDENCE OF SIMPLE GENETIC CONTROL

A. The genetic approach to biology

Very many genetic analyses have been undertaken in order to elucidate the life-cycle of an organism, the nature and behaviour of chromosomes, the mechanism of crossing-over, the relationship between the structure of a gene and that of its product, or some other facet of heredity itself. However, apart from being of interest for its own sake, genetic analysis also plays an extremely important role as an analytical tool in the elucidation of biological problems which in themselves have nothing to do with heredity. The power of the genetic approach lies in the nature of genes: the product of each gene has a specific role in the cell or organism; and each gene is mutable, with a low probability, to one of a series of alleles each producing an altered product. Thus two organisms differing in respect of a single mutation in one gene of the entire set of genes will in general differ in a single unit of function. Comparison of the two organisms by biochemical, physiological, morphological, or any other means may thus allow the recognition of a single unit process. If we are interested in a particular facet of an organism's function (for example a biosynthetic sequence), or structure (for example that of an organelle), or behaviour (for example its movement), we could hope to isolate organisms with mutations in all the genes "concerned" with the facet. Then, by comparing each mutant with the original organism (the "wild-type") and with other mutants, we could hope to identify each unit process and perhaps the ways in which the unit processes interlock. Although the isolation of mutants is fundamental to any genetic analysis, it is not in itself such an analysis. Why need we go further?

One answer to this question arises from the fact that an organism with an altered phenotype may well differ from the wild-type by more than a single mutation. Thus comparison of the two organisms would not reveal a unit process. However, genetic analysis can tell us that a single gene is involved if a "segregation" is observed when the mutant is crossed with the wild-type, as we shall see in a moment. Genetic analysis can also do much more. For example two mutants with the same phenotype can each be crossed with the wild-type and the mutation can be mapped to a position on the linkage map of the organism. If the two mutations map in unambiguously different positions, then we have identified two genes involved in controlling the function we are interested in. If the two mutations map in a similar position, crossing them together may yield rare wild-type recombinants: the two mutations must then be close but not identical. A complementation test between the two mutations can reveal whether they represent two different, but closely linked, genes or whether they represent changes in the same gene. A test of dominance between the wild-type and mutant alleles can reveal likely ways in which the normal gene operates. And so on. Such procedures will occupy us for the rest of this Chapter.

B. The recognition of segregation

In a eukaryotic microbe with a regular sexual cycle, evidence of single gene involvement in a mutant phenotype is very easily obtained by crossing mutant and wild-type and classifying a random sample of sexual progeny. A single gene difference between the parents is revealed by the finding of only the two parental phenotypes amongst the progeny, and in statistically equal numbers (assuming a haploid life-cycle). A departure from equality may simply reflect a viability disadvantage of one phenotype, usually the mutant or, if the ratio of the two phenotypes falls into a recognizable category, some alternative genetic situation may be revealed. For example the finding of one-quarter mutant and three-quarters wild-type could indicate that the mutant phenotype requires the simultaneous presence of two mutations, at unlinked loci, either of which alone gives a wild-type phenotype. The finding of new phenotypes in the progeny, not shared by the parents, would clearly indicate more than a single gene difference between the parents. If all the progeny had the same phenotype, cytoplasmic, rather than nuclear, determination might be revealed, and so on. In those eukaryotes in which tetrad analysis can be performed, single gene segregations can be particularly easily diagnosed, even in the presence of a fairly severe viability disadvantage (see page 58).

In prokaryotes, the occurrence of new phenotypes in the progeny will also reveal multiple gene differences between the parents. However, the converse, the recognition of single gene control, is not usually so straightforward because the sexual biologies of the organisms do not lead to a segregation of equal numbers of progeny bearing each allele. Acceptable evidence of single gene involvement in these organisms is usually provided only when the mutation is mapped; its segregation can then be followed in relation to those of other gene differences and can be seen to obey the rules of whatever sexual system is involved. Thus mapping itself assumes an analytical role in these organisms.

III. THE BASIS OF GENETIC MAPPING

A. The relationship between recombination frequency and genetic distance in eukaryotes

The concept of recombination of pairs of characters is fundamental to genetic mapping. Recombination can be illustrated by a simple example. Suppose two strains of a fungus differ in two well-defined pairs of contrasting characters; for example one has pigmented cells and grows in the absence of a particular growth factor such as thiamine (that is it is prototrophic), whereas the other strain has colourless cells and requires thiamine for growth (it is auxotrophic). When the two strains mate some, at least, of the sexually produced progeny will resemble one or other parent (that is they are "parental") in being pigmented and prototrophic or colourless and auxotrophic. However there will normally also be a proportion of progeny which show new combinations of the parental characters in being colourless prototrophs or pigmented auxotrophs. These "recombinants" arise as a result of the behaviour of chromosomes at meiosis, and their frequency has a characteristic value for a particular pair of characters, up to 50% of the total progeny.

It is easy to see why 50% of the progeny should be recombinants when the two character differences, pigmented (pig^+) versus colourless cells (pig^-) and prototrophy (thi^+) versus auxotrophy (thi^-) are determined by genes borne on different chromosome pairs. The parent strains carry haploid nuclei, each containing a unique set of chromosomes; suppose that chromosome 1 bears the locus concerned with the pigmentation difference and chromosome 2 that concerned in the auxotrophy for thiamine. A diploid zygote nucleus formed by sexual interaction of the two strains contains both sets of parental chromosomes and is heterozygous for both gene loci: pig^+/pig^-, thi^+/thi^-. The process of meiosis, which leads to the production of the haploid nuclei for the progeny generation, ensures the choosing of haploid sets of chromosomes at random from the diploid zygote nucleus; each must contain a chromosome 1 and a chromosome 2 and thus either pig^+ or pig^- and either thi^+ or thi^-. The combination of these alternative alleles at random therefore results in 50% of the progeny being parental (pig^+ thi^+ or pig^- thi^-) and 50% being recombinants (pig^+ thi^- or pig^- thi^+). Experimentally, the finding of 50% recombination, known as "no linkage", may thus indicate the carriage of the two loci under investigation on different chromosome pairs. However, this need not necessarily be the case, for the following reason.

As we have just seen recombination of loci on separate chromosomes results from the reassortment of whole chromosomes. In contrast, recombination of loci on the same chromosome requires a separate feature of the meiotic process: crossing-over. The relationship between the distance apart of the two loci and the probability of crossing-over between them provides the basis for genetic mapping. This procedure succeeds because the probability of crossing-over in a particular short segment of the chromosome happens to be low. Thus alleles at two loci not very far apart are rarely separated by crossing-over in a particular meiosis. The farther apart the two loci, the greater the probability that crossing-over will occur between them, resulting in recombination. The fact that, no matter how distant the two loci, the frequency of recombinants generated by crossing-over between them cannot normally exceed 50%, the same value as that for loci on separate chromosomes, is a curious biological coincidence. It stems from the fact that, at the time of crossing-over, the two chromosomes involved are double structures, consisting of two sub-units or chromatids; at any point of crossing-over, only one chromatid of each chromosome is involved in the exchange process. Thus a single cross-over results in the production of two recombinant chromatids, leaving two parental chromatids: that is 50% recombination occurs (Fig. 1).

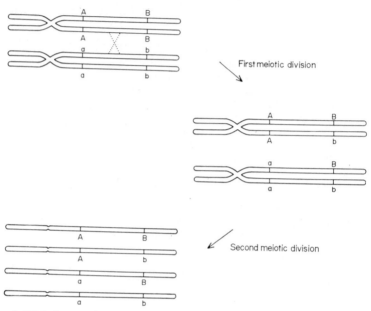

FIG. 1. Meiosis, crossing-over, recombination, and segregation. Two homologous chromosomes each have a centromere and two identified genetic loci (A/a and B/b). A single cross-over, involving one chromatid from each chromosome, occurs in the interval between the two loci. At the end of the first meiotic division, the two centromeres have led the chromosomes to separate nuclei; the alleles at the A/a locus (proximal to the cross-over with respect to the centromere) have segregated from one another at this first division, while those at B/b (distal to the cross-over) have not. At the end of the second-division, when meiosis is complete, the four original chromatids are in four separate nuclei (4 products of meiosis = a tetrad); the alleles at B/b have now segregated. Of the four meiotic products, two are parental in respect of the markers (AB and ab) and two are recombinant (Ab and aB): that is one cross-over leads to 50% recombination.

When we come to consider the occurrence of more than one cross-over in a segment of a chromosome between two loci, the situation clearly involves various combinations of the particular chromatids taking part in different cross-overs. Fig. 2 illustrates the case of a pair of cross-overs and shows that four combinations of chromatids (or "strands") are possible. They lead to various frequencies of recombinant production: 0, 50 or 100%. However, if we assume that the four possibilities occur at random (that is "chromatid interference" is absent) the average result is 50% recombination. The same average result can be shown to hold for three, four or indeed any number of cross-overs, provided always that there is no chromatid interference, an assumption that appears to be not very far

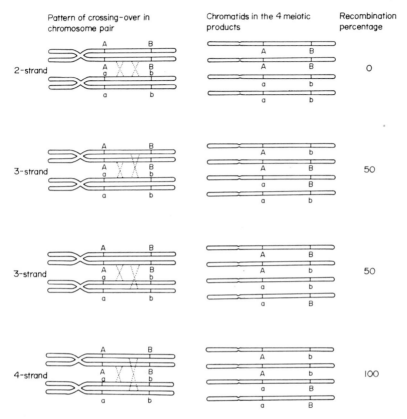

Fig. 2. The effect of various strand arrangements of two cross-overs on the proportion of recombinant products of meiosis. The four possible arrangements (two-strand, two possible three-strand, and four-strand) are illustrated.

from the truth (except in very short intervals: see later), although a slight excess of two-strand at the expense of four-strand exchanges has been reported in the few situations in which the problem has been investigated (Emerson, 1963).

We can now appreciate the relationship between the distance apart of two loci (assuming a uniform probability of crossing-over at all points along the chromosome) and the observed recombination percentage between them (Fig. 3). Leaving for a moment a consideration of the precise shape of the curve, we can recognize three general regions. At short distances (region A), there is near proportionality between distance and recombination percentage since in most meioses there is no cross-over between the loci (0% recombination), in most of the rest there is a single cross-over (50% recombination) with very few cases of multiple crossing-

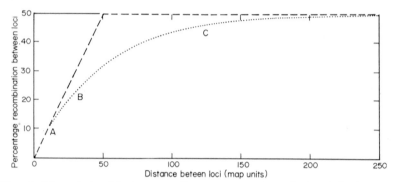

Fɪɢ. 3. The relationship between distance apart of two loci and the frequency of recombination between them. The lower curve shows the relationship assuming no interference (that is using the equation of Haldane, 1919). The upper curve shows the theoretical relationship assuming complete interference (that is no multiple crossing-over). Regions A, B, and C are referred to in the text.

over. At the opposite extreme (region C), there is at least one cross-over between the loci in almost every meiosis, so that increasing distance leads to hardly any increase in the observed recombination percentage. Region B represents a transition between the two extremes. Bearing in mind that recombination percentage is what is observed, and distance between the loci is what is deduced, we see that recombination percentage is a good measure of distance in region A, scarcely any measure in region C, and a poor measure in region B, unless of course the precise form of the curve in Fig. 3 were known or could be deduced.

The trouble is that the form of the curve depends on what assumptions are made regarding any possible influence of the occurrence of one cross-over on the probability of further cross-overs in the same interval, a phenomenon known as interference. Interference is certainly a real phenomenon, but it is known to vary between organisms, and between chromosome regions in the same organism. Thus searches for generalized "mapping functions", that is formulae describing the curve in Fig. 3, have proved largely sterile, in spite of a good deal of statistical ingenuity (Bailey, 1961; Walmsley, 1969). The classical formulation was by Haldane (1919), who assumed interference to be absent and that cross-overs were consequently distributed according to a Poisson function; the resulting relationship between distance and recombination is in fact that drawn in Fig. 3.

For all practical purposes, then, at least in eukaryotic microbes, we can simply make the following statements: a finding of approximately 50% recombination between two loci indicates either that they are on separate chromosomes or far apart on the same chromosome; the finding of a recombination percentage significantly less than 50 indicates location on the same chromosome (that is linkage); recombination percentages of (say) 15 or less are reasonably good measures of distance; whereas values in excess of 15 are underestimates of distance and are increasingly inaccurate as the values increase.

In all the above considerations, we have assumed cross-overs to occur with uniform probability at all points of the chromosome. In the few cases in eukaryotes where this assumption has been tested, it has been found wanting; cross-overs tend to be concentrated in certain regions and sparsely distributed in others. Thus we have to remember that, while recombination percentages are capable of giving a self-consistent, reproducible, and useful picture of the separation of loci on a linkage map, the actual spacing on the chromosome of which the linkage map is an abstraction may well be different.

B. The sequencing of loci on a linkage map: multi-factor crosses

Since, as we have just seen, recombination frequency is a function of genetic distance, even though it is not a linear function, it should clearly be possible to deduce the sequence of three or more loci on a linkage map by determining the recombination frequencies between each pair of loci; the different frequencies may derive from separate experiments, each involving two loci ("two-factor" or "two-point" crosses). For example suppose we measured, in separate experiments, the recombination percentages between loci a and b, b and c, and a and c, as x, y and z respectively and found that z was approximately the sum of x and y, then the order of the loci would be deduced as $a - b - c$. However, the measurement of any recombination percentage is subject to error, for example, due to the differential viability or difficulty of classification of particular genotypes of progeny, or because genetic or environmental factors are influencing the probability of crossing-over differentially in different crosses. A single cross involving three loci ("three-factor" or "three-point" cross) yields considerably more information than three crosses each involving a different pair of the loci, and in particular is much more reliable for the sequencing of loci. An example of such a "three-factor" cross is in Table I.

The results of this cross are idealized in that members of a complementary pair of genotypes have exactly equal frequencies: that is fluctuations due to statistical sampling error, which are bound to occur, and distortions due to a selective disadvantage of particular genotypes, which may occur to varying degrees, are ignored. The most important point to note in these data is that the sequence of the three loci can be deduced qualitatively by identifying the pair of complementary classes having the lowest frequencies, in this case AbC and aBc, since these must derive by crossing-over

D. A. HOPWOOD

TABLE I
Results of a hypothetical three-factor cross

	A	B	C	
Parental genotypes:	1	2		1 and 2 are map intervals
	a	b	c	

Progeny		Recombination
Genotype	Number	in interval
ABC abc	378 378 } 756	None
Abc aBC	42 42 } 84	1
ABc abC	72 72 } 144	2
AbC aBc	8 8 } 16	1 and 2
Total	1000	

Interval	Recombinants/ Total	Recombination percentage	Map distance (cM)†
$A/a - B/b$	84 + 16/1000	10 } 26	11·157 } 30·440‡
$B/b - C/c$	144 + 16/1000	16	19·283
$A/a - C/c$	84 + 144/1000	22·8	30·440‡

† Calculated from the recombination percentage by the formula of Haldane (1919).

‡ Note the additivity of cM units: the map distance calculated from the recombination percentage between A/a and C/c is equal to the sum of the two shorter map distances.

simultaneously in the two intervals. This kind of reasoning, that the progeny requiring most cross-overs have the lowest frequencies, is the basis of all deductions of gene sequence deriving, both in eukaryotes and prokaryotes, from recombination tests (that is excluding certain specialized procedures, not depending on recombination, which will be discussed when we consider mapping in particular prokaryotic organisms).

The data of Table I also provide a clear illustration of the consequences of the relationship between recombination and genetic distance discussed by reference to Fig. 3. Thus the recombination percentage calculated between loci A/a and C/c is 22·8, whereas the value for this distance obtained by summing the recombination percentages between A/a and B/b and between B/b and C/c is 10 + 16 = 26; clearly the latter is a better estimate for the genetic length $A/a - C/c$, although still an underestimate. The progressive underestimation of genetic distances of increasing length by considering recombination percentages is brought out clearly by the calculated values for map distance in Table I, assuming the absence of interference, that is using Haldane's mapping function: map distance $= -\frac{1}{2} \ln (1-2\theta)$, where θ is the recombination fraction. (The units of map distance obtained by applying this formula are known as centimorgans (cM), one centimorgan corresponding to 1% recombination when the effects of multiple cross-overs are allowed for.) Thus 10% recombination converts to only 11·2 cM, whereas 16% is increased to 19·3, and 22·8% to 30·4. The truly *additive* properties of map intervals in cM are also illustrated; the same value for the long interval is obtained by summing the two values for the short intervals or by calculation from the recombination percentage over the long interval. (Note also that interference between recombination in the two intervals in Table I can be seen to be absent, since the combined frequency of the "double recombination" classes, $AbC + aBc$ is 1·6%, exactly the product of the separate probabilities of recombination in the two intervals, 10% and 16%).

It should be emphasized that the brief outline up to this point of the bare essentials of genetic mapping in eukaryotes from recombination values does scant justice to the theoretical treatment of the subject that has taken place (Mather, 1951; Bailey, 1961), particularly of statistical procedures and experimental designs to overcome the errors introduced by differential recovery of particular genotypes of progeny. A basic concept in such an approach is the use of *sets* of crosses in which the same group of characters is studied; members of the set of crosses differ in the arrangement of the markers at each locus in the parents. For example a two-factor cross could involve parents AB and ab or alternatively Ab and aB; the crosses are said to differ in the coupling of the alleles, the former being described as a *coupling* or *cis* cross and the latter as *repulsion* or *trans*. The significance of using both kinds of cross is that a given genotype, say ab, is a member of a *parental* class in the first cross and of a *recombinant* class in the second, so that any factor, like reduced viability, which depresses the frequency of this class, will cause an overestimate or an underestimate respectively of the recombination frequency in the two crosses. A suitable statistical treatment will allow the calculation of an unbiased estimate of the recombination frequency from the combined data.

In most mapping studies with microbes, such precautions have not been taken very far, but this has not usually hampered achievement of the objective of the investigation. However, it is not hard to find examples where conclusions from a cross with one coupling arrangement of alleles could usefully have been checked by a reverse cross.

C. Linkage groups

We have seen how linkage between pairs of loci can be detected and measured and how the sequence of a set of linked genes can be determined. Such a set is called a *linkage group*. When the genetic analysis of an organism has reached a certain stage, the number of linkage groups is the same as the number of chromosomes in the haploid set for that organism. However, in an earlier stage of the investigation the number of linkage groups will usually at first fall short of the haploid chromosome number (when fewer genes than chromosomes have been studied) and may later *exceed* the haploid chromosome complement, before settling down to the final value. A temporary excess of linkage groups over chromosomes comes about because of the fact that most chromosomes are more than 50 map units long, and may be several hundreds of units. Thus several genes or groups of genes on the same chromosome may behave as separate linkage groups until other loci are detected which span the segments between them by breaking them into intervals of less than 50 units. Very many genes may have to be mapped before all the chromosomes of, say, a fungus are marked off into intervals of manageable proportions; this has not yet happened in either *Aspergillus nidulans* or *Saccharomyces cerevisiae*, in spite of the study of large numbers of mutations. This explains the usefulness of mitotic haploidization (see page 53) which allows the unambiguous assignment of genes to linkage groups, each representing a separate chromosome.

D. Genetic fine-structure

One of the success stories of microbial genetics was the deduction that a single gene, like the whole chromosome on a grosser scale, is a linear structure, containing many sites of mutation separable by recombination. Early studies in *Drosophila* foreshadowing this finding (Oliver, 1940) led to the demonstration by Pontecorvo and his colleagues (Roper, 1950; Pontecorvo, 1952; Pritchard, 1955) that a linear linkage map could be drawn to express the distribution of mutant sites within a single gene of *Aspergillus nidulans*. This finding was soon followed by the unambiguous demonstration by Benzer (1955) of a linear array of 8 sites, later increased to 308 (Benzer, 1961) in the r_{II} gene of phage T4, leading to the postulated correspondence between a specific linear DNA molecule and the linkage map of a gene, and eventually to demonstrations of the co-linearity of the linkage map and the polypeptide gene product (Sarabhai, Stretton, Brenner and Bolle, 1964; Yanofsky *et al.*, 1964).

Intragenic mapping depends on the same theoretical framework as chromosome mapping on a grosser scale; however the practical methodology is usually different since it has to overcome the problem of estimating the frequencies of very rare classes of recombinant progeny. In the phrase

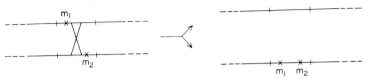

Fig. 4. A "normal" (reciprocal) cross-over between two mutants in the same gene would be expected to yield a double mutant recombinant as well as a wild-type recombinant. The boundaries of the gene are indicated. Only the two chromatids taking part in the cross-over are drawn, as single lines.

of Pontecorvo (1959) the "resolving power" of the genetic analysis has to be increased by techniques leading to the recognition of rare recombinants amongst large numbers of progeny of the parental classes. These techniques usually rely on the *selection* of particular classes of recombinants by their ability to grow on a medium on which the parentals cannot. For example, in a cross of two strains having mutations inactivating a particular gene responsible for a biosynthetic enzyme, and therefore being auxotrophic for the end-product of the biosynthetic pathway, rare prototrophic recombinants can be selected by inoculating large numbers of progeny to a medium lacking the growth factor in question. The total progeny population can be estimated by parallel platings, at considerably higher dilution, on a supplemented medium, and so the frequency of prototrophic recombinants can be calculated. This value would be expected to represent half the total recombinants, since an equal number of double mutant auxotrophs should be present (Fig. 4); however this is by no means always found to be true when put to the test by tetrad analysis or mitotic analysis (see later). By determining such prototroph frequencies for various pairs of mutations, a linkage map representing the relative spacing of sites within the gene can be built up (Fig. 5). Even more than in the case of mapping on a grosser scale, discrepancies between recombination percentages involving various pairs of mutant sites can arise (Fig. 5), and specific tests of sequence, involving further genetic "markers", are generally employed.

A common design of experiment to sequence sites within a gene uses "outside markers" on one or both sides of the gene. The principle involved is illustrated in Fig. 6. It depends on distinguishing the classes of progeny, with respect to the outside markers, attributable to various patterns of crossing-over in the outside regions, and choosing the sequence that gives the lowest frequencies to the multiple cross-over classes. In practice, the results of such tests are often not as unambiguous as could be expected, owing to the occurrence of "negative interference". This term describes the incidence, with a much higher frequency than in a random sample of progeny, of crossing-over in a chromosome region very close to a second

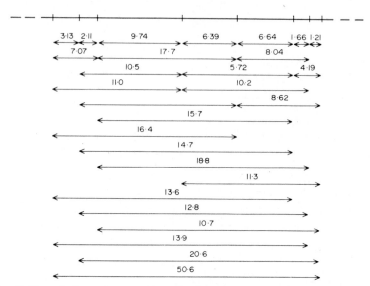

Fig. 5. Genetic fine-structure. Part of a fine-structure map of 8 mutations in the *ad*-7 gene of *Schizosaccharomyces pombe* (from Leupold, 1961). Each figure represents the frequency of selected ad^+ recombinants per million sexual progeny. Note that the order of mutant sites adopted leads to good additivity of recombination frequencies for most intervals, but with some exceptions, where recombination in a "longer" interval may actually be less frequent than in a constituent "shorter" interval.

region in which a cross-over has occurred. For example in Fig.6, the regions between *a* and m_1, and between m_2 and *b*, as measured in an unselected sample of progeny, are short; that is the proportion of progeny showing recombination in one or other of these regions is small, and the proportion showing recombination in both together is *very* small. However, it is often observed that, amongst the selected sample of progeny arising by crossing-over in the very short interval between the sites of m_1 and m_2, the proportion showing recombination in one or both of the outside regions is very much larger than in a random sample of all progeny; this effect may be so pronounced that the four classes of progeny in respect of the outside markers (ab^+, a^+b, a^+b^+, ab) have nearly equal frequencies. Under these conditions of "high negative interference", the sequencing of the sites m_1 and m_2 obviously breaks down. (As well as a general increase in crossing-over detected in short intervals, chromatid interference may also occur, i.e. there is an excess of involvement of particular chromatids in successive cross-overs, usually of exchanges involving the same two strands Whitehouse, 1963).)

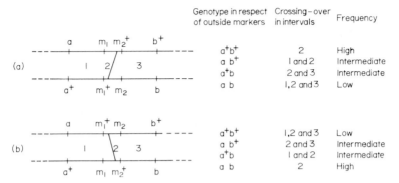

FIG. 6. Sequencing of mutations by the use of outside markers. Wild-type recombinants between two mutations in the same gene (m_1 and m_2) are selected and classified with respect to outside markers, a and b. The sequence shown is compatible with the observed frequencies of outside marker classes. Note that two crosses, with reversed coupling of outside markers, confirm the sequence.

As we saw earlier there is not always an accurate correspondence, at a gross level, between physical distance on a chromosome and the spacing of loci on a linkage map. The same appears even more true in the case of fine-structure mapping. The rules and irregularities of such effects must depend on the molecular details of the process of crossing-over, and their dependence on particular base sequences in the DNA, a field which has become highly specialized in recent years, without a clear picture of the mechanism of crossing-over having yet emerged (Whitehouse, 1963; Holliday, 1964a, b; Meselson, 1967b; Boon and Zinder, 1969; Holliday and Whitehouse, 1970; Whitehouse, 1970).

To summarize, it is possible to obtain very detailed maps showing the order of sites within a gene, provided enough checks are made for internal consistencies of the data. However, there is certainly no guarantee that the spacing of sites on the linkage map is a faithful representation of the distance between the sites in terms of nucleotide-pairs in the DNA; on the contrary, it can be anticipated that this will rarely be the case and that recombination frequencies derived from two-factor crosses may even bear little relationship to distance (Norkin, 1970).

E. Deletion mapping

One procedure for sequencing mutant sites over short intervals depends on the use of deletions—mutations which result from the loss of a length of genetic material, rather than the alteration of a single base-pair ("point mutation"). Such deletions may extend over parts of more than one adjacent gene, but are unlikely to cover a stretch of many genes since a long sequence of genes not containing one whose function is essential under all conditions (an "indispensible" gene) is unlikely; thus long deletions are usually lethal (in the organisms that we are considering, which spend a considerable part of the life-cycle in the haploid state). A first indication that a mutant carries a deletion is given when it never reverts back to the non-mutant condition, since the lost genetic material in a deletion cannot be regained, whereas a point mutation can be reversed.

Mapping depends on the fact that a set of deletions for a particular chromosomal region will include deletions overlapping in various combinations, and on the fact that two overlapping deletions cannot recombine to produce a non-mutant recombinant, whereas non-overlapping deletions can. For example, consider a set of deletions A, B, C. If A fails to give recombinants with B, and B fails to recombine with C, while A recombines with C, the order

is indicated. Having identified a set of overlapping deletions covering a region of the chromosome under investigation, they can be used in the sequencing of point mutations by a *qualitative* test. Each point mutation is crossed with the set of deletions. If a particular mutation gives recombinants with A and C but not B, then it lies in the region of B between its overlaps with A and C. If, on the other hand it gives recombinants with C but not with A or B, then it must lie in the overlap region of A and B, and so on. In this way a set of mutations can be assigned to particular segments of the map (five segments are defined by the three deletions under consideration) and more detailed mapping can proceed by pairwise crosses between mutants falling in the same segment, rather than with all the mutants. This method of mapping was made famous by Benzer (1961) in his classical study of the r_{II} gene of T4 phage, but less exhaustive studies have been made by the same technique, or modifications of it, in many other systems. (Although the method is qualitative in the sense that estimates of recombination *frequencies* are unnecessary, the limits of accuracy of the method obviously depend on the sensitivity of the test for recombination: failure to detect recombinants amongst a comparatively small sample of progeny is not proof of overlap of two deletions.)

IV. FUNCTIONAL GENETIC TESTS

A. Dominance tests

The concept of dominance is as old as modern genetics, having been clearly enunciated by Mendel; one character of a pair was said to be dominant to the other (the recessive character), when the first character was manifested by a hybrid. Later it was realized that the finding of dominance had analytical value in probing the action of genes. In very general terms, an allele producing an active product will normally be dominant to an allele producing no product or an inactive one, provided that gene dosage is not so critical that one copy of the active allele is insufficient for full function in a diploid cell. Thus the corresponding wild-type allele is usually dominant to an auxotrophic mutation resulting in an inactive biosynthetic enzyme. By the same token, two alleles, both producing active products that can be separately recognized—for example two electrophoretically distinct forms of a protein—will not show dominance.

The analytical power of dominance tests has been particularly clearly shown in studies of genetic regulation in micro-organisms. Thus the finding of certain dominant mutations affecting regulation in the *lac* operon of *E. coli* provided an important piece of evidence for the theory of negative regulation of the system (Jacob and Monod, 1961). Conversely, the dominance relationships of regulator gene mutations in the L-arabinose system of *E. coli* indicated positive control (activation) in that situation (Sheppard and Englesberg, 1966).

The principle of a dominance test is extremely simple: the two alleles whose relative dominance is to be tested are introduced into the same cell, and the phenotype of the cell in respect of the relevant characters is then determined. However, the way in which the operations are performed will obviously depend on the particular sexual biology of the organism, and the genetic constitution of the heterozygote can also be very different. In eukaryotic microbes we may be dealing with diploid nuclei, or with two genetically different kinds of haploid nuclei in a common cytoplasm; in the latter case we may have a variable ratio of the two kinds of nucleus, as in the heterokaryons of *Aspergillus nidulans* (page 65) or an exactly equal ratio, as in the dikaryons of *Coprinus lagopus* (page 73).

In viruses and prokaryotes, geneticists have made use of an even wider variety of systems leading to heterozygosity; for example mixed infection of a host cell by viruses carrying the two alleles whose dominance is to be tested (page 133); or in bacterial genetics, various situations leading to a partially diploid state, ranging from the rather stable F′ strains of *E. coli* (page 106), through the unstable heteroclones of *Streptomyces coelicolor* (page 124), to the highly unstable abortive transductants of *Salmonella typhimurium* infected by phage P22 (page 90), in which only one cell in the colony is a partial diploid. Some details of these systems for testing dominance will be described in later Sections of this Chapter.

B. Complementation tests

The clear interpretation of genetic complementation (Fincham, 1966) is much more recent than the understanding of dominance. The basis of a complementation test, like that of a dominance test, is to bring two copies of a particular stretch of genetic material together and to determine the phenotype of the resulting heterozygote. However, whereas in a dominance test we are normally dealing with a single mutant allele and the corresponding wild-type allele, in a complementation test we are looking for interaction between two mutations and deducing from the outcome of the test whether they are alleles of the same locus or of different loci. The power of such a test in defining the limits of a gene as a functional unit was shown by Benzer (1955), who coined the term "cistron" for the unit of function so defined. A new term was needed at the time, since the word "gene" had become ambiguous; however, in more recent usage, gene and cistron are essentially synonymous.

The basis of the complementation test is outlined in Fig. 7. It should be noted that, although the conclusion of allelism theoretically derives from the finding of a difference in phenotype when the two mutations are in the *cis* and *trans* configurations—wild-type in *cis*, mutant in *trans*—the test in the *cis* configuration, which is in the nature of a control, is rarely performed owing to the difficulty, in most systems, of preparing the double mutant strand. Note also that the interpretation of a complementation test requires prior information on the dominance relationships of the mutations concerned with respect to wild-type. The case illustrated is the simplest, in which each of the two mutations is recessive to the wild-type allele. If each mutation is dominant, and the mutations lead to identical phenotypes, a complementation test is not possible. However a difference in phenotype may allow an informative complementation test to be performed (e.g., Chater, 1970).

The finding of complementation between two mutations is not always proof of two separate loci, owing to the possible involvement of intragenic (inter-allelic) complementation. The features and interpretation of this phenomenon have been fully discussed (Fincham, 1966; Gillie, 1966). The chief criteria for the diagnosis of complementation as intragenic are, firstly, that the complementation is usually sub-optimal, the *trans* heterozygote not having a fully wild-type phenotype; and secondly that, although two mutations may complement one another, a third mutation (which is not a deletion) can normally be found that fails to complement each of the other two.

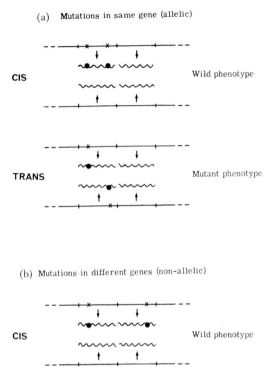

FIG. 7. The complementation test between recessive mutations. Two adjacent genes are indicated, in diploid condition. Two mutations, represented by ×, are shown either in the same gene (a) or in different genes (b), and alternatively in the *cis* or *trans* configuration. The products of each gene are shown, bearing alterations (●) at positions corresponding to the mutations. Note that, in all cases except that of two mutations in the same gene in the *trans* configuration, at least one unaltered product of each gene exists, giving a non-mutant (wild-type) phenotype. The finding of a mutant phenotype in the *trans* configuration leads to the conclusion of allelism.

Conversely, the finding of no complementation between a pair of mutations is not absolute proof that they are alleles of the same locus, because of the possibility that one is a *polar* mutation having an effect, not only on the function of its own cistron, but also of cistrons distal to it in the

4

direction of translation of the genetic message. Nonsense and frame-shift mutations can have this property, when the group of cistrons constitute an operon producing a polycistronic messenger RNA. When a fine-structure recombination study is combined with the complementation analysis, polar mutations can be recognized for what they are by being occasional non-complementing mutants *within* a sequence of other mutants all of which complement mutants in the adjacent cistron(s).

The practical basis of a complementation test is essentially the same as that of the dominance test discussed in the previous Section; again the details for particular organisms will be described later in this Chapter.

C. Epistasis

Epistasis is for non-allelic mutations what dominance is for alleles of the same locus. Like dominance it has considerable implications as an analytical tool. For example, suppose we have two auxotrophic mutants, x and y, defective in different genes involved in the biosynthesis of the same end-product. The mutants can be differentiated phenotypically, in that x accumulates substance X and y accumulates substance Y; X and Y may both be precursors of the end-product, or one or both may represent conversion products derived from precursors on the main pathway by side pathways. What is the sequence of action, in the wild-type, of the enzymes defective in x and y? One way to answer this question is to prepare a double mutant, combining the defects of x and y, and to see which compound it accumulates. If X is found, and not Y, we can assume that the enzyme defective in x operates before that defective in y, in the same pathway. The mutation in x is said to be epistatic to that in y.

In the study of the usual type of biosynthetic pathway, this kind of genetic test may not be very valuable, since more direct chemical methods may well be available. However, where the underlying chemistry is difficult or poorly developed, a study of epistatis may be revealing; it can indicate by purely genetic tests that mutations in two genes do in fact affect the same pathway. An example is the study of radiation-sensitive mutants of *E. coli* (Howard-Flanders, Boyce and Theriot, 1966) and *Streptomyces coelicolor* (Harold and Hopwood, 1970a).

V. GENETIC ANALYSIS IN EUKARYOTIC MICROBES

A. General considerations

Genetic analysis in eukaryotic microbes follows the classical patterns worked out for higher organisms much more closely than do the special procedures devised for bacteria which will be discussed in a later Section of this Chapter. One obvious factor of difference from higher organisms concerns the stages in the life cycle at which nuclear fusion and meiosis occur, which determine the relative lengths of time spent in the haploid and diploid states. In most eukaryotic microbes (some yeasts are exceptions), fusion of pairs of haploid nuclei to give diploid nuclei occurs immediately before meiosis, so that the diploid stage is very much circumscribed, although this generalization may occasionally be upset to give situations that can be exploited experimentally, as we shall see shortly in considering the parasexaul cycle. Another special feature of genetic analysis in many eukaryotic microbes is the technique of tetrad analysis: the genetic characterization of the four products of a particular meiosis. A third feature that characterizes the genetic manipulation of eukaryotic microbes is the variability of the entities harnessed for the study of complementation and dominance: heterokaryons, dikaryons, diploids or disomics.

1. *The parasexual cycle and its use in genetic analysis*

(a) *Nature of the parasexual cycle.* This is the name given by Pontecorvo and his colleagues (Pontecorvo *et al.*, 1953) to the sequence of events discovered in *Aspergillus nidulans*, and later in several other microbes, that leads to the same end result as sexual reproduction—that is the reassortment of genes both linked and unlinked—without the involvement of specialized organs of sexual reproduction or of a regular alternation of nuclear fusion and meiosis.

The mycelium of *A. nidulans* is septate and each compartment contains many nuclei. In a normal strain of the fungus, these nuclei are almost all haploid, and genetically identical. However, if two strains differing in one or more genetically determined characters grow in close contact, the hyphal fusions which are constantly occurring between hyphae that touch one another result in mycelia containing a mixture of genetically diverse nuclei: that is heterokaryons. The next step in the parasexual cycle is the occasional chance fusion of two haploid nuclei to give a diploid nucleus, which, if the haploid nuclei were non-identical, will be heterozygous for those genes differentiating the parent haploid strains. Such diploid nuclei will divide to give clones of diploid descendants. By various visual and selective techniques, cells carrying diploid nuclei can be recognized and experimentally isolated to give rise to pure diploid, heterozygous cultures.

The propagation of such diploid strains gives rise to sufficient numbers of diploid nuclei for the chance occurrence of the remaining two components of the parasexual cycle: mitotic crossing-over and haploidization. The former must occur if recombination of linked genes is to result; haploidization alone leads to the reassortment of whole chromosomes and so of unlinked genes only. As such it is an extremely useful device for the definition of linkage groups since even distant genes show complete linkage

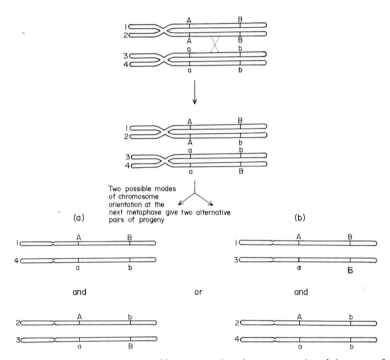

Fig. 8. Mitotic crossing-over and homozygosity. A cross-over involving two of the four chromatids of a homologous pair of chromosomes leads to a rearrangement of alleles between the two chromosomes. At the next mitotic division, the chromosomes can orientate themselves in two possible ways, leading to two possible modes of chromatid separation: (a) 1 and 4 separating from 2 and 3, or (b) 1 and 3 separating from 2 and 4. One of these two possible modes, (b), leads to homozygosity of the products of mitosis for alleles at any locus (B/b) distal to the position of mitotic crossing-over.

during haploidization. The process of haploidization in the parasexual cycle appears to be an imprecise one and to reflect accidents of mitosis leading to the production of nuclei with chromosome complements ranging between the diploid and haploid numbers. Those nuclei with unbalanced chromosomal constitutions are at a severe disadvantage compared with the balanced haploid nuclei, which therefore predominate amongst the segregants. The frequency of haploidization is greatly increased by treatment with agents such as p-fluorophenylalanine, which probably act by interfering with the mitotic spindle.

Mitotic crossing-over has similar characteristics to crossing-over at meiosis, in involving one chromatid only of the two present in each participating chromosome, but the frequency of such exchanges is low. Thus more than one simultaneous cross-over in a given chromosome pair is very unlikely (except in short intervals: see later). Mitotic crossing-over, as the name implies, is not associated with reduction in the chromosome number, so that the result of a mitosis in which a cross-over has occurred is a pair of diploid nuclei with some rearrangement of alleles between homologous chromosomes. The genetic constitution of the daughter nuclei depends on the relative orientation at the mitotic metaphase of the two homologous chromosomes that were involved in crossing-over. There are two possibilities (Fig. 8). One of them (Fig. 8(a)) results in no loss of heterozygosity of the daughter nuclei, only reassortment of markers between homologous chromosomes; the other (Fig. 8b) leads to homozygosity for any markers that were previously heterozygous, provided that they lie between the point of crossing-over and the end of the chromosome in a distal direction (that is away from the centromere).

To complete the parasexual cycle, haploidization suffered by any of the descendant nuclei after mitotic crossing-over gives rise to haploids with new combinations of linked markers.

(b) *Occurrence of the parasexual cycle.* The parasexual cycle was first characterized in *A. nidulans*, a fungus with a normal sexual cycle which could be used to confirm the validity of conclusions of genetic analysis derived from the parasexual cycle. From this species it was extended to related ascomycetes with no sexual cycle, such as *Aspergillus niger* (Pontecorvo, Roper and Forbes, 1953) and *Penicillium chrysogenum* (Pontecorvo and Sermonti, 1954) and to other fungi, both filamentous and yeast-like (see later).

(c) *Genetic analysis by the parasexual cycle.* The parasexual cycle is of prime importance in the genetic analysis of imperfect fungi (fungi lacking a regular sexual cycle), since it provides a means of gene mapping, both long-range and fine. In organisms having a regular sexual cycle, parts of the parasexual cycle are invaluable in providing systems for carrying out tests of dominance and complementation (using the heterokaryons and diploids). Mitotic crossing-over provides a means of determining centromere position on a linkage map, while haploidization provides a method for the definition of linkage groups.

i. *The definition of linkage groups by haploidization.* In the recognition of genes borne on the same chromosome, advantage is taken of the consequences of mitotic haploidization for the recombination of linked and unlinked genes. Suppose we use two haploid strains of genotype *ABC* and *abc* to produce a heterozygous diploid strain; let the loci *A/a* and *B/b* be borne on the same chromosome and *C/c* on a different chromosome.

When haploidization of this diploid occurs, and assuming, as will usually be the case, that mitotic crossing-over has not resulted in a rearrangement of alleles in the diploid before haploidization, then only four classes of haploid will arise, and in approximately equal numbers: *ABC*, *ABc*, *abC* and *abc*. That is, linked genes show no reassortment, whereas unlinked genes show 50% recombination. Thus genes can readily be recognized as being on the same chromosome, even if they are near the ends of a long chromosome and consequently would show very nearly 50% recombination in an ordinary meiotic analysis. (There are in fact several genes in *A. nidulans* which have been assigned to one of the eight linkage groups by mitotic haploidization, yet they show no linkage with any of the other markers of that linkage group in a sexual analysis, and this in spite of the fact that the linkage map of this fungus is comparatively well marked (Dorn, 1967); evidently at least some of the chromosomes are very long in terms of the probability of crossing-over.)

ii. *Centromere mapping and gross mapping by mitotic crossing-over.* The consequences of mitotic crossing-over on the homozygosity of markers previously heterozygous have been described (Fig. 8b). The important point is that the centromere provides a barrier to homozygosity since all daughter nuclei must inevitably contain a descendant of *both* centromeres present in the nucleus of the previous cell-generation: the centromere must remain "heterozygous". Since crossing-over between the centromere and a particular locus gives the opportunity for markers at that locus to become homozygous, the frequency of homozygosity for a locus will be directly related (assuming no multiple crossing-over) to its distance from the centromere in terms of the probability of crossing-over. Thus, in principle, the frequency of homozygosity of each of a series of linked loci could be measured and the centromere then inserted on the linkage map in the position fitting best its distance from each locus.

In practice this method would be tedious because of the difficulty of measuring accurately the frequency of homozygosity of each locus. This is due partly to the low frequency of mitotic crossing-over, which means that mitotic segregants have to be identified selectively or by visual examination, and partly to the fact that each mitotic segregant found in a diploid colony does not represent an independent cross-over event; instead crossing-over gives rise to *clones* of segregants, of varying size depending on the timing of segregation. Thus in practice the map interval containing the centromere is found by a qualitative consideration of the classes of homozygous segregants arising in a multi-factor cross.

Let us consider the example in Table II where we have five linked heterozygous loci in the parent diploid prepared from two haploid strains of genotypes *ABCDE* and *abcde*. Suppose for the sake of argument that markers

TABLE II

A hypothetical example of mitotic recombination analysis showing the effect of selecting segregants homozygous for an interstitial (*b*) or terminal (*e*) recessive marker

	A	B		C	D	E
Parental marker arrangement:	1	2	3	4	5	
	a	*b*	*c*	*d*	*e*	

Phenotypes of progeny selected for homozygosity of

	b			*e*	
Type	% Frequency	Cross-over in interval	Type	% Frequency	Cross-over in interval
abCDE	100	2	*ABcde*	30	3
			ABCde	50	4
			ABCDe	20	5

at two of the loci are selectable when homozygous, say *b* and *e*. *b* might be a recessive mutation conferring resistance to a particular growth inhibitor such as acriflavine; thus the heterozygote, *B/b*, will be sensitive, but mitotic segregants homozygous for *b* will be resistant and selectable by plating the diploid on a medium containing acriflavine. Perhaps *e* is a recessive mutation causing altered conidial colouration, say yellow instead of the wild-type green; thus clones of mitotic segregants homozygous for *e* can be recognized visually as yellow spots on the green diploid colony.

The experiment would consist of isolating a sample of mitotic segregants by each of the two methods, selecting *b/b* and *e/e* respectively, with all the other markers non-selected. The selected segregants would then be classified in respect of the non-selected markers. Suppose the results in Table II were obtained. The result of selecting segregants homozygous for *b*, namely that all segregants have become homozygous for *a*, but none are homozygous for *c*, *d* or *e* indicates that the locus of *a* must be distal to that of *b* while the other three loci are either very close to the centromere or on the opposite side of it from *b* and *a*. The selection of homozygotes for *e* indicates that *e* is a terminal marker, since no other marker becomes homozygous whenever *e* is homozygous. The pattern of segregation at the *c* and *d* loci, namely that *d* can be homozygous without *c*, but the converse does not occur, indicates the order: centromere–*c*–*d*–*e* for these markers. Combining this conclusion with the result of the *b* selection indicates the overall sequence: *a*–*b*–centromere–*c*–*d*–*e*. The relative frequencies of the classes homozygous for *c*, *d*, *e*; *d*, *e*; and *e* alone, namely 30, 50 and 20%, reflect the relative probabilities of mitotic crossing-over in map intervals 3, 4 and 5 respectively. This conclusion brings out a practical feature of mitotic segregant selection for gross mapping: selection of a terminal marker such as *e* allows mapping of a whole chromosome arm, whereas selection of an interstitial marker such as *b* allows no estimate of crossing-over distal to this locus to be made. Thus terminal "selectors" are the most useful (Pontecorvo and Käfer, 1958).

Further discussion of the methodology of mitotic crossing-over analysis in *A. nidulans* is on page 64.

iii. *Fine-structure analysis by mitotic crossing-over.* Some fine-structure studies have been made in *A. nidulans* by mitotic crossing-over (see page 65) but the purpose has in general been the study of the nature of crossing-over itself rather than the mapping of an extensive series of sites within a gene. An advantage of mitotic analysis here is the possibility it offers of recovering both strands involved in a cross-over, allowing a test of the production of reciprocal recombinant genotypes. In yeast, on the other hand, the finding of X-ray induced mitotic crossing-over has provided a good method of intragenic mapping (see page 71) which has proved particularly useful in the analysis of the sequence of amino-acid substitutions in the cytochrome c gene (Parker and Sherman, 1969).

iv. *Dominance and complementation studies.* In most fungi, fusion of haploid nuclei to give diploids in the sexual cycle immediately precedes meiosis, so that heterozygosity is not sufficiently prolonged for functional genetic tests to be performed. It is therefore fortunate that the parasexual cycle lends itself to the problem. In filamentous fungi such as *A. nidulans*, complementation and dominance may be studied in two ways: using the heterokaryons and diploid strains. Heterokaryons have the advantage of simplicity, whereas the preparation of a diploid strain heterozygous for the appropriate markers is more laborious. The disadvantage of heterokaryons in this species is their irregular nuclear constitution, such that one genotype of haploid nucleus may greatly outnumber the other in many parts of the mycelium. Thus a negative result of a complementation test can reflect the absence of a suitably balanced heterokaryon. Use of a heterozygous diploid strain overcomes this difficulty, as well as the possibility that two mutants may fail to complement one another in a heterokaryon, not because they are allelic, but because their products are confined to the nucleus. This situation, while a theoretical possibility, has been elusive to prove with certainty (Pontecorvo, 1963).

The possibility of using heterokaryons in the study of complementation and dominance is not confined to fungi with a complete parasexual cycle. Thus in *Neurospora crassa*, heterokaryons are routinely used for this purpose, the only other possible situation being provided by the so-called "pseudowild" strains (see page 69), which are not complete diploids but aneuploid strains with a single chromosome present as an extra copy; that is in disomic condition.

In basidiomycetes, such as *Coprinus lagopus*, a regular dikaryotic stage precedes sexual reproduction, and this stage is harnessed as a means of carrying out functional genetic tests. Since regular dikaryotization occurs spontaneously between strains of different mating-type, there is no need to select for establishment and maintenance of the heterozygous condition by the use of "forcing markers" as in the case of the heterokaryons of ascomycetes.

2. *Tetrad analysis*

Tetrad analysis amongst eukaryotic microbes is particularly associated with fungi, but has also been usefully applied in *Chlamydomonas*. Amongst fungi, filamentous ascomycetes (*Neurospora, Aspergillus, Bombardia, Ascobolus, Sordaria*), yeast-like ascomycetes (*Saccharomyces, Schizosaccharomyces*), heterobasidiomycetes (*Ustilago*), and homobasidiomycetes (*Coprinus, Schizophyllum*) have been studied.

Tetrad analysis provides particularly clear evidence for the involvement of a single pair of alleles in the determination of a character difference, as well as a technique for mapping the centromere. In addition, tetrad analysis yields information on the process of crossing-over, particularly on the involvement of chromatids in single and multiple cross-overs (chromatid interference; Emerson, 1963) and the nature of crossing-over itself (e.g., Holliday, 1964; Whitehouse, 1963; Stahl, 1969; Holliday and Whitehouse, 1970); however a discussion of these phenomena is outside the scope of this Chapter.

(a) *Segregation of alleles.* An advantage of tetrads over random meiotic products in the recognition of single gene segregations lies in the *exact* equality, in a tetrad, of meiotic products bearing each member of a pair of alleles, compared with their *statistical* equality amongst random products. This feature is advantageous particularly when the phenotypic difference involved is slight, or subject to strong environmental modification, as in the case of the colonial mutants of *Venturia inaequalis* studied by Keitt and Longford (1941), or when one genotype is characterized by severe inviability or complete lethality. The latter is revealed, in a tetrad, by inviability, or

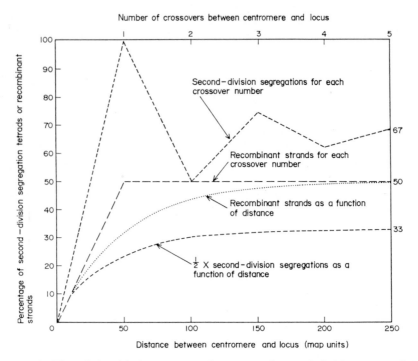

Fig. 9. The relationship between crossing-over and second-division segregation frequency and between distance and second-division segregation frequency. Note that, with increasing numbers of cross-overs in the interval between a locus and the centromere, the value of the second-division segregation frequency shows a damped oscillation around 67%. This means that the curve relating distance of the locus from the centromere with half the second-division segregation frequency rises asymptotically to 33%. This curve follows that relating distance and recombination frequency (reproduced from Fig. 3) over its early part, but soon falls below the latter curve, which rises asymptotically to 50%.

even a failure to develop, on the part of half the members of each tetrad. Thus the normal 2 : 2 ratio is replaced by a 2 : 0 ratio (4 : 0 instead of 4 : 4 in those ascomycetes in which a mitotic division normally follows meiosis in the tetrad). A departure, *by many or all tetrads in a cross*, from the expected 2 : 2 ratio would be indicative of a situation more complex than segregation of a single pair of alleles (see Emerson, 1963, p. 169). On the other hand, *occasional* "aberrant" tetrads showing 3 : 1 or 4 : 0 tetrads (or 5 : 3 ratios in 8-spored asci) do not contradict the simple hypothesis, but rather reflect the "gene conversion" characteristic of the crossing-over process (Holliday, 1964; Whitehouse, 1963).

(b) *Centromere mapping.* The ability of tetrads to yield information on centromere position derives from the influence of crossing-over on the timing of segregation of a pair of alleles: that is, whether this occurs at the first or second meiotic division. Recalling the behaviour of centromeres in a normal meiosis, we remember that the centromeres of paired homologous chromosomes separate from one another during the first division; at the second division, each centromere "splits" and carries a chromatid to either pole. The consequences of this behaviour are that, in the absence of crossing-over between a particular locus and the centromere, alleles at that locus are separated from one another, along with the centromeres of homologous chromosomes, at the first division; in contrast, crossing-over between the locus and the centromere leads to second-division segregation (Fig. 1).

Leaving aside for a moment how second-division segregation may be experimentally recognized, we see that the proportion of tetrads in which a pair of alleles segregates at the second division is related to the probability of crossing-over between the locus and the centromere. This relationship (Fig. 9) is somewhat more complex than that between crossing-over and recombination of alleles at two linked loci (page 37). As we have seen, no crossing-over between centromere and locus leads to no second-division segregation, while one cross-over leads to 100% second-division segregation. It is easy to calculate that two cross-overs lead, in the absence of chromatid interference, to 50% second-division segregation (2- and 4-strand doubles cause first-division segregation, while 3-strand doubles cause second-division segregation), and so on for higher numbers of cross-overs (Fig. 9). The consequence of this is that the curve relating distance to second-division segregation frequency rises asymptotically to 67%. As in the case of the curve relating distance to recombination percentage (pages 37–38), the relationship approximates to a linear one only in its early part.

Since one cross-over between centromere and locus leads to 100% second-division segregation, but only 50% recombination, any value of the second-division segregation percentage has to be halved to give the equivalent recombination percentage, with a maximum value of 33%. Thus, if locus *A/a* shows 20% second-division segregation, locus *B/b* 10%, and the recombination percentage between *A/a* and *B/b* is about 15%, the following linkage map may be drawn:

$$A/a\text{————centromere————}B/b$$
$$10\phantom{\text{————centromere——}}5$$
$$\text{————————}15\text{————————}$$

(Note that, since recombination between two points on the chromosome rises asymptotically to 50%, whereas half the value of the second-division

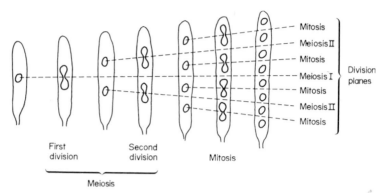

FIG. 10. Recognition of division planes in an ordered ascus. The diagram illustrates how, in an ordered 8-spored ascus, the plane of each nuclear division in the origin of the 8 ascospores can be recognized.

segregation frequency rises asymptotically to 33%, the two methods of estimating distance are equivalent only over short distances.)

The recognition of first- or second-division segregation tetrads in respect of a particular pair of alleles depends on the tetrads being "ordered". This situation occurs particularly in certain ascomycetes such as *Neurospora* and *Sordaria* where the developing ascus is so narrow that nuclear divisions occur exclusively in the long axis of the ascus with no overlapping of division spindles, and no passing of daughter nuclei takes place. The result is that first- and second-division planes can be recognized (Fig. 10), and the well-known first- and second-division patterns of segregation observed, either directly in the ascus (Fig. 11), when visually recognizable markers are involved, or else after separate culture of the spores.

An unordered tetrad, such as those found in *Aspergillus*, *Coprinus*, *Chlamydomonas*, and most strains of *Saccharomyces*, is one in which the meiotic products become jumbled. In this case, the second-division segregation frequency of a particular locus cannot be directly measured; however, it can be calculated, provided at least two other loci are segregating in the same cross (Perkins, 1949). The loci are considered in pairs, the tetrads being classified in respect of each pair into the three possible classes: parental ditype (PD), when each parental combination is represented twice; non-parental ditype (NPD), when each *recombinant* combination is represented twice; and tetratype (T), when the four possible genotypes, two parental and two recombinant, are each represented once. The proportion of T tetrads is the critical parameter, related to the second-division segregation frequencies of the two loci. When the three pairs are considered, one has three simultaneous equations, with three unknowns, the three second-division segregation frequencies (Whitehouse, 1957).

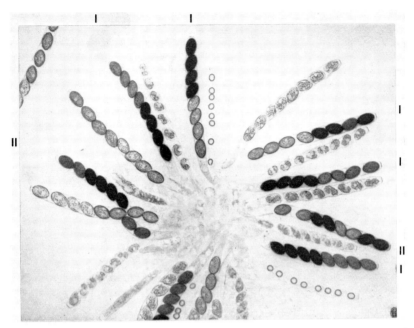

FIG. 11. Ordered asci of *Sordaria brevicollis*. The asci arose from a cross of a pale-spored mutant with the dark-spored wild-type. Each *mature* ascus (many are immature) contains four pale and four dark ascospores. First-division segregation (I) and second-division segregation (II) spore arrangements are shown.

(c) *Linkage estimation.* The estimation of linkage between loci from unordered tetrad data also depends on the classification of the tetrads into PD, NPD, and T classes. Perkins (1949) derived the relationship between map distance and the fraction of the total tetrads in each class, assuming no interference and ignoring cases of more than two cross-overs in the map interval concerned, as 50[(fraction of T tetrads) +6 (fraction of NPD tetrads)]. The relationship will, of course, be a good measure of map distance only over comparatively short intervals. This method of mapping is used routinely in those organisms in which the analysis of unordered tetrads is the best technique, for example in *Saccharomyces* (page 70) and *Chlamydomonas* (page 77).

B. Genetic analysis in particular organisms

1. *Aspergillus nidulans*

(a) *Mapping*

i. *By the sexual cycle.* The classical work on the genetics of this fungus was described by Pontecorvo *et al.* (1953). A recent summary is given by Sermonti (1969).

A. nidulans is a homothallic ascomycete with 8-spored, unordered, asci. Homothallism, which might have been regarded as a disadvantage in leading to difficulty in producing pure populations of "hybrid" progeny, was turned to good account by Pontecorvo and his collaborators. Homothallism confers the advantages of crossability of any two strains, with no mating-type limitation, and also of genetic homogeneity of the stocks, which were all derived from a single haploid conidium; thus difficulties stemming from genetic inhomologies of the kind encountered in *Neurospora crassa* (Frost, 1962) were avoided, at least until chromosomal rearrangements arose in the stock of cultures (Käfer, 1965). Controlled crosses were realized by choosing the ascospores contained in single crossed "perithecia" (strictly cleistothecia) as the source of sexual progeny, following the discovery that a whole perithecium, containing many thousands of ascospores, is normally either crossed or selfed (Pontecorvo *et al.*, 1953). Crossed perithecia were recognized because they gave rise to genetically heterogeneous progeny in respect of one pair of characters, usually involving distinct colonial colouration. Thus a series of spore suspensions, each derived from one perithecium, were tested for segregation of this marker, and those giving a positive result were then analysed further in the knowledge that they contained hybrid progeny. Alternatively, ascospores could be harvested from a collection of random perithecia and recombinants in respect of a particular pair of markers *selected* from the mixed population of "crossed" and "selfed" ascospores as in bacterial genetics (see page 82). The segregation of non-selected markers among these selected samples of progeny could then be studied.

Crosses are usually performed simply by mixing dense conidial suspensions of the two strains to be crossed on an agar medium in Petri dishes and allowing perithecia to develop. Pontecorvo *et al.* (1953) found that the proportion of hybrid perithecia arising in a cross was very variable and did not appear to be increased by taking the precaution of selecting heterokaryotic growth (see below) for the production of perithecia. Instead, they found that particular combinations of strains were characterized by high proportions of crossed perithecia, and others by low proportions ("relative heterothallism"). The mature perithecia are collected and, before bursting them to liberate the ascospores, they are washed free of attached conidia that would contribute an excess of spurious "parental" progeny.

Samples of the ascospore suspension are plated, either on a non-selective medium, on which all classes of progeny can grow, or on a medium selective for a particular combination of markers. The former procedure is usually adopted when long-range mapping is the objective. Samples of the resulting colonies are classified directly for visually distinct markers, such as conidial colour or colonial morphology (e.g. Clutterbuck, 1969), or sub-cultured to test-media in the case of nutritional, enzymological or resistance markers. Selective plating is normal in the case of fine mapping, as between sites in the same gene. Fine-structure maps have been constructed from two-point data representing the frequency of selected recombinants between a pair of close sites, compared with the frequency of recombination in a standard map interval, the ordering of the close sites being achieved by following the segregation of one or more outside markers (Pritchard, 1955, 1960; Sidiqi, 1962).

Tetrad analysis has been comparatively little exploited in *A. nidulans*, partly perhaps because of the comparative difficulty of isolating separately the ascospores from individual asci (a micromanipulator is needed) and because of the unordered nature of the asci, which renders the calculation of centromere distances dependent on the simultaneous segregation of at least three loci (Whitehouse, 1957). In view of the fact that mapping and centromere location are more easily performed by meiotic and mitotic recombination analysis respectively, the chief interest of tetrad analysis has been in the study of abnormal segregation and of interference (Strickland, 1958a, 1958b). The absence of autonomous mutations affecting ascospore colour, in spite of a search for them (Apirion, 1963), has meant that no information on gene segregation has been obtained by the direct observation of intact asci, as in some other fungi such as *Sordaria* and *Ascobolus* (page 69).

ii. *By the parasexual cycle.* The first step in the parasexual cycle is the establishment of heterokaryosis. In *A. nidulans* heterokaryons are invariably "forced"; that is the two strains to be fused each have a nutritional requirement, in addition to the marker or markers under investigation, and "balanced" (that is prototrophic) heterokaryons are *selected* by mixing dense conidial suspensions of the two strains on a medium lacking these two nutrients. Heterokaryotic hyphae then grow out from the area of inoculation and can be subcultured by the transfer of portions of the advancing hyphal front. Diploid strains are normally selected by plating conidia from the heterokaryon on the same selective medium; since conidia are normally uninucleate, only those containing a diploid nucleus can give rise to a growing colony. Alternatively, diploid sectors can be recognized visually in the heterokaryotic growth by using conidial pigment mutants. For example a heterokaryon between white and yellow mutant strains produces a mixture of white and yellow conidia, since the colour of a conidium reflects the genotype of its nucleus, whereas diploid conidia have the wild-type, green, colour.

Another method of selecting diploids depends on the occasional formation of diploid ascospores (Pritchard, 1953; Arditti and Strigini, 1962). Colonies arising from these can be recovered by selecting two closely linked markers, one from each parent, since haploid recombinant spores carrying both the markers are comparatively rare.

Diploid strains can easily be distinguished from haploids by the greater diameter of their conidia (Pontecorvo *et al.*, 1953) or, more laboriously, by their DNA content (Heagy and Roper, 1952).

The first step in gross mapping by the parasexual cycle is the assignment of a new mutation to its linkage group by haploidization (see page 53). This procedure depends on the efficient recognition and recovery of haploid segregants from the diploid strain. Initially haploids were recovered by "double selection", for two unlinked recessive markers simultaneously (Käfer, 1958; Forbes, 1959): for example acriflavine-resistance and suppression of an adenine-requirement. The idea was that expression of one such marker at a time would occur commonly as a result of mitotic crossing-over, so that the still diploid products of this segregation would outnumber and mask the haploids, whereas simultaneous expression of two such markers would occur with considerable frequency only by haploidization. More recently, with the discovery of the haploidizing effect of *p*-fluorophenylalanine, haploids have been selected simply by inoculating the diploid culture in defined areas on a medium containing this compound and picking the vigorously growing haploid sectors that grow out (McCully and Forbes, 1965).

The assignment of a new marker to its linkage group in *A. nidulans* has been rationalized by the use of haploid "master strains" bearing a marker on each linkage group (McCully and Forbes, 1965). A diploid between a strain bearing the new marker and a master strain is made, and a small sample of haploid segregants is selected and classified in respect of all markers. Amongst this sample, recombinants in respect of the new marker and markers on all but one of the linkage groups should occur; the remaining linkage group is that to which the new marker belongs.

Gross mapping within a linkage group by mitotic recombination depends on the recovery of the products of mitotic crossing-over. This, as we saw on page 55, depends on the selection or visual recognition of clones that have become homozygous for a particular marker, and their testing for homozygosity of other markers in the same linkage group. This can be done directly when the recessive allele at each locus is in the *cis* configuration with the homozygous marker that is selected, since homozygosity at each locus then leads to a phenotypic difference from the parent heterozygote. If not, the genotype of a selected diploid can be determined by selecting haploid segregants from it, when the two frequent classes of haploid in respect of the linkage group in question define the genotype of the diploid. As discussed on page 56, selective markers for the recognition of mitotic segregants are most efficient when terminally located. Articles on the general methodology of mitotic analysis are by Pontecorvo and Käfer (1958), Käfer (1958), Pritchard (1963) and Roper (1966).

Fine genetic analysis by mitotic recombination has been described by Pritchard (1955) and Putrament (1964). Diploids between pairs of allelic mutants (adenine or p-aminobenzoic acid-requiring respectively) were made; these, of course, required adenine or p-aminobenzoic acid. Diploid mitotic recombinants were then selected on unsupplemented medium, and the genotype of each was determined by haploidization as just outlined. Mitotic crossing-over in short intervals is characterized by high negative interference between multiple cross-overs, comparable to that in meiotic crossing-over, as well as non-reciprocal events, a feature revealed when both strands involved in a particular mitotic cross-over are recovered.

(b) *Testing for complementation and dominance.* This makes use of either heterokaryons or diploids. The use of heterokaryons for qualitative complementation tests (e.g., Foley, Giles and Roberts, 1965; Roberts, 1967) is clearly less time-consuming, if somewhat less certain, than the use of diploids (e.g., Foley, Giles and Roberts, 1965; Clutterbuck, 1969). For quantitative tests of dominance or complementation (e.g., Cove, 1970), diploids are required because heterokaryons have variable ratios of each component haploid genotype, and are unstable, constantly breaking down again to homokaryons.

2. Aspergillus niger *and other "imperfect" fungi*

A. niger was the first fungus lacking a sexual cycle in which the parasexual cycle was experimentally harnessed as a means of achieving genetic recombination (Pontecorvo, Roper and Forbes, 1953). The techniques worked out for *A. nidulans*, in which genetic conclusions from the parasexual cycle could be checked against the results of sexual analysis, were applied to *A. niger* by Lhoas (1967) to give a linkage map. A modification was necessitated by the lack of suitable distal markers for the selection of diploid recombinant strains from the synthesized diploids. Thus mitotic analysis of the kind used in *A. nidulans* to determine map order and distance was not practicable. However, in *A. niger* the frequencies of haploidization and mitotic crossing-over were so much higher than in *A. nidulans* that they frequently occurred together in the same nuclear lineage, leading to the segregation of haploids recombinant for linked markers. The frequencies of various marker combinations amongst a sample of such haploids provided the basis for mapping. Thus the end result was as if a kind of "irregular meiosis" had occurred.

Penicillium chrysogenum is another "imperfect" fungus in which the parasexual cycle has been exploited, this time with the motive of opening the way to a genetic approach to strain improvement in this industrially important organism (Pontecorvo and Sermonti, 1954; Sermonti, 1957, 1959). The methodology employed was essentially the same as that worked out for *A. nidulans*, with slight modifications in the procedure used in the isolation of heterokaryons as the first step in the parasexual cycle. A recent account of the techniques of *Penicillium* genetics is by Sermonti (1969).

The parasexual cycle has also been studied, at least in a rudimentary fashion, in several other filamentous ascomycetes. In *Fusarium oxysporum*, Buxton (1956) successfully selected stable prototrophic strains from nutritionally balanced heterokaryons prepared in the normal way. Conidial diameter confirmed the diploid nature of these strains, which gave rise to variants in respect of pathogenicity.

Hastie (1962) found that balanced heterokaryons of *Verticillium albo-atrum* were highly unstable; sub-culture from hyphal tips, a method that allows the transfer of heterokaryons in most other species, often failed to give rise to heterokaryotic growth. However, heterozygous, presumably diploid, strains could be isolated from the unstable heterokaryons. These diploids, also, were unusually unstable, giving rise to segregants; some of these were diploid and others haploid (Hastie, 1964), providing information for the assignment of markers to linkage groups.

An interesting technique for recovering the reciprocal products of mitotic crossing-over, when formed, was later described by Hastie (1967). This depended on the fact that each uninucleate side-branch of the conidiophore (phialide) gives rise to a small group of conidia, each containing one product of successive mitotic division of the phialide nucleus, and the members of each group could be isolated and characterized genetically.

Other studies of parasexuality in imperfect fungi, such as *Cephalosporium mycophilum* (Tuveson and Coy, 1961), *Aspergillus fumigatus* (Strömnaes and Garber, 1963). *Aspergillus oryzae* and *Aspergillus sojoe* (Ishitani, Ikeda and Sakaguchi, 1956) have been reviewed by Sermonti (1969), as well as those in ascomycetes with a sexual stage, such as *Cochliobolus sativus* (Tinline, 1962) and *Emericellopsis* spp. (Fantini, 1962).

3. *Neurospora crassa*

This fungus has been used more in genetic studies than any other, and a large literature exists. Much of this is listed in a monograph by Bachmann and Strickland (1965). Later references are to be found in the *Neurospora* Newsletter produced at regular intervals by B. J. Bachmann, School of Medicine, Yale University, New Haven, U.S.A.

(a) *Mapping*. *N. crassa* is a heterothallic ascomycete with 8-spored, ordered asci. Heterothallism is controlled by a single pair of mating-type alleles,

A and *a*. This means that a given strain is capable of crossing, on average, with half of a random collection of strains. It also means that the crossable stocks in laboratory use could have arisen from a minimum of two original cultures; in fact several independent wild-types have contributed to the stocks, with undesirable consequences on the heterogeneity of linkage values for the same map interval derived from non-isogenic pairs of strains (Frost, 1961).

No parasexual cycle has been described for this fungus, in spite of a search for it (Case and Giles, 1962).

Although strains of each mating-type can function as both male and female parents, crosses are often made by treating one strain as female and the other as male. This is achieved by inoculating the female parent first, and only when protoperithecia are produced is the other strain added as a conidial suspension, which then fertilizes the protoperithecia of the female parent.

Mapping is carried out either by tetrad analysis or by the study of random ascospores. Ordered tetrad analysis dominated the early work on gene location in *Neurospora*; much of this is summarized by Barratt, Newmeyer, Perkins and Garnjobst (1954). Many of the data consisted of the equivalent of two-point analysis, that is the centromere distances of individual loci. Such data suffered from the drawbacks of other two-point analyses, which were exacerbated by the genetic heterogeneity of the stocks, which caused great variability in recombination values from cross to cross.

Asci of *N. crassa* are large enough for dissection by hand-held mounted needles, and ordered data are obtained by removing the spores one at a time (or in mitotic pairs) for individual culture after the temperature-shock at about 60°C which is required to cause efficient germination of the ascospores. (This treatment has the advantage of killing most contaminating asexual conidia that may be accidentally included with an ascospore.) Alternatively the asci may be analysed in unordered fashion, making use of the fact that the entire contents of individual asci tend to be shot out in groups which can be collected on an agar surface placed below mature perithecia (Strickland, 1960); the members of each group can then be separated and cultured separately. The spores from sequentially maturing asci of the same perithecium can even be collected in an automatic device (Strickland and Thorpe, 1963).

Random spore analysis was originally carried out by collecting ascospores that had been spontaneously shot from perithecia on to the wall of the culture tube or lid of a Petri dish, spreading them on a block of agar and picking off each one with a needle to a separate tube for temperature-shocking and culture. An example of an extensive random spore analysis is by Perkins (1959). Other studies have been performed by collecting

ascospores *en masse* from Petri dish lids, and suspending them in 0·1%
agar (a viscous solution that minimizes problems of settling out of the
heavy spores). The spores were then counted in a haemocytometer, diluted
and plated, after temperature-shocking the suspension; plating was usually
in soft-agar overlays on basal layers of medium (Newmeyer, 1954). This
approach has been used in some gross mapping studies, but has been
adopted particularly in fine mapping (e.g., Murray, 1963; Stadler and
Towe, 1963; Smith, 1965; Fincham and Pateman, 1967; Stadler and
Kariya, 1969) when selection has been made for rare recombinants between
closely linked alleles. Outside, non-selected, markers have usually been
used to assist in the difficult problem of arriving at a consistent linear
order for the closely linked mutant sites.

At least one study of recombination in short intervals has been by
selective tetrad analysis; Stadler and Towe (1963) collected unordered
ascospore groups by the method of Strickland (1960) from a cross between
allelic *cys* mutations, incubated them and examined the groups for germinating *cys+* recombinant spores. The spores in those groups revealing at
least one *cys+* spore were then analysed individually.

In the absence of a parasexual cycle with haploidization of diploids, the
assignment of a new marker to its linkage group is not so straightforward
as in *A. nidulans*. The problem has been alleviated by the synthesis of
tester strains (called "*alcoy*" strains) containing three reciprocal trans-
locations (interchanges) involving linkage groups I and II, IV and V, and
III and VI; the remaining linkage group, VII, is unaffected (Perkins,
Newmeyer, Taylor and Bennett, 1969). Three "visible" markers are incorporated, one near the interchange point of each reciprocal translocation.
An *alcoy* tester of each of the two mating-types is available, so that any
strain bearing a new marker can be crossed to *alcoy*. When this is done, a
high proportion (about 7/8) of the resulting ascospores are inviable and
colourless because they lack certain chromosomal regions. Such spores
contain most of the products of crossing-over between the "normal" and
translocated chromosomes. Therefore, when the minority of viable black ascospores is germinated and classified, close linkage of any new marker is shown
with the marker identifying one pair of translocated chromosomes, if the new
marker is on one or other of these chromosomes. Markers showing 50%
recombination with all three *alcoy* markers may be in linkage group VII, or
occasionally in one of the translocated linkage groups but far from the interchange point. Further "follow-up" testers are available to investigate further
the result indicated in the preliminary test with *alcoy* (Perkins *et al.*, 1969).

(b) *Complementation*. Most studies of complementation in *N. crassa* have
made use of heterokaryons (e.g., Ishikawa, 1962; Gross, 1962). Pairwise

tests of a group of mutants may be made on solid medium, when dense patches of conidia of each strain are mixed on selective agar medium and examined for growth due to the formation of complementing heterokaryons. Alternatively, tests can be made by mixing conidia in liquid selective medium (de Serres, 1962). Unfortunately a negative result is not always meaningful since it can reflect a failure to establish a heterokaryon. In this case nutritionally unrelated markers, one in each strain of the pair, can be used to establish a heterokaryon. In certain cases (Ishikawa, 1965), complementation was not detected in some allelic combinations when tested in heterokaryons, but was found by the study of pseudowild-type strains, that is aneuploid strains containing an extra copy of the chromosome bearing the locus in question (Pittenger, 1954). Such strains arise during ascospore production at a low frequency, presumably as a result of occasional non-disjunction at nuclear division.

4. *Other ascomycetes*

(a) *Sordaria.* At least three species of this genus have been used in genetic work: *Sordaria brevicollis, Sordaria fimicola* and *Sordaria macrospora.* The first is heterothallic, with control by two mating-type alleles, and the other two are homothallic. Most of the interest in these organisms has centred round the information obtainable on crossing-over, gene conversion and post-meiotic segregation by the visual recognition of asci showing abnormal segregation in respect of spore-colour markers. These fungi have 8-spored, ordered asci, although in *S. brevicollis,* an appreciable proportion of the asci show disturbances in spore sequence attributable to spindle overlap at the second meiotic division (Chen and Olive, 1965). Techniques of genetic analysis are generalized, most linkage studies having involved ordered tetrad analysis. Some references are: *S. brevicollis,* Chen (1965); *S. fimicola,* Olive (1956; Kitani and Olive, 1969); *S. macrospora,* Esser and Straub (1958).

(b) *Ascobolus immersus.* As in the case of *Sordaria,* studies of *A. immersus,* which is heterothallic with 8-spored unordered asci, have largely concentrated on the segregation of spore-colour markers (e.g., Lissouba *et al.,* 1962; Emerson and Yu-Sun, 1967; Baranowska, 1970). Mapping has been by the analysis of unordered tetrad data by the method outlined on page 61 (Yu-Sun, 1964).

5. *Yeasts*

Two species have been the subject of considerable genetic analysis: the classical yeast, *Saccharomyces cerevisiae*; and the fission-yeast *Schizosaccharomyces pombe.*

(a) *Saccharomyces cerevisiae*. The genetics of this yeast has recently been reviewed in an excellent lucid chapter by Mortimer and Hawthorne (1969), and only a brief summary need be made here. The organism is basically heterothallic, with two mating-type alleles *a* and α, although certain strains carry a mutation that overrides this heterothallism to make the strain homothallic. Asci containing four spores are produced, and these tetrads may be ordered or unordered depending on the strain. The haploid spores emerging from a tetrad in a heterothallic strain give rise to clones of haploid vegetative cells, and the cells in such cultures will mate with cells of opposite mating-type to produce stable diploid vegetative cells. Only under defined cultural conditions do these cells differentiate into asci containing tetrads of spores.

Matings are usually made by mixing haploids of opposite mating-type, and carrying complementary auxotrophic markers, so that the diploid products of fusion can be selected on unsupplemented medium.

i. *Long range mapping*. Tetrad analysis provides the best general method of mapping because of problems associated with the isolation of random spores in suitable numbers, although this *can* be done (Mortimer and Hawthorne, 1969, p. 396). The practical details of tetrad analysis are described by Mortimer and Hawthorne (1969, p. 397). Ordered tetrads, in those strains that produce them, differ from those of most other ascomycetes in that alternate, not adjacent, spores contain the products of each second division of meiosis; presumably the spindles at the second division of meiosis regularly overlap one another. However, there is usually a small proportion of asci in which nuclei have slipped past one another, giving spurious spore sequences. Nevertheless it is possible to use the classes of asci whose frequencies are not modified by nuclear passing to estimate the frequency of second-division segregation, and so the centromere distance, of a marker. Having identified markers close to their centromere, a cross incorporating at least one such marker, even if it produces completely unordered asci, can be analysed directly; the "centromere marker" can be used to put the four spores in order. Thus the centromere distance of each new marker can be estimated.

For the recognition and estimation of linkage between pairs of markers, Perkins' formula is used (see page 61).

Spontaneous mitotic crossing-over occurs very rarely in yeast diploids, but its frequency can be very greatly increased (as in other organisms) by chemical mutagens and radiations. Spontaneous mitotic segregants can be selected, as in *A. nidulans*, by markers such as those conferring recessive resistance, but after induction by mutagens, such selective markers are not essential. As in *A. nidulans*, the probability of multiple mitotic crossing-over is low, so that markers on the same chromosome arm normally segregate together, providing useful information on gene location. Mitotic haploidization as a means of defining linkage groups has not been developed; *p*-fluorophenylalanine causes the production of various aneuploid strains, but not true haploids (Emeis, 1966; Strömnaes, 1968). This factor presumably accounts for continued uncertainty over the number of linkage groups in spite of over 100 genes having been mapped (Mortimer and Hawthorne, 1669).

ii. *Fine mapping.* Owing to the difficulties of isolating large populations of random meiotic products, two-point meiotic fine-structure mapping has made use of random whole asci. For example, Esposito (1968) crossed *adenine*-8 mutants in pairs to produce adenine-requiring diploid cultures. These were then caused to sporulate, after samples had been plated on adenineless medium to estimate the "background" of adenine-independent cells in each diploid culture. Samples of the sporulated cultures were then examined in a haemocytometer to estimate the proportion of asci in the culture (which usually ranged from 40–60%, the remainder being unsporulated diploid cells) and also plated on medium with and without adenine. Those asci in which intragenic crossing-over had produced at least one adenine-independent spore gave rise to colonies on the adenineless medium, while the total population was measured on the medium with adenine. The unit of mapping was prototrophs/10^5 asci, calculated as (prototrophs/10^5 plating units after sporulation −prototrophs/10^5 plating units before sporulation) × 1/proportion of asci after sporulation.

Individual tetrads in crosses between alleles have been studied by Fogel and Hurst (1967), using a selective technique for the visual identification of asci in which intragenic recombination had generated at least one prototrophic spore. Asci from diploids heterozygous for two *histidine* alleles were plated on histidineless medium and examined for the presence of spores beginning to bud. The proportion of such asci out of the total examined was recorded, and each such ascus was dissected and analysed.

A method of fine mapping by mitotic recombination depends on the finding that X-irradiation greatly increases the frequency of intragenic recombinant production in a diploid heterozygous for a pair of alleles, and that a consistent map emerges when the frequencies of induced recombinant production between various pairs of alleles are put together (Manney and Mortimer, 1964; Manney, 1964). The technique, applicable to auxotrophic mutations, is to plot prototroph frequency (determined by plating samples of the diploid on selective and non-selective medium) as a function of X-ray dose, for each diploid containing a particular pair of alleles. From the curves, a mapping function of prototrophs/10^8 survivors/roentgen is obtained. This technique has been used for several genes (e.g., Fink, 1966; Esposito, 1968; Parker and Sherman, 1969). More recently, Snow and

Korch (1970) found that alkylating agents like methyl methanesulphonate could be used instead of X-rays in such studies.

iii. *Complementation and dominance tests.* Functional genetic tests make use of the stable diploids that are so easily available in yeast. Complementation tests between large numbers of auxotrophs can be made by growing a set of mutant haploid strains of mating type *a* in parallel stripes on one plate and a set of mutants of mating type α on another; cross-replicating the two plates to a third non-selective plate for mating to occur; and then replicating the resulting patches of diploids to a selective medium to test for complementation (e.g., Dorfman, 1964; Fink, 1966).

iv. *Non-Mendelian inheritance.* Yeast is one of the microbes, along with *Chlamydomonas* and *Paramecium* (see later), that lends itself to the study of what used to be called cytoplasmic inheritance, but which is probably better referred to as non-Mendelian in view of the fact that organelle "chromosomes" may well turn out to be involved. The classical studies in this field involved respiratory-deficient, or "petite colony" strains (Ephrussi, 1953). Although the phenotype of some such strains (nuclear petites) was determined by Mendelian genes, in other cases all members of a tetrad produced by crossing petite and normal strains were normal; this provided a good example of the use of tetrad analysis in demonstrating very directly the existence of cytoplasmic determinants. Much work, implicating loss of or changes in the mitochondrial DNA in the determination of cytoplasmic petites, is now being done (e.g., Williamson, 1970).

In addition to petite, other phenotypes such as resistance to "bacterial type" antibiotics have been associated with mitochondrial mutations (Wilkie, Saunders and Linnane, 1967; Thomas and Wilkie, 1968); the use of such mutations promises to reveal an analysable genetic system (perhaps analogous with that of the chloroplast in *Chlamydomonas*: see page 78) involving mitochondrial "mating" (Slonimski *et al.*, 1970). It is too early to discuss in detail the features of such a system.

Apart from the characters fairly clearly associated with the mitochondria, other cytoplasmic determinants have been recognized in yeast. Somers and Bevan (1969) described phenotypes (killer and neutral) dependent on the presence of "allelic" cytoplasmic determinants, the lack of which gave rise to a third phenotype, "sensitive". At least one nuclear gene was required for the maintenance of the cytoplasmic determinants.

(b) *Schizosaccharomyces pombe.* This yeast has been exploited for genetical studies particularly by U. Leupold and his collaborators (Leupold, 1970). His strains are either heterothallic, when they carry the mating type alleles h^+ or h^-, or homothallic, when they carry a third allele, h^{90}. In mapping studies, heterothallic haploids are usually grown up, then mixed under conditions favouring mating, to produce diploid cells. These zygotes undergo meiosis to yield linear four-spored asci. Tetrad analysis can be carried out by ascus dissection. Random spore analysis is done by harvesting ascospores from mass mating mixtures. Usually vegetative cells are killed by alcohol treatment so that progeny can be analysed non-selectively. Alternatively (Leupold, 1958) the presence of viable vegetative cells in the ascospore suspension is overcome by plating on media selective for recombinants, as in *A. nidulans* (see page 62). Extensive fine-structure mapping by two-point selections has been carried out in this way (e.g., Leupold, 1961; Leupold and Gutz, 1964; Barben, 1966).

Complementation studies (e.g., Leupold and Gutz, 1965) make use of stable diploid strains produced by fusing combinations of haploid strains of opposite mating-type (Leupold, 1970).

Diploid strains undergo mitotic crossing-over (Ali, 1967), and can also be haploidized by the use of *p*-fluorophenylalanine (Gutz, 1966). Such studies have recently been exploited in the definition of several new linkage groups (Flores da Cunha, 1970).

6. *Filamentous basidiomycetes*

(a) *Coprinus lagopus*. This fungus is a higher basidiomycete, producing unordered tetrads of basidiospores. Heterothallism is controlled by two mating-type loci, A and B, each represented by a series of multiple alleles; fertile matings occur between haploid strains bearing different alleles at each locus. Linkage data have been obtained by both random basidiospore analysis and tetrad analysis (Lewis, 1961; Day and Anderson, 1961; Moore, 1967). Moore (1967) provided examples of the usefulness of White-house's (1957) technique for the calculation of centromere linkages from unordered tetrad data when three markers are segregating simultaneously.

The normal method for carrying out functional genetic tests (e.g., Lewis, 1961; Morgan, 1966) makes use of the dikaryons formed when two mono-karyotic mycelia of different mating-type (at both loci A and B) meet and fuse. The dikaryotic mycelium can be distinguished macroscopically (by a particular colonial morphology due to the angle of branching of the mycelium) and microscopically (by the presence of clamp connections). Since the dikaryotic mycelium, once formed, is very stable, no "forcing" markers are required to select or maintain it; dikaryons can be synthesized on non-selective medium. Dikaryons are indeed so stable that the experimental re-isolation of their monokaryotic components requires a special technique; the germination of specialized uninucleate asexual spores (chlamydospores) borne on the dikaryon (Lewis, 1961).

The complete parasexual cycle does not appear to have been developed in *C. lagopus*, but stable diploid strains have been isolated (Casselton, 1965; Casselton and Lewis, 1966). Use was made of the fact that heterokaryons will form between two strains having the same allele at the A mating-type locus; in such "common A" heterokaryons the normal strict relationship between the two kinds of nucleus breaks down. Prototrophic diploid strains were obtained by harvesting uninucleate oidia from such a heterokaryon synthesized between complementary auxotrophs and plating them on selective medium.

Day and Roberts (1969) investigated the possibility that cases of complementation in heterozygous diploids but not in the corresponding heterokaryons may exist, and if so may implicate nuclear-localized gene products (see pages 56–57). *Coprinus* should provide ideal material for such studies since diploids can be compared with dikaryons consisting uniformly of cells with one nucleus of each genotype, rather than with the unstable and irregular heterokaryons of ascomycetes. In their preliminary study, Day and Roberts (1969) in fact found no example of a difference in complementation between diploid and dikaryon.

(b) *Coprinus radiatus*. The genetics of this species has been developed more recently than that of *C. lagopus*. Details of the methodology for this species were described in a thesis by Prévost (1962); they closely resemble those employed for *C. lagopus*. Random spore and tetrad analysis for gross mapping (Prévost, 1962), fine structure analysis by two-point crosses, and complementation by the use of dikaryons (Gans and Masson, 1969) have been carried out. In addition, a parasexual cycle has been described (Prud'homme, 1970). Special techiques for the isolation of diploids were needed because of the absence, from this species, of a uninucleate asexual spore stage corresponding to the oidia of *C. lagopus* (or the conidia of *A. nidulans*). Diploids were isolated as vigorously growing sectors from heterokaryons, which were either inately unstable, or else were rendered unstable by culture in an unaerated, acidic, liquid medium. The diploids were relatively stable, but they did give rise to haploids, by progressive loss of chromosomes at detectable rates. However mitotic crossing-over was not detected with certainty.

(c) *Schizophyllum commune*. The emphasis in genetic work with this fungus has been primarily on the genetic and physiological control of mating behaviour (e.g., Raper and Raper, 1966; Koltin and Raper, 1967; Raper and Raudaskoski, 1968; Koltin and Flexer, 1969; Wang and Raper, 1969) and recombination (Simchen, 1967; Stamberg, 1968), rather than with studies that might have called for the development of specialized genetic tests. The life-cycle is basically the same as in *Coprinus*, and mapping has been carried out by random basidiospore analysis (Raper and Miles, 1958).

7. *Ustilago*

(a) *Ustilago maydis*. Genetic analysis of this smut fungus was pioneered by Perkins (1949) and has been developed and exploited particularly by Holliday (1961a). Two mating-type loci, one with two and the other with multiple alleles, control the ability of haploid strains to fuse and give rise to the sexual cycle in the host plant, maize. A dikaryotic, obligately parasitic mycelium is produced when compatible haploid strains are inoculated together, and this eventually gives rise to diploid fusion cells (brandspores) in galls produced in the tissues of the host plant. A brandspore germinates to produce an elongated "promycelium" consisting of a linear tetrad of four cells, each of which then buds off a basidiospore that proceeds to divide to produce a clone of haploid, uninucleate cells ("sporidia"). Sporidia do not fuse with each other, even if they differ in mating-type, unless inoculated into plants, so that pure recombinant clones can easily be established and maintained on artificial media.

Long-range mapping in *U. maydis* has been carried out by random spore analysis, tetrad analysis, and mitotic crossing-over, but detailed knowledge of the linkage map is still comparatively scarce; rather little systematic mapping has been done. Random spore analysis depends on making a suspension of brandspores from galls and freeing it from viable vegetative cells by treatment with copper sulphate (Perkins, 1949; Holliday, 1961a). The brandspores are then plated out on non-selective medium and allowed to give rise to small colonies of progeny sporidia. These are collected and replated on non-selective medium to give individual clones of meiotic products, a sample of which is classified. The method works only when there is no bias against particular progeny genotypes due to selective brandspore germination or differential multiplication of different sporidial genotypes. Provided that alleles at each locus segregate 1 : 1 and complementary genotypes of progeny also have equal frequencies, linkage data are reasonably reliable (Holliday, 1961a, 1962).

Although the tetrads of *U. maydis* are fundamentally ordered, morphologically irregular germination of the brandspores and precocious division of the first formed basidiospores before the tetrad is complete, combined with the small size of the structures, make ordered tetrad analysis unprofitable (Holliday, 1961a). Instead, brandspores have been allowed to germinate to give colonies containing the mixed, unordered, descendants of the tetrad. These colonies can be caught at a young stage and respread to give rise, after further incubation, to groups of colonies;

amongst these the two (in the case of PD or NPD tetrads) or four (in the case of T tetrads) possible genotypes for each pair of markers can be identified. Alternatively, when linkage between a particular pair of auxotrophic markers is to be tested, the brandspore colonies are allowed to grow larger, and then classified for the presence of prototrophic sporidia; this can be done either by replica-plating to selective medium or resuspension and plating on selective medium. Both approaches to tetrad analysis are subject to bias due to selective recovery of different genotypes (Holliday, 1961a), but can yield reliable linkage data when adequate checks are included.

Vegetative diploids are isolated by plating infected plant gall tissue, before mature brandspores are produced, on medium selective for recombinants. Diploid clones then grow out, presumably derived from diploid pre-brandspore cells (Holliday, 1961b). The diploid clones are very stable, and have not been found to undergo mitotic haploidization. They undergo spontaneous mitotic crossing-over very rarely, but the frequency is greatly increased by irradiation and other mutagenic treatments (Holliday, 1961b, 1962, 1964a, 1965; Esposito and Holliday, 1964). Mitotic segregants have not usually been selected, as they are in *A. nidulans* (page 64); instead irradiated cells were plated on non-selective medium and the resulting colonies replicated to selective medium to identify those that had become auxotrophic or contained auxotrophic sectors. These could then be isolated and analysed, to give information on linkage in each chromosome arm. As in other systems multiple mitotic crossing-over was rare, so that the order of loci, even in linkage groups that were very long meiotically, was readily deduced.

Complementation and dominance tests are performed on the diploid clones (e.g. Lewis and Fincham, 1970). Diploids between allelic auxotrophic mutations are not always easy to make, probably because of a shortage of the required growth factor in the inoculated plant (Holliday, 1961a). This could sometimes, but not always, be overcome by adding the growth factor with the inoculum.

(b) *Ustilago violacea*. Genetic work with this species was begun in an attempt to develop a complete parasexual cycle (Day and Jones, 1968) which, as we have just seen, was not found in *U. maydis*. Sporidia of *U. violacea*, in contrast to those of *U. maydis*, will fuse with sporidia of opposite mating-type in culture to yield dikaryotic cells. These grow well only if inoculated to a host plant. However if dikaryons formed between complementary auxotrophs on agar are replated on selective medium, vegetative diploids can be selected. These were studied by mitotic analysis (Day and Jones, 1968, 1969; Day, 1971). The diploids were very stable, but could be induced to undergo mitotic crossing-over by ultraviolet irradiation, and mitotic haploidization by treatment with p-fluorophenyl-alanine. Complementation was studied in the diploids, and linkage groups were defined by haploidization (Day and Jones, 1969). Meiotic analysis was by the use of random meiotic products, as described for *U. maydis* (see page 75).

A complication in the parasexual cycle as developed in *U. violacea* (Day and Jones, 1969; Day, 1971) was the finding that two of the linkage groups usually remained disomic after treatment with p-fluorophenyl-alanine; true haploids were rarely produced.

8. *Chlamydomonas reinhardii*

Although at least two other species of *Chlamydomonas* (*C. eugametos* and *C. moewusi*) have been studied from a genetical point of view, most recent work has concentrated on *C. reinhardii*. A clear outline of the life-cycle and genetic analysis of this species is given by Levine and Ebersold (1958), and in more detail by Levine and Ebersold (1960) and Ebersold, Levine, Levine and Olmsted (1962). Haploid, uninucleate vegetative cells of this green alga are of two mating-types, + and −. Clones can be grown indefinitely in the light on a simple nitrogen-containing medium. The cells are converted into gametes by a period in nitrogen-free medium. Mixtures of gametes of opposite mating-type give rise to zygotes; these need about a day in the light and a further period in the dark of at least 5 days for matura-tion. When mature, the zygotes can be germinated at will by incubation in the light, when an unordered tetrad of four cells is produced. These then proceed to divide to give rise to clones of haploid vegetative cells.

(a) *Mapping.* Owing to the difficulty of obtaining unbiased populations of random meiotic products, nearly all mapping has been done by the analysis of unordered tetrads. The zygotes are arranged in single rows on agar by means of a fine hand-held glass needle. Contamination by unmated vegetative cells is eliminated by briefly inverting the Petri dish over chloro-form, which kills vegetative cells but not zygotes. After overnight incubation, each zygote has produced 4 or 8 products and these are separated and allowed to develop into colonies which are characterized. Linkage and centromere locations are indicated by classification of the tetrads into PD, NPD and T types for each pair of markers (see pages 60–61).

Instead of "dissecting" individual tetrads, the zygotes (again after chloroform treatment) can be plated out and allowed to develop into colonies each consisting of the mixed descendants of a tetrad. The tetrad colonies can then be replicated (using rough filter paper) to test for a particular class of progeny, usually prototrophs when the cross involves two auxotrophs. This method, similar to that used for yeast (see page 71), is efficient for the detection of close linkage (e.g. Warr *et al.*, 1966).

(b) *Complementation and dominance tests*. A complete parasexual cycle does not appear to have been described in *C. reinhardii*, but vegetative diploids have been isolated (Ebersold, 1967). Gametes of two auxotrophic strains were mated and plated on a selective medium in the light. Under these conditions, some of the fusion cells did not mature into zygospores, as they would have done if one day of light had been followed by the normal dark maturation treatment, but developed instead into prototrophic diploid colonies. The cells of these colonies were larger than haploid cells; strains could also be diagnosed as diploid by the occasional presence of tri- or quadriflagellate cells (Starling, 1969). Diploid strains have been invaluable in allowing tests of dominance and complementation (Starling, 1969).

Vegetative diploids of *C. reinhardii* are very stable, and have not been found to undergo mitotic haploidization. However mitotic crossing-over has recently been studied, particularly after ultraviolet irradiation (Martinek, Ebersold and Nakamura, 1970) and we can expect this to be increasingly used as a genetic tool.

(c) *Non-Mendelian inheritance*. A particular interest of *Chlamydomonas* is its usefulness in the study of non-Mendelian inheritance, a topic developed particularly by R. Sager and her collaborators.

Non-Mendelian genes are recognized because, when two strains differing in such a marker are crossed, all four members of the tetrad normally inherit the marker of the + mating-type ("female") parent. However, occasionally the descendants of a zygote show segregation, and this can be made to be the rule if the "female" parent is irradiated with a small dose of ultraviolet light (Sager and Ramanis, 1967). Such "bi-parental" zygotes are the tools for the study of the segregation of non-Mendelian genes, which occurs during the mitotic divisions following the meiosis that produced the tetrad. The technique consists of separating the successive products of these cell-divisions and then, when enough divisions have been followed (say meiosis followed by two mitoses, giving 16 cells), each cell is grown into a clone; this is classified as being homozygous or heterozygous for each marker. A recent paper by Sager and Ramanis (1970) describes some of the striking conclusions of such analysis, which appears to have analogies with mitotic analysis of nuclear genes; different genes become homozygous at different mitotic divisions, and a gradient, perhaps indicating the existence of a "centromere-equivalent" at one end of the linkage group, is observed. The working hypothesis is that the non-Mendelian linkage group so defined corresponds to the chloroplast DNA.

9. *Ciliates*

Two ciliates have been the subject of considerable genetic experimentation: *Paramecium aurelia* and *Tetrahymena pyriformis*. The interest of these organisms to geneticists has been focused particularly on problems of genetic control by micro- and macro-nuclei (Nanney, 1968; Gibson, 1970a), on organelle genetics (Gibson, 1970b), and on aspects of the self-determination of cell-surface structures (Sonneborn, 1970). The formal genetics of the organisms, in terms of mapping and complementation studies, has been very little developed up to the present.

(a) *Paramecium aurelia*. Basic methods for the culture and genetic manipulation of the organism are in an article by Sonneborn (1950), and a review of the basis and results of genetic studies in a monograph by Beale (1954). Vegetative cultures of *P. aurelia* are diploid. Cultures of suitable mating-type can be crossed and, as a result of complex nuclear changes (see Beale, 1954, pp. 23–31), each member of a conjugating pair comes to be diploid and heterozygous in respect of any alleles differentiating the parents. Such individuals and their descendants produced by vegetative fission will not mate until the process of autogamy (Beale, 1954, pp. 31–33) has occurred. Autogamy can be induced at will by suitable nutritional adjustment and results in the organisms becoming diploid but homozygous. Homozygosity occurs at random for one or other of the alleles originally heterozygous. This arises as a result of a meiotic division, followed by a mitosis and fusion of *identical* haploid products. Thus, in terms of genetic analysis, conjugation produces clones of heterozygous individuals that can be used to test complementation and dominance, while autogamy allows the study of the segregation of alleles from such heterozygotes in ratios equivalent to those in the haploid meiotic products of other eukaryote microbes. Many studies of dominance and Mendelian segregation have been made, but only recently has linkage been detected (Beisson and Rossignol, 1969). This study made use of pairs of temperature-sensitive mutations (*ts*). The phenotype of the progeny of conjugation (F_1) was wild-type, indicating complementation of the two mutations. In an example of non-linkage, one-quarter of the individuals produced by autogamy from the F_1 (the F_2 generation) were recombinants with wild-type phenotype, while three-quarters were temperature-sensitive (one-quarter each of the two parental *ts* mutations and one-quarter of the double mutant recombinant). Linkage in a cross of a different pair of *ts* mutations was indicated by a proportion of ts^+ recombinants much smaller than one-quarter.

(b) *Tetrahymena pyriformis*. Genetic studies with this organism have recently been reviewed by Allen and Gibson (1970) and little detail will be mentioned here. The system has many similarities with that of *P. aurelia*, but some differences. Mating-type expression appears much more complex, and nuclear behaviour at conjugation less predictable. Conjugation occurs more or less as in *P. aurelia*, but autogamy does not; heterozygous individuals produced by conjugation can be crossed with each other to produce an F_2 generation or backcrossed to individuals bearing recessive markers. Thus, in a two-factor cross, the familiar ratios of diploid genetics are seen: for dominant markers at a pair of unlinked loci, 9 : 3 : 3 : 1 in a cross of two heterozygotes or 1 : 1 : 1 : 1 for the four phenotypes in a backcross. A shortage of the two recombinant classes in one backcross has indicated a linked pair of loci (Allen, 1964).

The consequences of conjugation followed by autogamy in *P. aurelia* are mimicked in *T. pyriformis* by the process of "genomic exclusion", which amounts to two rounds of mating. The first round leads to heterozygosity of markers differentiating the parents, and the second to homozygosity for one or other allele, at random. This leads, therefore, to ratios *equivalent* to those in the haploid meiotic products of other eukaryote microbes. By suitable techniques, the genomic exclusion pathway can be chosen by the experimenter, as an alternative to the conjugational route, with useful applications in genetic manipulation of the organism, particularly the preparation of homozygous individuals (Allen and Gibson, 1970).

10. *Slime moulds*

The genetic analysis of slime moulds is still in its infancy. However, since both groups of slime moulds, the Myxomycetes (true, acellular slime moulds) and the Acrasiales (cellular slime moulds), have been the subject of considerable interest in relation to morphogenesis and colonial interaction, a brief mention is warranted here. *Physarum polycephalum* is an example of the first group, and *Dictyostelium discoideum* of the second.

(a) *Physarum polycephalum.* Spores produced on fruit bodies hatch into haploid amoebae of four alternative mating-types (Dee, 1966a), two on any one fruit body, which can be cultured indefinitely (on bacteria). When amoebae of two different mating-types are mixed they fuse to produce a coenocytic plasmodium containing diploid nuclei. The plasmodium can also be grown indefinitely or, under the right cultural conditions, can be made to produce fruit bodies in which meiosis precedes haploid spore production (Dee, 1962). Different plasmodia will coalesce, provided that they are of suitable genotype with respect to the alleles at a specific locus distinct from that controlling mating-type (Poulter and Dee, 1968). Genetic analysis consists of the classification of the amoebae ensuing from a fruit body in respect of the markers differentiating the amoebae that formed the plasmodium on which the fruit body arose. Unfortunately there is so far a dearth of markers, only three unlinked loci having been identified controlling characters of the amoebae (Dee, 1962, 1966b; Dee and Poulter, 1970): mating-type and resistance to emetine and actidione.

The isolation of mutations affecting the plasmodium has hitherto been hampered by its diploid nature and obligate hybrid origin from amoebae of the two different mating-types. With the recent identification of a homothallic strain (Wheals, 1970) whose haploid amoebae can fuse with each other to produce homozygous diploid plasmodia it should be possible to isolate recessive plasmodial mutations by mutagenizing amoebae and allowing surviving clones to form plasmodia. The homothallic strain crosses preferentially with strains of the four standard mating-types, so that genetic analysis of the mutations, once identified in the homothallic strain, should not present problems due to selfing.

(b) *Dictyostelium discoideum*. Haploid amoebae of this organism can be cultured indefinitely in the free state, or alternatively, by restricting the supply of food, can be induced to aggregate. Within such aggregates (pseudoplasmodia), some of the haploid cells differentiate into asexual spores. It has been possible to isolate diploid strains by forming mixed aggregates between strains of amoebae bearing markers affecting growth rate or temperature-sensitivity (Loomis and Ashworth, 1969; Sinha and Ashworth, 1969; Loomis, 1969). As in some imperfect fungi (see pages 65–66) these diploids are highly unstable, rapidly reverting by chromosome loss to the haploid condition (Sinha and Ashworth, 1969). Thus part of a parasexual cycle has been described, which should lead to successful genetic analysis in this organism in which a great deal of work has already been done on colonial behaviour and morphogenesis.

VI. GENETIC ANALYSIS IN PROKARYOTES

A. General considerations

One of the most fascinating general observations to emerge from the study of bacterial genetics is the diversity of phenomena leading to heterozygosity, and so to gene reassortment, to be found amongst prokaryotes. The overwhelming proportion of research effort has been concentrated on a very few systems, which have therefore been developed in great detail, while almost every study on a different system has revealed new phenomena, or at least important variants of known ones. Consequently many new phenomena must remain to be discovered. In view of this diversity, minute practical details developed for well-studied systems may be inapplicable in a new situation; on the other hand a survey of the range of practical approaches and analytical procedures that have already yielded genetic data in particular situations could speed success in a new system. This survey is what I shall try to supply in the next few Sections of this Chapter.

5

The existence of three diverse modes of transfer of genetic material between bacterial cells was established by 1952, and since then genetic transfer in new systems has fallen under one or other of the existing headings: conjugation, transduction and transformation. It is likely that most or all new discoveries will continue to do so, since it is hard to see how genes could be transferred other than by some form of cellular contact (conjugation), by a virus (transduction) or as naked nucleic acid (transformation). Nevertheless these categories may well become more heterogeneous, particularly the first, where it is already clear that the conjugation systems of *Escherichia coli* and *Streptomyces coelicolor*, for example, are not closely similar (Hopwood, 1967, 1971).

Two features that are common to all bacterial systems in which genetic recombination has so far been studied have important consequences for genetic analysis. The first concerns the relative rarity of genetic recombination in bacterial cultures; thus cultures in which recombinants are to be sought, and in which their frequencies are to be estimated, usually contain an overwhelming majority of individuals derived from the parents by simple fission and therefore of parental genotype. This means that recombinants in general have to be *selected* from the total population, in much the same way as rare recombinant classes arising by crossing-over in short segments are selected from the total *sexual* progeny in eukaryote systems (see the earlier discussion of genetic fine-structure). This means that in any bacterial cross, with the exception of a few special cases (see page 94), some of the character differences must be "selective markers": that is alleles conferring nutritional independence for a particular growth factor, resistance to an inhibitor, etc. These markers serve to identify recombinants on a medium lacking a growth factor, containing an inhibitor, etc.; the recombinants may then be classified in respect of other, "non-selected" markers. The latter may be basically incapable of use as selected markers, for example most character differences concerned with morphology or pigmentation, or they may be further selectable markers which are treated as non-selected by adding the relevant growth factor (in the case of an auxotrophic character) or omitting an inhibitor (in the case of a resistance), etc.

The second hallmark of bacterial recombination systems is the involvement of "merodiploids"; that is, entities containing the complete genetic complement of one parent, but only an incomplete complement from the other. Thus only a portion of the genetic material can be heterozygous in any particular merodiploid. The amount varies from system to system. In the case of transduction it is most circumscribed, since only a small length of DNA can be accommodated in a transducing phage particle; the larger the phage head, the longer the piece of DNA. In the case of transformation, the length of DNA is also quite short, but much more variable, and dependent on the conditions used in preparing the DNA, which determine its molecular size. Only in the case of conjugation can the proportion of heterozygous genes extend, as a limiting case, to the full complement.

Merodiploidy imposes limitations on the detection of linkage by measuring recombination frequencies since, if two loci are not transferred into the same merodiploid, recombination between them cannot be studied in a meaningful way. On the other hand linkage tests, either qualitative or quantitative, have been devised that depend on merodiploidy itself and are largely or partially independent of the estimation of recombination frequencies between pairs of loci. These tests were developed for specific organisms; mapping by gradient of transmission and by time-of-entry have been largely confined to *E. coli*, although more recently applied to *Salmonella typhimurium* and *Pseudomonas aeruginosa*, and will be described in the Section dealing with *E. coli*. However, some prior remarks on those features of the sexual biology of the organisms that make such mapping procedures possible may be appropriate. Mapping by co-transduction and co-transformation have been more widely applied, and, although the practical details will be outlined in Sections dealing with *S. typhimurium* and *Bacillus subtilis* respectively, some general remarks on the conception of the genetic tests may be useful here.

1. *Fertility types and conjugational mapping*

In nearly all bacteria in which conjugational ability has been demonstrated, it has been found to depend on "sex factors" capable of existing, at least for part of the time, as "extrachromosomal" genetic entities, better regarded as small, supernumerary chromosomes called plasmids (Novick, 1969; Richmond, 1970). Possession of an appropriate sex factor makes a strain "male" in the sense that it can transfer pieces of chromosome into cells of a "female" strain differing from it only in lacking the sex factor. Thus genetic transfer is *"one-way"* from *"donors"* to *"recipients"*. (The exceptions, *Streptomyces coelicolor* (Hopwood, 1967) and *Rhizobium lupini* (Heumann, 1968), which so far give no evidence of such sex factors, nevertheless show sexual variants involving, at least in part, chromosomal genes (Hopwood, Harold, Vivian and Ferguson, 1969; Vivian and Hopwood, 1970; Heumann, 1970), and also showing the alternative roles of donor or recipient.) In several of these systems, genetic analysis began before the roles of the sex factors were clear, and the early crosses involved exclusively male strains. Genetic analysis was greatly simplified when male × female crosses could be made and analysed in terms of the linked transfer of genes from male to female, as we shall see in later Sections of this Chapter. A particular flavour and versatility is given to genetic experimentation with these bacteria by the possibility of interconverting male and female strains by loss or gain of the appropriate sex factor.

2. *Transduction and co-transduction*

A very useful general article on transduction is by Hartman (1963), who discussed criteria for the recognition of transduction, in addition to methods of genetic analysis dependent upon it. A useful review of transduction *mechanisms* is by Ozeki and Ikeda (1968).

(a) *General and special transduction.* Two phenomenologically and experimentally distinct classes of transduction have been recognized: general and special. Phages capable of the former can transfer any gene from a strain on which they have previously been propagated (the donor) to a second strain (the recipient); phages promoting special transduction transfer only circumscribed groups of genes adjacent to the chromosomal attachment sites of the phages. While some phages would appear to promote exclusively special transduction, for example the *E. coli* phage λ, general transducing phages have in some cases been found to promote special transduction. For example the classical general transducing phage P22 of *S. typhimurium* (Zinder and Lederberg, 1952) has more recently been found to cause special transduction of the *pro* region (Smith-Keary, 1966), while the general transducing phage, P1, of *E. coli* can promote special transduction of the *lac* region (Luria, Adams and Ting, 1960).

Recognition of special transduction, in the presence of general transduction, depends on the phenomenon of high-frequency transduction (HFT). A phage promoting special transduction, when grown lytically on the donor, transduces with a low frequency not very different from that characteristic of general transduction. On the other hand the same phage derived by induction from a donor strain lysogenic for it can transduce with a very much higher frequency; the explanation is presumably that the prophage, in the lysogenic state, is already joined to the bacterial genes adjacent to its attachment site on the bacterial chromosome, and when it detaches from the chromosome by "looping-out" (Campbell, 1962), these bacterial genes have a high probability of inclusion in a phage coat. As can be imagined, the methodologies of special and general transduction are rather different, but only general transduction will be dealt with in this Chapter.

(b) *Co-transduction.* Mapping by general transduction depends on the concept of co-transduction: since only genes rather close on the bacterial chromosome (not more than one phage genome's equivalent of DNA apart) can be carried by the same transducing phage, joint transfer, that is co-transduction, is in itself diagnostic of close linkage. Co-transduction is detected by adding phages grown on a donor strain bearing a selectable allele to a recipient strain bearing the corresponding counter-selected allele, and testing for the transfer of another marker along with the first. Suppose we have two strains *thi arg⁺* and *thi⁺ arg* (*thi* = thiamine requirement; *arg* = arginine requirement). Phages can be grown on the first strain, which is thus chosen as the donor, and added to the second (the recipient), which is then plated on a medium lacking arginine but containing thiamine: selection is thus made for transfer of the *arg⁺* allele from the donor to the recipient, and its inheritance by recombinant progeny of the merodiploid, while *thi⁺/thi* is unselected. A sample of the *arg⁺* progeny growing on the selective plates is then classified as *thi⁺* or *thi*. If, say, 60% are *thi*, the frequency of co-transduction of the *arg⁺* and *thi* markers is said to be 60%.

Clearly the co-transduction frequency will depend on the distance apart of the two loci; however, the precise relationship between the two is not simple. Sufficient evidence of this is provided by the not uncommon finding (Eggertsson and Adelberg, 1965; McFall, 1967; Taylor and Trotter, 1967) that the co-transduction frequency between two loci is not the same in reciprocal crosses. In the example above, reciprocal crosses would consist of selecting for transfer of *arg⁺* from strain 1 to strain 2, with *thi⁺/thi⁻* non-selected, and alternatively selecting for transfer of *thi⁺* from strain 2 to strain 1, with *arg⁺/arg* non-selected.

Two conditions must be satisfied if both donor markers, selected and non-selected, are to be inherited by the same recombinant: (a) both markers must be included on the same fragment of bacterial chromosome contained in the transducing phage; (2) both markers must be incorporated, by appropriately positioned cross-overs, into a recombinant chromosome. The primary reason for the finding of unequal co-transduction frequencies in reciprocal crosses, just referred to, may well be the non-random inclusion of donor markers in transducing fragments, for which there is a good deal of evidence. One aspect of the problem concerns the overall probability of inclusion of single markers in such fragments; the second concerns the probability of inclusion of particular combinations of linked markers. There has been considerable discussion of the latter aspect since Ozeki's much-quoted paper (Ozeki, 1959), which proposed that the ends of generalized transducing fragments of *S. typhimurium* carried by phage P22 occurred at a limited number of fixed positions. This conclusion was later shown to be an over-simplification (Pearce and Stocker, 1965; Roth and Hartman, 1965). The most likely conclusion is that, although ends can occur in very many positions, the probability of occurrence at certain positions is much higher than at others; this would appear to be true also for phage P1 of *E. coli* (McFall, 1967), and no doubt in other systems.

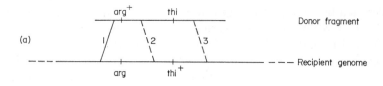

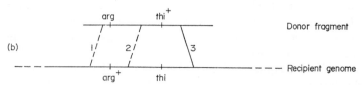

Fig. 12. The effect of non-random transducing fragments on the incorporation of non-selected markers into recombinant genomes. (a) and (b) represent reciprocal crosses. The transducing fragment introduced into a merodiploid is assumed to have ends in fixed positions, resulting in regions (1 and 3) of different length on either side of the pair of loci under consideration. In (a), selection is made for the allele *arg*+ from the donor; thus crossing-over in interval 1 is obligate, while the second cross-over can occur in either 2 or 3. Crossing-over in 3 leads to co-transduction of the donor allele *thi* with the selected *arg*+, while crossing-over in 2 leads to incorporation of *thi*+ instead of *thi* and so to no co-transduction. In (b), selection is for *thi*+, with obligate crossing-over in 3 and the second cross-over in 1 or 2; 1 leads to co-transduction and 2 does not.

This feature could affect co-transduction frequencies in two ways: firstly via the joint introduction of the two markers into the same merodiploid, and secondly via crossing-over within the merodiploid to produce recombinants.

On the first point, the frequencies of three classes of transducing fragment affect co-transduction of two loci, *thi* and *arg*: those carrying locus *thi* alone, those carrying locus *arg* alone, and those carrying loci *thi* and *arg* together. Suppose the frequencies of the first and third classes of fragments are considerable, but that of the second is much lower, then the co-transduction frequency determined by selecting at locus *thi* will be lower than that determined by selecting at *arg*, since many of the fragments carrying *thi* will not carry *arg*, but most of those carrying *arg* will also carry *thi*.

The effect of non-random transducing fragments on recombination in the merodiploid can be seen in Fig. 12. When selection is for *arg*+ (Fig. 12(a)), crossing-over to incorporate this marker into a recombinant must occur in interval 1, and in either 2 or 3: the co-transduction frequency will be the ratio of crossing-over in 3 to that in 2 or 3, expressed as a percentage. On the other hand, when selection is for *thi*+ (Fig. 12(b)), the co-trans-

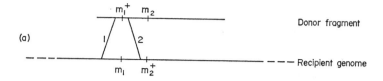

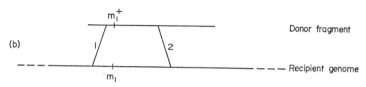

FIG. 13. Recognition of linkage between two mutations (m_1 and m_2) by trans-duction of a recipient auxotroph alternatively by means of phages grown on a second auxotroph or on wild-type. In (a), the auxotrophic recipient carrying mutation m_1 was transduced by phages grown on a second auxotroph, carrying m_2. In order to produce prototrophic recombinants, crossing-over in interval 1 has to be accompanied by crossing-over in the short interval 2 between the two mutations. In (b), transduction is by phages grown on a wild-type donor; in this case the second cross-over can occur anywhere in the long interval 2 between m_1 and the right-hand end of the transducing fragment. Consequently the proto-troph frequency is much higher in (b) than in (a).

duction frequency will be the ratio of crossing-over in 1 to that in 1 or 2. Clearly, if the average lengths of intervals 1 and 3 differ, as they may well do if the transducing fragments have non-random ends, the estimated co-transduction frequency will differ in the two experiments.

The moral from these considerations is that the measurement of co-transduction frequencies is not always straightforward, so that precise gene sequencing by co-transduction frequencies alone may be subject to considerable error, just as mapping by two-point data in other systems (see pages 39 and 45) may be ambiguous. Nevertheless this is a valuable method for arriving at preliminary information on gene sequence, and moreover the finding of a measurable co-transduction frequency is con-clusive evidence for rather close linkage; the closeness will depend on the maximum size of the transducing fragments, which will be longer the larger the head of the particular phage concerned: for example P22 of *S. typhimurium* can transduce of the order of 1% of the whole chromosome, whereas the very large phage PBS1 of *B. subtilis* can carry up to 10% (Dubnau *et al.*, 1967). However, the latter is probably unusually large. Thus transduction alone is not capable, at least in the early stages of genetic analysis of an organism, of yielding a complete linkage map, but only a

TABLE III

Two-factor transduction crosses between *cys* mutants of *Salmonella typhimurium* (from Clowes, 1958)

Recipient bacteria carrying *cys* mutation number	Phages grown on *cys* mutant number								
	A3	A20	B4	B10	B18	B24	C7	C38	D23
A3	**0**	**0**	157	185	287	141	47	205	155
A20	**0**	**0**	171	247	206	120	52	62	125
B4	102	53	**0**	**0·5**	**1·8**	**0·9**	61	24	75
B10	52	32	**0**	**0**	**4·2**	**1·6**	40	51	59
B18	88	132	**3**	**0·6**	**0**	**0·4**	102	89	45
B24	75	46	**3**	**1·5**	**0·5**	**0**	56	93	126
C7	133	148	131	68	94	129	**0**	**0·1**	10*
C38	152	147	146	105	150	133	**0·9**	**0**	5*
D23	59	119	113	83	91	40	6·3*	3·6*	**0**

Each figure represents the value:

$$\frac{\text{number of } cys^+ \text{ transductants with phage grown on } cys \text{ mutant donor}}{\text{number of } cys^+ \text{ transductants with phage grown on } cys^+ \text{ donor}} \times 100$$

Figures in bold type indicate co-transduction of the *cys* mutations in donor and recipient. Note that the figures marked with an asterisk are rather larger than the others that indicate co-transduction. This reflects linkage of genes *cysC* and *cysD*, but with lower co-transduction frequencies than for mutations within the same gene, e.g. A3 × A20, B4 × B10, etc.

series of small fragmentary linkage groups which may later be joined by conjugational studies (as happened in *S. typhimurium*: page 95), or by exhaustive transduction tests, as in *E. coli* (page 105).

An alternative approach to the detection of linkage between two markers, when both are selectable, is to estimate the frequency of transductants when selection is made for both markers, one in the donor and one in the recipient, and compare this with the frequency when selection is made for a donor marker only. This method has been used in many studies of linkage between a set of mutations, each conferring auxotrophy for the same compound. Phages are grown on one mutant and used to transduce a second; the frequency of transductants observed is compared with that when phages grown on the wild-type are used to transduce the same mutant. If the first frequency is appreciably less than the second, the two mutations can be deduced to be linked (Fig. 13). An example of such a study is that of Clowes (1958): Table III.

A possible source of error in such studies is the innate variation in the frequency with which particular markers are transduced, in part due to effects ("marker effects") of particular mutations on the probability of

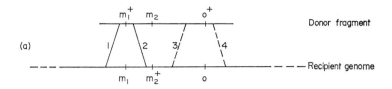

(a)

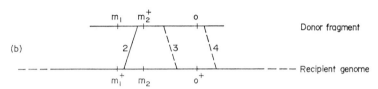

(b)

Fig. 14. Marker sequencing by a three-factor transduction cross involving an outside marker. (a) and (b) are reciprocal crosses between two closely linked mutations, m_1 and m_2. In each cross, m_1^+ and m_2^+ are selected, and the outside marker o/o^+ is non-selected. In (a), cross-overs in 1 and 2 are obligate; the result is incorporation of the recipient non-selected allele (o), *unless* two extra cross-overs, in 3 and 4, occur. In (b), crossing-over in 2 is obligate, with a second cross-over in 3 or 4, leading to incorporation of o^+ or o respectively. Thus in (a), the great majority of selected recombinants have the recipient outside marker, whereas in (b), there is a fair segregation of donor and recipient alleles. If the alternative sequence of m_1 and m_2 had been adopted, the expectations in (a) and (b) would have been reversed.

crossing-over in their immediate neighbourhood (Hartman, Loper and Šerman, 1960; Balbinder, 1962). However a matrix of crosses involving many possible combinations of a set of mutations will usually give sufficiently self-consistent data to overcome fluctuations in individual crosses.

In view of all these factors, sequencing is best done by crosses involving three or more markers, one of which is selected from the donor while the others are non-selected (Fig. 14). A typical analysis of this type is by Demerec, Goldman and Lahr (1958).

A mathematical treatment for the calculation of map distances from three-point transduction data when each of the three donor markers can be selected in turn was provided by Wu (1966). He assumed random transducing fragments, an assumption that seemed to be justified for transduction of his markers by P1 in *E. coli*, as judged by the good fit of his calculated values for the frequencies of each class of progeny with those observed. However, this was possibly a fortunate coincidence over the particular map intervals considered since, as mentioned earlier (page 85), there is some evidence for non-random ends even with P1.

(c) *Functional genetic tests by transduction.* In many genetic systems based on general transduction, no means have been discovered to test dominance or complementation, since heterozygosity is not prolonged enough for functional tests to be performed. Exceptions are provided by those systems in which "abortive transduction" occurs, as first described for the *S. typhimurium*–phage P22 system (Stocker, Zinder and Lederberg, 1953; Ozeki, 1956). Abortive transduction occurs when a donor chromosomal fragment introduced into the recipient cell fails to undergo crossing-over with the recipient chromosome. The fragment is unable to reproduce itself, but it is not destroyed, and is inherited by one of the pair of cells produced when the originally transduced cell divides. This process is repeated at each cell division, and so if the original cell gives rise to a colony, *one* cell in the colony at any time harbours the donor fragment, and is thus heterozygous for any genes on the fragment that differentiated the donor and recipient. If two complementary auxotrophic mutations are involved, complementation within one cell of the colony may be sufficient for limited cellular division of the colony on a selective medium. Thus a complementation (or dominance) test involves detecting the colonies due to abortive transduction, which are minute in comparison with those of the normal (complete) transductants arising on the same selective plates as a result of integration of the donor fragment by crossing-over. The test certainly needs practice to perform accurately, but it has been used successfully in some genetic studies of particular biosynthetic pathways in *S. typhimurium* (Hartman, Hartman and Šerman, 1960; Margolin, 1963).

Special transduction systems allow complementation and dominance tests to be made by virtue of the fact that comparatively stable heterozygotes can arise following introduction of the transducing fragment into a recipient cell and its integration into the host chromosome (e.g., Buttin 1963, 1968).

(d) *Practical methodology of transduction.* Much of the methodology of transduction depends on the general techniques of phage work (Adams, 1959). These include methods of preparing phage stocks, freeing them from viable bacterial cells and cell debris, determining their titre in terms of plaque-forming units, etc. Particular points of technique involve the use of conditions favouring the lytic cycle of a temperate phage when transducing phages are to be prepared on the donor strain, and conditions favouring lysogenization and survival of the recipient during transduction. (For details of phage techniques see Billing (this Series, Vol. 3B) and Kay (Volume 7A).

(e) *Transduction by virulent phages.* One can imagine that a temperate transducing phage may not always be available for a bacterial strain in which a transduction system would be desirable. A possible solution may lie in the use of a virulent phage, taking encouragement from the successful transduction by virulent T1 phage reported by Drexler (1970) for *E. coli*. The problem is that a transductant arising as a result of transfer of a piece of bacterial chromosome by such a phage would normally be killed by virulent phages released from other cells in the recipient population infected with viable phages; this would be a problem even if the multiplicity of infection were low enough for the transduced cell itself to escape simultaneous infection by a viable phage. Drexler overcame this problem by using *amber* phage mutants grown on permissive donors to transduce non-permissive recipients; thus phage reproduction in the recipient culture could not occur. This approach would not be available in the early stages of genetic analysis of a new species, but could perhaps be applied if and when known nonsense mutants, or other conditional lethals, became available.

3. *Transformation and co-transformation*

(a) *Transformation in genetic analysis.* The classical studies of transformation were made with nutritionally exacting bacteria, the pneumococcus and *Haemophilus influenzae*. Thus, although the principles of genetic mapping by transformation were worked out in these bacteria, and some studies were made of particular short chromosome segments concerned with capsular synthesis (Jackson, 1962; Mills and Smith, 1962) or antibiotic resistance (Goodgal, 1961), progress in the overall genetic description of these species was slow. Only with the discovery of transformation in a nutritionally non-exacting species, *Bacillus subtilis*, was it possible to use the full range of nutritional genetic markers and to arrive at a reasonable development of the linkage map (Dubnau *et al.*, 1967). This is not to minimize the importance of the studies on the pneumococcus and *H. influenzae* since the unique contribution of transformation has been the possibility it offers for investigating *in vivo* the results of *in vitro* manipulation of DNA (e.g., Fox, 1962; Lerman, 1963); however, such studies are outside the scope of this Chapter, in which we shall deal with transformation as a mapping procedure, and therefore primarily in relation to *B. subtilis*.

One defect of transformation in genetic analysis is its current inability to provide the basis for complementation and dominance studies, since it does not lead to heterozygosity prolonged enough to be harnessed to these ends.

A phenomenon of the utmost importance in the practical methodology of transformation is *competence* (see Hayes, 1968: pp. 583–591; Tomasz, 1969): that is the state of a recipient culture in which cells can be effectively transformed. The exact nature of competence is still poorly understood;

moreover the proportion of competent cells in a culture at the peak of competence, and the growth conditions required to bring about competence, vary widely with different species. Consequently, in a fruitless search for transformation in a new system, one may be left with the frustrating feeling that failure to demonstrate transformation may well have been due to a failure to find the right conditions to achieve competence in the recipient rather than to an innate inability of the cells to be transformed.

(b) *Co-transformation*. The normal criterion for linkage in transformation systems is co-transformation, a concept analogous to co-transduction. A fairly high frequency of co-transformation is unambiguous evidence for linkage, although the precise relationship between co-transformation frequency and distance is not simple. However, the interpretation of low frequencies of co-transformation is more difficult than in the case of co-transduction. The reason is that, whereas two independent transducing fragments are not normally introduced into the same cell, it is perfectly possible for a recipient cell to pick up more than one separate molecule of transforming DNA. Thus the mere inheritance, by a recombinant, of two donor markers introduced by transformation is not sufficient evidence of close linkage, since they could have been introduced on two separate DNA molecules deriving from well separated regions of the donor chromosome. (This can be directly demonstrated by mixing transforming DNA derived from two separate donor strains, each bearing a separate marker: such mixed DNA can transform a recipient strain in respect of both markers (Goodgal, 1961; Nester, Schafer and Lederberg, 1963).)

The problem is resolved by considering the frequency of joint transformation of the two markers in relation to their frequencies of independent transformation, at different concentrations of transforming DNA. The proportion of recipient cells transformed for a given marker is proportional to the DNA concentration, over a suitable range up to saturation. The same is true of the co-transformation of two markers borne on the same DNA molecule. However, if transformation of the two markers results independently, by the incorporation of two separate DNA molecules, the proportion of joint transductants increases as a higher order of the DNA concentration. Another way of analysing the data is to consider the proportion of transformants in respect of a first marker which are co-transformed for a second marker, as a function of DNA concentration (Fig. 15). In the case of linked markers, this proportion remains more or less constant, whereas with unlinked markers the proportion falls as the DNA concentration is reduced.

For gene sequencing, two-point selections involving pairs of auxo-trophic mutations, analogous to those mentioned for transduction (page

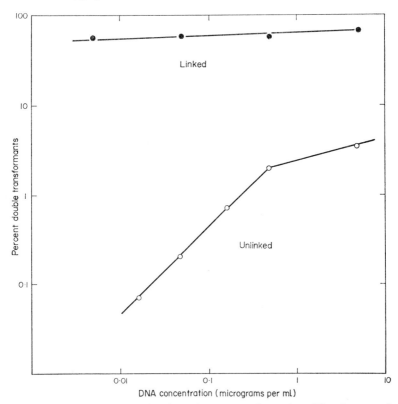

FIG. 15. The recognition of linkage by co-transformation. The data are from Kelly and Pritchard (1965). In each experiment, transformants inheriting ind^+ from the donor were selected and classified in respect of a non-selected donor marker, met^+ (closed symbols) or tyr^+ (open symbols). In the case of *linked* markers, the proportion of double transformants (that is the proportion of ind^+ that are also met^+) is almost independent of DNA concentration, whereas for *unlinked* markers (ind^+ and tyr^+), the proportion of double transformants falls rapidly with decreasing DNA concentration.

88), have been performed (Dubnau *et al.*, 1967), as well as sequencing by three-point analysis (Nester, Schafer and Lederberg, 1963).

(c) *Gross mapping by transformation.* As in the case of transduction, the normal genetic analysis by transformation leads to a fragmentary linkage map consisting of a number of independent short linkage groups. The limits of the separate groups are less clearly defined, owing to the much greater variability in the size of transforming DNA fragments, as compared with transducing fragments; it is possible to increase the length, and so the ability of the analysis to detect linkage over longer intervals, by very gentle handling of the transforming DNA during is extraction and purification, resulting in a higher average molecular weight of the DNA (Kelly and Pritchard, 1965).

Ingenious means for harnessing transformation to yield information on gene location over the whole linkage map of the organism were found by Yoshikawa and Sueoka (1963a, 1963b) for *B. subtilis*. These methods (marker frequency analysis and density transfer) have so far been confined to particular strains of this species, and so, although they have general biological interest, they will be deferred until the genetics of *B. subtilis* is described in more detail (pages 114–115).

4. *Non-selective analysis*

As we have seen, bacterial crosses have characteristically to be analysed selectively: the medium on which recombinants are recovered has to select against one (in the case of transduction and transformation) or both (in the case of conjugation) parental classes. A few exceptions to this state of affairs have been described. In *E. coli*, advantage has been taken of the high rate of mating between male and female cells to recognize mating pairs microscopically in a mixed culture of an Hfr with an F⁻ strain, and separate them by means of a micromanipulator (Lederberg, 1957; Anderson, 1958). Individual pairs were then allowed to segregate progeny for a few cell generations before these were separated, cultured individually, and characterized in respect of all markers that differentiated the parent strains. The outcome was not so much a technique of genetic analysis, but rather an insight into the events immediately following gene transfer.

In *Streptomyces coelicolor*, the discovery of a genetic system controlling fertility has allowed the non-selective analysis of crosses involving strains of two different fertility types, NF (normal fertility) and UF (ultra-fertility) (Hopwood, Harold, Vivian and Ferguson, 1969). Such crosses yield up to 100% of sexually produced progeny, so that all classes of progeny, even those having all of the nutritional requirements of both parental strains, can be recovered and their frequencies estimated by plating on a non-selective medium (see pages 124–125).

B. Genetic analysis in particular organisms

1. Escherichia coli, Salmonella typhimurium *and other Enteric bacteria*

Although genetic analysis of *E. coli* and *S. typhimurium* started along quite separate paths, the methodology applicable to the two species has more recently converged. Moreover the two organisms are clearly quite closely related; they will mate with each other, provided suitable fertility types are used, and their linkage maps are almost identical (Sanderson,

1970). At first, genetic analysis in *E. coli* made use exclusively of gene transfer by the sex factor (F) in conjugation; this system permits gross mapping, and led to the finding that the linkage map of *E. coli* is a single circle (Jacob and Wollman, 1958). Conjugation also permitted fine-structure mapping, but with the later discovery of a generalized transduction system, using phage P1 (Lennox, 1955), this largely replaced conjugation as a fine mapping tool. Genetic analysis of *S. typhimurium*, on the other hand, began with generalized transduction (Zinder and Lederberg, 1952), which led to very detailed knowledge of gene locations within a number of small separate linkage groups; however this method was not applicable to the problem of joining these separate linkage groups into a gross map. The latter became possible with the development of a conjugation system, at first depending on transfer by colicinogenic (*col*) sex-factors (Smith and Stocker, 1962); however, this system was rather inconvenient, partly due to the fact that males were efficient donors only when they harboured a recently transferred *col* factor, a situation later explained by the natural repression of sex-factor activity (Meynell, Meynell and Datta, 1968); and partly due to the absence of donors analogous to *E. coli* Hfr strains. Conjugation studies in *S. typhimurium* became much more feasible when the *E. coli* sex-factor, F, was transferred by conjugation to the new host, and allowed to integrate to give Hfr donors; a complete linkage map was then obtained (Sanderson and Demerec, 1965).

Thus today, both gross and fine mapping, by conjugation and generalized transduction respectively, are very effectively performed in *E. coli* and *S. typhimurium*, and their circular linkage maps, each with more than 250 genes (Sanderson, 1970; Taylor, 1970) are the best known of any micro-organisms with the exception of some phages. In *E. coli*, the linkage map is so thickly marked, and the length of DNA carried by the generalized transducing phage P1 is not too short, so that almost any new gene can be found to be co-transduced with a known marker; however, conjugation analysis has not been superseded, since it still provides a good indication of the approximate position of a new gene, and hence of the known markers that are likely candidates for co-transduction studies. Moreover, a few segments of the map are still comparatively empty (Taylor, 1970) and un-bridged by transduction analysis.

Complementation and dominance studies, also, began in quite different ways in these two bacteria, but later converged. At first the only hetero-zygous stage in the *E. coli* system was the newly fertilized female cell, or zygote; however, heterozygosity was lost very quickly as integration of parts of the incoming male chromosome by crossing-over took place, so that functional tests only of rapidly expressed genes could be made (Pardee, Jacob and Monod, 1959). Later, abortive transduction by phage P1 became available (Gross and Englesberg, 1959) but does not appear to have been widely used. A great advance in the experimental genetics of E. coli was the discovery and understanding of the F′ strains (Jacob and Adelberg, 1959), which were found to be comparatively stable partial diploids, and these are now the normal basis of functional genetic tests in E. coli. In S. typhimurium, on the other hand, early dominance and complementation studies had to make use exclusively of abortive transductants. More recently, with the discovery of sex-factor transfer from E. coli to S. typhimurium, F′ factors also have been transferred, and a number of successful functional tests have been performed on the resulting "hybrid" partial diploids, as well as on pure Salmonella merodiploids containing Salmonella F′ elements (see page 109).

(a) Genetic analysis in E. coli

i. *Conjugational analysis in* E. coli *K12.* A very useful account of the methodology of conjugational analysis is by Hayes, Jacob and Wollman (1963). The general biology of conjugation is fully discussed by Hayes (1968) and, earlier, by Jacob and Wollman (1961). The mechanisms involved in conjugation are reviewed by Curtiss (1969).

Mating. Although, as we shall see, the conjugation system may be analysed in more than one way, the basic methodology of mating is rather uniform. Since mating is virtually always performed between an Hfr (high-frequency recombination) male donor and an F⁻ female recipient (rather than in the much less fertile F⁺ × F⁺ or F⁺ × F⁻ combinations) a first prerequisite is the availability of strains that are not only genetically complementary but also of suitable sex (see Hayes, Jacob and Wollman, 1963, p. 140). In isolating new mutations, the possible interconversions of the sexual types has to be borne in mind. In a sense, the F⁺ is the key type, since it can lose the sex-factor to become F⁻, and integrate it to become Hfr. The frequency of loss of the sex factor from F⁺, but not Hfr cells, is greatly increased by growth in sub-lethal concentrations of acridine dyes (Hirota, 1960), providing a useful routine method of preparing female strains. Hfr donors are isolated from F⁺ strains either by replica plating (Jacob and Wollman, 1956), or by a modification of the fluctuation test of Luria and Delbrück (1943). In this method, samples of a number of small sub-cultures of the F⁺ strain are tested separately for their ability to mate efficiently with an F⁻ tester strain; having identified a culture with high donorability, and therefore containing an appreciable proportion of Hfr cells, these can be identified by replica plating (Jacob and Wollman, 1956). F⁻ strains can be converted to F⁺ males by infectious transfer of the sex factor. The progeny of Hfr × F⁻ matings are almost all F⁻, since the sex factor is transferred only with the terminus of the chromosome.

Diagnosis of maleness can be done by testing for donorability, either by a rapid replica plating test (Hayes, Jacob and Wollman, 1963, p. 140) or by mating in broth. An alternative very useful test is for sensitivity to a male-specific phage such as MS2, or, if an F*lac* sex factor is to be transferred to a *lac*⁻ recipient, by testing for *lac*⁺ cells bearing recipient chromosomal markers.

Occasionally, use can be made of the fact that male cells (either Hfr or F⁺) can be converted to the female phenotype (so called "phenocopies": Lederberg, Cavalli and Lederberg, 1952) by conditions that remove the sexual appendages, called sex-pili: either incubation of a stationary phase culture with aeration, or by treatment with periodate (Sneath and Lederberg, 1961). Such emasculated cells will then accept Hfr mates, and crosses that would otherwise be sterile, or nearly so, can be performed (e.g., Clark, 1963).

Male *E. coli* cells mate efficiently only when they are actively growing in the exponential phase. The cell density in the mating mixture has to be high if contacts between cells are to be frequent; routinely, conditions are arranged so that nearly every male rapidly finds a mate, the females being in an excess of about tenfold or more; recombinant frequencies are related to the input of males, the excess of frustrated females being ignored. To achieve the above requirements, stationary (overnight) broth cultures of the two parental strains are diluted, say fiftyfold, into fresh broth and grown with shaking, or other means to ensure aeration, usually at 37°C, until the cell density approaches about $1–3 \times 10^8$. If one strain is seen to be growing much faster than the other, it can be held back by plunging in an ice bucket, and then re-started after an appropriate delay, provided enough time is allowed for the cells to begin exponential growth again before mating. Mating is started by pipetting, say, one-tenth volume of the male cells into the female culture and mixing *rapidly*; thereafter the mating mixture must be handled gently to avoid forcibly separating the paired cells, and not chilled, if reproducible results are to be obtained, since the rate of chromosome transfer is very sensitive to temperature. However, the culture must remain aerated if the males are not to run out of energy; a satisfactory solution is to incubate about 2 ml of mixed culture in a wide (25 mm), inclined test-tube in a *gently* reciprocating water bath, or about 10 ml in a 125 ml Erlenmeyer flask.

Subsequent handling of the experiment depends on the type of analysis being performed: time-of-entry, transfer gradient, or recombinational analysis.

Time-of-entry analysis. Most commonly, mapping is by this method, and for this the famous interrupted mating technique of Wollman and Jacob (Wollman, Jacob and Hayes, 1956) is used. The principle is to shear the mating pairs apart at various time intervals after mixing, and plate on a series of media each selecting a different male marker. Each marker, lying at a fixed distance from the "origin", or leading end of the donated Hfr chromosome, enters the female cells at a characteristic time; since not all males find their partner immediately the two cultures are mixed, it is the *earliest* time at which recombinants inheriting a particular marker can be selected that is important.

The practical details of an interrupted mating experiment vary somewhat in different laboratories, but the common features are: the withdrawal of small samples from the mating mixture (without chilling it) at defined times after mixing the parent strains; separating the mating pairs, usually by mechanical agitation; diluting the samples in order to achieve a suitable colony count on the plates on a series of media all counter-selecting a common male marker known to be transferred after the markers to be mapped, and each selecting a different male marker. The numbers of colonies on each medium at each time interval are counted after suitable incubation of the plates, and plotted against the time interval to give time-of-entry curves as in Fig. 16.

Originally, mating cells were separated in a Waring-type blendor or alternatively by killing the male cells with T6 phage, to which the female was chosen to be resistant (Hayes, 1957); more recently, various mechanical shaking devices have been used. A particularly useful technique was described by Low and Wood (1965), who found that reciprocal shaking in their apparatus for only 5 seconds was enough to shear the mating pairs. In their technique, samples of cells from the mating mixture were diluted by adding them to aliquots of soft agar which were shaken and then poured over base-layers of appropriate selective medium, so that withdrawal, dilution, shaking and plating could be completed in half a minute or so, allowing one person to sample the mating mixture at least every minute over the critical time period and thus determine accurate times of entry.

The estimation of accurate times of entry is not always entirely straightforward. When relatively distant markers are studied and samples are taken relatively infrequently (Fig. 16(a)), each entry curve rises almost linearly from the base line, and the linear part of the curve, before the plateau is reached, can simply be extrapolated to the time axis. However, when more frequent samples are taken, it is regularly observed that the curve has a non-linear portion over the first couple of minutes of entry of each marker (Fig. 16(b)). This is handled differently by different authors; some ignore it and extrapolate the linear portion, while others take account of the lag part of the curve (e.g., Taylor and Trotter in Fig. (16b)). An additional complication arises with closely linked markers near the origin, and can lead to a "pseudoinversion" in marker sequence (Glansdorff,

1967). The explanation is probably that to be mentioned shortly in connection with transfer gradient analysis—that the frequency of incorporation of a donor marker very near the origin may be depressed and thus the time-of-entry curve of a slightly later marker may actually appear to start earlier.

The map constructed in this way in terms of time units has been shown to bear a close relationship with the physical distribution of genes on the chromosome in terms of DNA. This was demonstrated by the experiments of Fuerst, Jacob and Wollman (1956) in which P^{32} labelled Hfr cells were stored in liquid nitrogen for varying periods before mating with female cells. Decay in the efficiency of transfer of each male marker was directly related to its map distance from the origin. The total length of the map, obtained by summing various shorter intervals turns out to be some 90 min (Taylor, 1970).

Transfer gradient analysis. This method of analysis depends, not on the artificial interruption of mating, but on the fact that spontaneous rupture of the chromosome occurs with a certain probability which is more or less constant with time. Thus the earlier a marker is transferred, the larger the proportion of zygotes that it enters and the higher the frequency of recombinants inheriting that marker. This is shown by the successively lower plateaux on the time-of-entry curves for later markers in interrupted mating experiments (Fig. 16(a)). A mating mixture is prepared as above and allowed to mate for a considerable time (say 90 min) before dilution and plating on a series of selective media as before. The colony counts for the various markers are related to their distance from the origin, so that when an experiment incorporates several known markers, and an unknown, the position of the new marker can be obtained by plotting the colony counts of the known markers against their map position and interpolating the new marker. A complication that may arise is the finding of a reduced transfer frequency for a very early marker (one transferred in the first 4–5 min of mating: Low, 1965). This effect is probably due to the requirement of a cross-over between the origin and a marker in order to integrate the donor allele into a recombinant; if this distance is short, such crossing-over has a reduced chance of occurring. The consequence is that the transfer frequency of a new marker may sometimes give an ambiguous indication of its map position. The solution is to use a second Hfr strain with its origin at a different position.

The main function of marker frequency analysis in *E. coli* is not to provide a definitive map location for a new marker, which is much more accurately done by time-of-entry, but to give a preliminary indication from which a choice of a suitable Hfr strain for accurate mapping can be made.

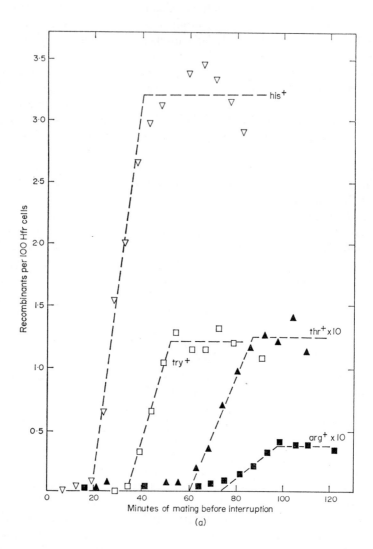

Fig. 16. Time-of-entry curves for donor markers in interrupted mating experiments between Hfr donors and F⁻ recipients of *Escherichia coli* K12. (a) Data of Low and Wood (1965); (b) data of Taylor and Trotter (1967). In both experiments the male (Hfr) strain was streptomycin-sensitive (str^s) and the female (F⁻) streptomycin-resistant (str^r). Recombinants inheriting str^r from the female and one of

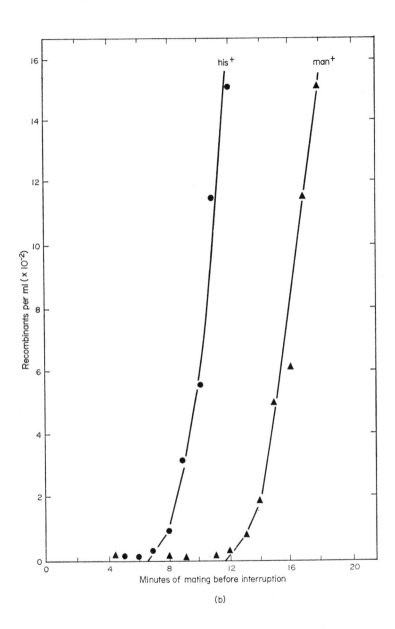

(b)

several Hfr markers were selected. In (a), note the decreasing height of the plateau for Hfr markers with progressively later entry times, due to spontaneous chromosome rupture. (Note the expanded scale for the last two markers.) In (b), note the concave initial region of the entry curves.

Recombination analysis. Mapping by recombination analysis in *E. coli* provides one of the closest analogies currently available to genetic analysis in a eukaryote. The mating is made as before, and allowed to procede uninterrupted. However, at the end of the mating period, samples of the mixture are plated on a medium selective for a particular donor marker transferred comparatively late (that is a distal marker); the medium is non-selective for other markers, between the selected marker and the origin (that is more proximal markers). An even more distal marker is used to counterselect the Hfr parent. When recombinant colonies appear, a sample of them is characterized in respect of the non-selected markers, usually by picking them off in a regular array to a "master" or "patch" plate and replicating this plate with velvet to a series of media each diagnostic for one non-selected marker. In this way the genotype of a sample of recombinants is determined, in respect of the series of non-selected markers.

Published data for medium-range mapping obtained by this method of analysis are not very numerous, probably because time-of-entry provides a more accurate measure of map distance between most pairs of loci owing to its linear relationship with distance (see above), whereas recombination values are subject to errors including those of the general kind due to multiple crossing-over discussed in Section II of this Chapter. However, in the case of certain unselectable markers, this form of analysis is useful (Cooper, Hirshfield and Maas, 1969).

One of the most extensive sets of data of this kind was published by Jacob and Wollman (1961 : Table 35), and others were provided by De Haan and Verhoef (1966: Table 3). The briefer data of De Haan and Verhoef are analysed in Table IV, and we see that they resemble data that could have been obtained in a eukaryote, except for the constraints imposed by merozygosis, which introduce two further map intervals, x and y, extraneous to the markers.

We see from the table that we cannot consider the four pairs of complementary genotypes in respect of the three non-selected markers simply in terms of crossing-over in the intervals (2 and 3) between them, since various patterns of crossing-over in the remaining intervals are also required to produce them. One approach, ignoring interference, would be to estimate the *relative* lengths of intervals 1, 2 and 3 from the frequencies of the three simple classes of progeny, those arising by crossing-over in intervals x, 1; x, 2; and x, 3. This would involve a minimum of assumptions but would use only part of the data. Instead, Verhoef and De Haan (1966) and De Haan and Verhoef (1966) provided a mathematical treatment for the calculation of relative map distances from the whole data. These distances, which reflected closely the ratio of the simple cross-over classes, were shown to be consistent with those based on time-of-entry studies so

TABLE IV
Recombination analysis in *Escherichia coli* K12 (from de Haan and Verhoef, 1966)

	100	73	69	56	
	ade^+	thr^+	leu^+	pro^+	

$$-- \!\!\! -\!\!\!-\!\!\!-\!\!\!\triangle\!\!\!-\!\!\!-\!\!\!-\!\!\!|\!\!\!-\!\!\!-\!\!\!|\!\!\!-\!\!\!-\!\!\!|\!\!\!-\!\!\!\longrightarrow \quad \text{Donor chromosome}$$

x 1 2 3 y

$$---\triangle\!\!\!-\!\!\!-\!\!\!|\!\!\!-\!\!\!-\!\!\!|\!\!\!-\!\!\!-\!\!\!|\!\!\!-\!\!\!-\!\!\!|\!\!\!-\!\!\!-- \text{Recipient chromosome}$$

str^r	*ade*	*thr*	*leu*	*pro*	
100	0	27	31	44	100

Non-selected marker combination			Cross-overs in intervals	Number of progeny	Cross-overs in respect of intervals 2 and 3
thr^+	leu^+	pro^+	x, y	664 ⎫ 906	None
thr	*leu*	*pro*	x, 1	242 ⎭	
thr^+	*leu*	*pro*	x, 2	66 ⎫ 89	2
thr	leu^+	pro^+	x, 1, 2, y	23 ⎭	
thr^+	leu^+	*pro*	x, 3	337 ⎫ 462	3
thr	*leu*	pro^+	x, 1, 3, y	125 ⎭	
thr^+	*leu*	pro^+	x, 2, 3, y	38 ⎫ 55	2 and 3
thr	leu^+	*pro*	x, 1, 2, 3	17 ⎭	
		Total		1512	

A sample of 1512 recombinants inheriting ade^+ from the Hfr (donor) and str^r from the F⁻ (recipient) was classified in respect of the three non-selected markers. The diagram at the top of the table gives the percentage frequency of each allele in this sample, while the table gives the number of recombinants of each genotype and the intervals in which crossing-over was required to produce it. (Triangles indicate the selected alleles).

that, provided one map interval was known in minutes, the others could be calculated from it. This treatment succeeded even for data from an interstrain cross, *E. coli* K-12 Hfr × *E. coli* B F⁻, in which the frequency of incorporation of donor markers was very different from that in a K-12 × K-12 cross.

Recombination analysis combined with interrupted mating can be used to sequence two markers whose times of entry in a normal interrupted mating experiment are indistinguishable (Wann, Mahajan and Wood, 1970). The approach is to study the two markers in the same cross, selecting each in turn, with the other unselected, after interrupting the mating at various time intervals; the order of markers is deduced from the asymmetry in the inheritance of the non-selected marker. In the example given by Wann et al. (1970), when the distal marker, $purF^+$, was selected, the non-selected donor marker, $aroC^+$, was inherited by 95% of the selected progeny, no matter at what time the mating was interrupted; on the other hand when the proximal marker, $aroC^+$, was selected, the non-selected donor marker, $purF^+$, had a frequency of only 50% amongst progeny resulting from interruption at the earliest times, rising to 95% only after longer periods of mating. This asymmetry is due to the fact that, when the distal marker is selected, the population of merozygotes yielding progeny is homogeneous in respect of the region between the selected marker and the origin, for all interruption times; this region includes the non-selected marker. On the other hand, when the proximal marker is selected, the length of duplicated chromosome distal to it, which includes the non-selected locus, will vary with the time of mating; at early times the donor non-selected marker will not have entered all zygotes or, when it has entered, the length of donor chromosome beyond it, in which crossing-over must occur if the donor non-selected marker is to be inherited, will be short.

A new approach to the problem of measuring short map intervals accurately in physical terms was made by Wann et al. (1970). As we have seen, time-of-entry analysis gives estimates of physical distance, but is impracticable for markers separated by less than about 1 min in entry time. Co-transduction is a good technique for the study of linkages over shorter distances but gives values (see pages 57–61) which are not simply related to physical distance, unless assumptions are made about the randomness of transducing fragments (Wu, 1966). There is a general correlation between co-transduction frequencies and differences in entry time, as shown by the values collected by Taylor and Trotter (1967), but a considerable scatter is observed, presumably largely reflecting the difficulties of accurate measurement. Wann et al. (1970) used recombinational analysis, selecting the distal donor marker of a closely linked pair of markers. They measured the frequency of reassortment with the donor non-selected marker, after various periods of irradiation of the zygotes with u.v. or X-rays immediately after interruption of mating, which was done soon after entry of the selected marker. The effect of irradiation was to increase the frequency of crossing-over between the selected and unselected markers, presumably by the production of lesions randomly distributed along the chromosome. From the observed recombination frequencies, a "radiation crossing-over coefficient", was obtained, graphically, and this appeared to be a good measure of genetic distance for short intervals (down to about 0·1 min of entry time).

Recombination analysis has been used for studies of genetic fine structure on a number of occasions. For example Jacob and Wollman (1961, pp. 228–230), in an investigation of medium-fine structure, studied the *lac* region. In this case selection was made for a donor marker distal to the *lac* region, and that region was studied non-selectively in the way just outlined. On a finer scale, selected recombinant frequencies in two-point crosses have been used to construct intragenic maps (Garen, 1960; Jacob and Wollman, 1961, pp. 230–232). However, in most recent studies of genetic fine structure in *E. coli*, general transduction by phage P1 has taken the place of conjugational analysis.

ii. *Transduction in* E. coli. General transduction by phage P1 has become the method of choice, both for fine-structure analysis of short chromosomal regions and even for the routine mapping of new markers, now that nearly all of the linkage map of the organism is closely enough marked for co-transduction of a new marker with a previously mapped marker to be almost certainly obtainable. Transduction studies in *E. coli* K12 make use of a mutant strain P1kc (Lennox, 1955), of a temperate *Shigella* phage; this mutant was selected to give better plating efficiencies and higher titres on *E. coli* K12 strains than the original wild-type P1 phage.

The practical methodology of transduction with P1 varies somewhat in different laboratories. For example transducing phages can be grown on the donor strain on plates inoculated with a top-layer containing bacteria and enough phage particles to give confluent lysis of the whole culture (Adams, 1959); phages are then eluted from the top layer by suspending it in a suitable buffer or liquid medium and removing the debris by low speed centrifugation (e.g., Arber, 1960). Alternatively, the transducing phages can be prepared by adding phages to a growing donor culture in liquid medium, incubating it until massive lysis occurs, as judged by an obvious loss of turbidity, and spinning off the bacterial debris at low speed (e.g., McFall, 1964). In either case, any bacteria in the phage preparation are killed by means of chloroform (Luria, Adams and Ting, 1960), which is removed by incubating samples of the phage suspension briefly before use. Transduction is performed by adding the transducing phage suspension at a suitable multiplicity of infection to samples of the recipient culture, incubating for some minutes to allow phage adsorption, and plating on suitable selective media (e.g., Taylor and Trotter, 1967).

Linkage tests based on co-transduction of the kinds discussed earlier (page 88) have been used. For example selections of prototrophs in pairwise crosses of auxotrophs were made by Yanofsky *et al.* (1964); in this case the frequencies of tryptophan-independent recombinants in pairwise crosses of tryptophan auxotrophs were normalized by reference to the frequency of histidine-independent transductants in the same cross, which always involved a *his+* donor and a *his−* recipient. Other experimental designs have also been used, including the analysis of multi-point crosses (see for example Taylor and Trotter, 1967, and the references in that review).

iii. *Complementation and dominance tests*. As mentioned above (page 96), functional genetic tests in *E. coli* are routinely made with the partially diploid F′ strains harbouring sex-factors that have incorporated various regions of the bacterial chromosome (Jacob and Adelberg, 1959). Such "substituted sex-factors" appear to arise when the normal process of detachment of the sex-factor from the chromosome of an Hfr strain by crossing-over occurs inaccurately, with the result that a piece of bacterial chromosome adjacent to the site of sex-factor integration, and of variable length, is incorporated into a plasmid (Hayes, 1968, pp. 794–797). Thus any Hfr culture no doubt contains a small minority of cells harbouring F′ factors carrying various lengths of bacterial chromosome from near the site of insertion of the sex-factor in that particular Hfr strain.

The original method of isolating pure clones carrying such F′ factors was to mate an Hfr culture with an F− culture for a rather brief period, interrupt the mating, and plate on a medium selecting recombinants inheriting a donor marker that would be transferred by a normal Hfr cell only after a prolonged period of mating; that is a late, if possible terminal, marker. Any colonies developing on the selective plates would be likely to be those of F′ strains, since a cell in the Hfr culture harbouring an F′ factor could transfer early a normally late marker (Jacob and Adelberg, 1959). Low (1968) discovered a more efficient method of isolating a wide variety of F′ strains which depended on his finding that, on mating an Hfr strain with a particular kind of recombination-deficient (*recA−*) F−, no normal recombinants were produced. Instead colonies developed of at least two kinds of plasmid-carrying strains: normal F′ strains, recognized by being males (that is genetic donors, sensitive to male-specific phages), and strains carrying defective sex-factors that were neither normal males nor normal females (they were resistant to both male- and female-specific phages).

F′ strains vary considerably in stability. In general, the shorter the piece of bacterial chromosome incorporated into the sex-factor, the less prone is the plasmid to loss during growth of the culture. There are reports of partially diploid *E. coli* strains in which about one third of the chromosome is duplicated, but it is not certain whether they represent typical F′ strains or partial diploids of some other kind (Clark, Maas and Low, 1969).

(b) *Genetic analysis in* Salmonella typhimurium

i. *Conjugational analysis*. Although, as we saw above, long-range mapping was first done on a considerable scale by the use of conjugation mediated by colicinogenic factors (Smith and Stocker, 1962), conjugation became much more convenient with the use of Hfr strains arising after the transfer of the F factor from *E. coli* to *Salmonella*; this transfer was first achieved by Zinder (1960). Matings of such *Salmonella* Hfr strains with *Salmonella* F⁻ strains have been analysed both by time-of-entry studies in interrupted mating (Sanderson and Demerec, 1955) and by recombination analysis of uninterrupted matings. It was found that efficient mating did not occur in liquid media. For interrupted mating a technique was therefore devised in which the mixed culture was drawn on to a nitrocellulose filter, from which the cells were gently resuspended, after allowing a short period for the establishment of effective mating contacts (Sanderson and Demerec, 1965). Interruption was then achieved by mechanical agitation. For uninterrupted mating, the donor and recipient bacteria were mixed and allowed to mate on the selective plates and the resulting colonies were analysed in respect of non-selected marker combinations (Sanderson, 1967). This form of analysis is more useful in *S. typhimurium* than in *E. coli* because of the greater difficulties of time-of-entry studies.

ii. *General transduction by phage P22*. A full and clear account of one version of the practical methodology of transduction is given by Smith (1961). Transducing phages were prepared in broth, starting with a low multiplicity of phage to exponential-phase bacteria (0·05–0·1), in order to favour the lytic cycle. In this version of the technique, the resulting phages were concentrated by ultracentrifugation, and resuspension in buffer; this procedure was adopted to avoid the loss of viability of phages stored in the lysis medium, so that the same phage preparation could be used repeatedly over a considerable period. It also produced a phage suspension concentrated enough for effective transduction even in conditions of inherently low frequency, such as in crosses of the methionine mutants studied by Smith, or in spot-tests of transduction on plates. Transduction was effected by adding phages, at a multiplicity of 5 to minimize lysis, to overnight (stationary phase) recipient cultures.

In other versions of the technique, (e.g., Smith-Keary, 1960), transducing phages are prepared from confluently lysed top-layer cultures, as in the case of transduction of *E. coli* by P1 (see page 105).

A consequence of the fact that a multiplicity of about 5 phages per bacterium is required to favour lysogenization and so to give efficient transduction, is that most of the transductant colonies that arise are lysogenized by non-transducing phages in the preparation. This means that such strains cannot be used *as donors* for further transduction studies, since they are immune to P22 infection. A solution to this problem has been to use a non-lysogenizing mutant of P22, called L4 (Smith and Levine, 1967); general transduction by this mutant appears to be essentially the same as by the P22 wild-type (Kuo and Stocker, 1969; Levinthal and Simon, 1969). Another solution was to use a virulent mutant of P22, heavily irradiating the transducing phage preparation before adding it to the recipient (Goldschmidt and Landman, 1962).

In the case of *S. typhimurium*, the linkage map derived from transduction analysis alone would still consist of many isolated small linkage groups (Sanderson, 1970). This contrasts with the situation in *E. coli* (Taylor, 1970) in which, as pointed out previously, the great majority of the map can be covered by transduction. The greater size of the P1 phage head as compared with that of P22 is doubtless one reason for this difference. Another possible cause is a greater tendency for transducing fragments carried by P22 to arise by non-random breakage of the chromosome in particular positions (see page 85), so that even rather close genes may fail to show measurable co-transduction. Consequently mapping in *S. typhimurium* still depends on a suitable combination of conjugation and transduction analysis; conjugation serves to join short linkage segments mapped in detail by transduction, and also to orientate such segments with respect to the whole linkage group (e.g., Roth and Sanderson, 1966).

iii. *Genetic mapping using hybrid* Salmonella *strains.* By conjugal transfer, it has been possible to prepare hybrid strains containing mainly *S. typhimurium* chromosome but with various segments substituted by *E. coli* or alternatively *Salmonella montevideo*; for the preparation of *E. coli/S. typhimurium* hybrids, particular *S. typhimurium* recipient substrains, probably with altered restriction-modification properties, are best used if the foreign DNA is to be successfully introduced (Okada, Watanabe and Miyake, 1968). The hybrid strains have then been used as transductional donors to *S. typhimurium* recipients; Demerec and Ohta (1964) found that markers carried on the "foreign" part of the donor chromosome (that is the *E. coli* portion) were transduced at very low frequencies compared with those carried on *S. typhimurium* regions. This finding was harnessed as a qualitative gene-sequencing technique (that is it did not depend on the quantitative estimation of recombinant frequencies) by preparing a series of hybrids, each selected for inheritance of a particular *E. coli* marker (for example *cys+*). These hybrids were used in turn as donors in transduction to a series of *S. typhimurium* recipients carrying various recessive markers in the region of the locus of the marker derived from *E. coli*. The frequency of transduction of the latter marker was always low. Various of the other markers were also transduced with low frequencies from particular hybrid donors; in other words they too had come originally from *E. coli*, whereas markers transduced with high frequency resided on *S. typhimurium* chromosome. The possible patterns of markers showing such low frequency transduction provided evidence on gene sequence (Table V), on the reasonable assumption that each hybrid would contain a single continuous insertion of *E. coli* chromosome rather than several discontinuous stretches. There are analogies with gene sequencing by crossing to deletion mutants (page 45). Further studies of this kind were made by Ino and Demerec (1968), and on *S. montevideo* × *S. typhimurium* hybrids by Glatzer, Labrie and Armstrong (1966).

An ingenious study of the map location of ribosomal protein cistrons by O'Neil, Baron and Sypherd (1969) utilized *E. coli* × *Salmonella typhosa* hybrids, the latter species being selected because the electrophoretic pattern of its ribosomal proteins differs in several bands from that of *E. coli* K-12. The procedure was to select several hybrids, all inheriting a particular locus (the *strA* gene implicated in ribosomal protein structure) from *E. coli*, together with various lengths of *E. coli* chromosome extending from the *strA* locus. On comparing the electrophoretic pattern of ribosomal proteins of each hybrid with those of the parent strains, a comparatively short chromosomal region was implicated as the site of many of the ribosomal protein cistrons.

iv. *Complementation and dominance tests.* These, as we have seen, originally made use exclusively of abortive transduction; references to the methodology of such studies have already been given (page 90).

F′ strains of *S. typhimurium* used more recently in functional genetic tests have been of at least two types, harbouring plasmids containing respectively *E. coli* or *S. typhimurium* chromosomal segments; the sex factor itself has always been that originating in *E. coli* K12. Strains harbouring F′ factors containing *S. typhimurium* chromosomal material (F′T factors) have been prepared by the same technique of early inheritance of a late marker (see page 106) originally used for *E. coli* (Sanderson and Hall, 1970), starting with *S. typhimurium* Hfr strains. However, the existence of certain large regions of the *Salmonella* chromosome devoid of Hfr integration sites (Sanderson, 1970) means that F′T factors covering these regions are unavailable. Examples of dominance tests using F′ and F′T factors are the studies of Fink and Roth (1968) and Chater (1970).

(c) *Transductional analysis in* Proteus mirabilis. A temperate phage promoting general transduction in this bacterium has been harnessed by Prozesky (1968) for linkage analysis, particularly of the genes controlling

TABLE V

Gene sequencing by the study of the numbers of transductants in respect of markers in *Salmonella typhimurium* recipient strains using phages grown on hybrid *Escherichia coli* × *S. typhimurium* strains
(from Demerec and Ohta, 1964)

Donor strains	Recipient strains carrying these markers															
							Mutations in *cysC, D, H, I, J* cluster									
	cysG	*metC*	*argE*	*serA*	*lys*	*argB*	537	1021	519	75	68	538	*phe*	*tyr*	*purG*	*gly*
S. typhimurium wild-type control	1260	555	485	446	960	750	1400	377	1000	1170	1890	1130	3200	2800	2600	3800
Various *E. coli* × *S. typhimurium* hybrids	599	732	732	205	619	305	**11**	**4**	..	..	..	**5**	736	1362	2840	1336
	451	1470	385	282	**5**	0	0	..	**1**	..	..	0	3000	3000	2200	3000
	1430	506	326	1	**3**	0	0	..	**0**	..	..	2	3000	3000	2600	3000
	541	893	**96**	1	**4**	0	0	**0**	**0**	**0**	0	0	0	**2**	**25**	1277
	..	6000	0	1	0	0	0	..	**0**	..	..	2	1500	4800	..	2750
	..	104	0	6	**19**	1	0	..	..	..	0	0	**12**	**14**	**56**	1480
	590	3	0	0	**5**	1	0	**1**	**1**	**4**	**3**	1	137	198	567	1663

Bold figures are values significantly reduced below wild-type levels, indicating that the marker resided on *E. coli* genetic material. The sequences of such low values in the various donors indicate the following map-order of loci: *cysG–metC–argE–serA*–(*lys, argB*)–(*cys C, D, H, I, J*)–(*phe, tyr, purG*)–*gly*. Sequences within parentheses are unresolved.

arginine biosynthesis. The practical methodology and techniques of genetic analysis are basically those developed for *E. coli* and *S. typhimurium*. A slight complication was the origin, with a rather high frequency, of virulent mutants of the temperate transducing phage; these mutants had the effect of significantly reducing the yields of transductants when the frequency of virulent mutants rose above 1×10^4 per ml so that the phage had to be constantly purified. No abortive transductants were detected.

(d) *Conjugational analysis in* Pasteurella *and* Shigella. Although members of these genera have not been shown to undergo spontaneous conjugation, just as in the case of *Salmonella* it has been possible to transfer to them the sex factor F of *E. coli*; this has then promoted conjugation in the new host. Schneider and Falcow (1964) isolated Hfr strains of *Shigella flexneri* after transfer to them of the sex factor from an *E. coli* Hfr strain, while Lawton, Morris and Burrows (1968) transferred a sex factor from an *E. coli* F' *lac* strain to *Pasteurella pseudotuberculosis*.

(e) *Other intergeneric transfers*. Baron, Gemski, Johnson and Wohlhieter (1968) have recently reviewed some aspects of the intergeneric transfer of plasmids between *E. coli* and its relatives. Such transfers are often not associated with the recombination of chromosomal genes, presumably because of lack of sufficient homology between the genomes of the various hosts.

2. Pseudomonas aeruginosa *and* Ps. putida

The methodology of genetic analysis in these organisms has much in common with that in the enteric bacteria.

Genetic recombination in *Pseudomonas aeruginosa* was discovered comparatively early in the history of bacterial genetics (Holloway, 1955), but the development of good systems of genetic analysis is of quite recent origin (Holloway, 1969). Currently, two pseudomonads are the subject of considerable research effort: *Ps. aeruginosa* and *Ps. putida*. *Ps. aeruginosa* has a sex-factor mediated conjugation system allowing gross mapping. In both species, generalized transducing phages have been harnessed for finer mapping: phage F116, transferring some 1–2% of the genome of *Ps. aeruginosa*, and Pf16, transferring up to 5% of the *Ps. putida* genome (Holloway, 1969). There appears to be no shortage of pseudomonad phages, so that the isolation of phages capable of transducing even larger genome segments would seem to be a distinct possibility. On the debit side, there is at present no general system in either species for carrying out functional genetic tests. However, the recent discovery that a defective phage transducing the mandelate region of *Ps. putida* (Chakrabarty and

Gunsalus, 1969a) can function as a sex factor (Chakrabarty and Gunsalus, 1969b) promises to lead to a partial diploid situation (Chakrabarty and Gunsalus, 1970) that may be harnessed for such tests.

(a) *Conjugation.* The conjugation system in *Ps. aeruginosa* promoted by a sex-factor, FP, has some similarities with the F-factor mediated system in *E. coli*, but also shows some differences. In particular, no high-efficiency donor exactly comparable to *E. coli* Hfr strains has been isolated, but the male strains (FP+) do not behave as classical F+ strains either. In FP+ × FP⁻ matings, one does not observe *random* transfer of chromosome segments from FP+ to FP⁻: certain donor markers are invariably inherited with high frequencies by recombinants, and others with low frequencies. Moreover, markers enter zygotes at characteristic times, so that mapping by time-of-entry succeeds; evidently one or a small number of origins of chromosome transfer predominate. Perhaps the situation is at least partly analogous with that in certain F+ *E. coli* strains, in which the sex-factor integrates, unstably, at one (Richter, 1961) or two (Low, 1967) preferred chromosomal sites (sex-factor affinity loci), rather than in the more random, but usually more stable, manner typical of most F+ strains.

Experimental interconversion of fertility types is not yet so straight-forward as in *E. coli*; for example acridines are not readily effective in eliminating the FP factor (Holloway, 1969). However, the finding of mercury-resistance associated with the presence of FP (Loutit, 1970) should facilitate experiments on elimination and transfer of the sex factor.

Several recent papers have described genetic analysis of *Ps. aeruginosa* by conjugation, using procedures representing modifications of the stand-ard methods evolved for *E. coli*; practical details are in papers by Loutit and Marinus (1968), Loutit (1969) and Stanisich and Holloway (1969). Loutit (1969) and Loutit and Marinus (1968) have described the results of the three mapping procedures previously employed for *E. coli*: gradient of transmission in uninterrupted matings, recombinant analysis, and time-of-entry in interrupted matings. Recombinant analysis was done by select-ing each marker in turn and measuring the frequency of the non-selected marker(s) in each cross. Usually a big difference was found in reciprocal selections and the proximal marker of a pair was identified by having the higher frequency when unselected. For example, suppose the merozygotes resulting from mating have the following constitution (o is the origin):

$$
\begin{array}{c}
a^+ \quad b^+ \\
\text{o} \,\underline{\text{————} \,\text{– – –}} \\
a \quad b
\end{array}
$$

The frequency of a^+ recombinants amongst those selected to carry b^+

will be higher than the frequency of b^+ amongst a^+ recombinants. This is because all merozygotes containing the distal marker b^+ must also contain the proximal marker a^+, but the converse is not true.

As in *E. coli*, time-of-entry measurements were only possible with useful accuracy for relatively early markers, in *Ps. aeruginosa* for those entering up to about 25 min after mixing the parent strains.

In these studies, each experiment normally provided evidence on only a pair of markers. This approach had the advantage that multiply marked strains did not have to be prepared: instead a series of mutant derivatives of a strain bearing a particular marker were isolated, each serving to map a different new mutation in relation to a common marker. However, the disadvantage was that of all two-point mapping procedures: evidence of order was derived from the comparison of map distances from different experiments with different strains, subject to possibly different causes of bias. In these data, some marker sequences based on time-of-entry were in fact later revised in the light of the results of recombinant analysis. Another ambiguity in the time-of-entry data, revealed by the recombinant analysis, was due to a peculiarity of the sexual system: two separate linkage groups were transferred, so that certain markers at similar distances from the origin of their linkage groups had similar times of entry, but were in fact unlinked.

(b) *Transduction.* A growing number of publications deal with transduction studies on *Ps. aeruginosa*, each aimed at the elucidation of the linkage relations of a particular phenotypic class of mutations, and using the generalized transducing phage F116 (Holloway, Monk, Hodgins and Fargie, 1962). The methods employed have been largely those already used for *E. coli* and *S. typhimurium*: transducing phages were prepared from confluently lysed top-layers and mixed at suitable ratios and cell densities with stationary phase recipient cells. After a period of incubation for phage adsorption, samples were plated on selective media. Co-transduction frequencies were estimated by standard methods. Some examples of such studies are: Mee and Lee (1967), Kemp and Hegeman (1968), Marinus and Loutit (1969).

With the discovery of generalized transducing phages for *Ps. putida* (Chakrabarty, Gunsalus and Gunsalus, 1968), this species is also now open to genetic analysis (Wheelis and Stanier, 1970). Again the methodology is fairly standard.

3. *Bacillus subtilis*

Genetic mapping in *B. subtilis* currently depends on four techniques: co-transduction analysis by the general transducing phage PBS1; co-trans-

6

formation studies; and marker frequency and density transfer combined with transformation. The value of these techniques is shown by the studies summarized by Dubnau *et al.* (1967).

As in other bacteria, transduction studies alone led to the identification of a number of isolated linkage groups. In *B. subtilis* these each represent a larger proportion of the whole linkage map than usual because of the comparatively very large fragments of DNA carried by the PBS1 phage. Dubnau *et al.* (1967) estimated that it can carry nearly 10% of the chromosome of *B. subtilis*: ten times the length carried by P1 for *E. coli*.

The methodology of transduction by PBS1 was described by Takahashi (1961, 1963) and this procedure has been followed in later investigations (Dubnau *et al.*, 1967). A slight complication is that only flagellated cells will adsorb the phage; several strains were found to be non-transducible because they had become non-motile (Dubnau *et al.*, 1967).

The size of the chromosomal fragments carried by PBS1 is considerably greater than that of transforming DNA prepared in the normal way, although, as pointed out earlier (page 93), DNA prepared by very gentle means had a larger average molecular weight and could detect linkages over longer distances by co-transformation. However the only way in which the separate linkage groups detected by co-transduction or co-transformation have been joined to form a continuous linkage map is by the use of the two methods provided by Yoshikawa and Sueoka (1963a, 1963b), marker frequency analysis and density transfer.

The first method (Yoshikawa and Sueoka, 1963a) depends on the reasoning that, in an exponentially growing culture, and assuming that the chromosome begins replicating at a fixed point, there would be twice as many copies of a gene lying at the beginning of the chromosome than of a gene at the end. A preparation of transforming DNA from such a culture should therefore give twice as many transformants for the early marker as for the late one, with intermediate values for intermediate markers, assuming all markers to have the same innate probability of transformation. The latter assumption is certainly not true, owing to "marker effects" of the kind referred to earlier in the discussion of transduction (Lacks, 1966); however, Yoshikawa and Sueoka (1963a) overcame this difficulty by using for comparison with the DNA from the log phase culture, DNA prepared from a stationary phase culture, in which all markers should have had the same frequency if DNA replication ceased at the end of the chromosome in such a culture. They found that the ratio of transformants given by the two DNA preparations, exponential : stationary, varied for different markers, *between the expected limits of 2 and* 1. A genetic map was then constructed, using the transformation ratio as the unit of distance, and this map has tended to be confirmed by later studies by the same authors and others (Dubnau *et al.*, 1967).

The second method, that of density transfer (Yoshikawa and Sueoka, 1963b), depended on the availability of cultures of the donor strain synchronized in respect of DNA replication. This was readily achieved, in the *B. subtilis* strain W23, since, as we have just seen, the cells in a stationary phase culture cease DNA replication at a fixed point; when growth re-starts the chromosomes replicate in a synchronous fashion. In the method of Yoshikawa and Sueoka (1963b), the cells were previously grown in a heavy medium, containing D_2O and N^{15}, and were brought out of stationary phase in a light medium, with H_2O and N^{14}. At various time intervals, the DNA was extracted and separated by caesium chloride density gradient centrifugation into heavy (that is not yet replicated) and half-heavy-half-light (that is semi-conservatively replicated) fractions, which were used separately to transform recipients in respect of various markers. The expectation was that markers near the beginning of the chromosome would rapidly be transferred to the replicated fraction of the DNA, and markers further from the beginning would be transferred at characteristically later times. Thus the unit of mapping would be minutes at which the transfer took place. As in the case of the marker frequency method, linkages established by this procedure were later confirmed and extended (Dubnau *et al.*, 1967).

Many strains used in genetic analysis of *B. subtilis* derive from a different wild-type, 168, in which growth to stationary phase does not synchronize the culture in respect of chromosome replication. It has been possible to adapt the density transfer method to such strains by using previously density labelled *spores* as the inoculum for growth in light medium (D. Dubnau, pers. comm.), since chromosome replication is synchronous in this strain in spore germination.

A later modification of the density transfer technique was made by Oishi and Sueoka (1965), who substituted 5-bromouracil for the much more expensive D_2O–N^{15} density label.

In *B. subtilis*, the proportion of competent cells in a culture showing maximum overall competence does not reach 100%, as it does in some other species. There is evidence that competence is associated with a particular stage in the growth cycle, and this fact was harnessed by Cahn and Fox (1968) and by Hadden and Nester (1968) to separate the competent cells from the rest of the population. This was achieved by centrifugation in a gradient of Renografin, when the cells separated into two bands, the lighter containing the competent cells.

It has recently been found (Ephrati-Elizur, 1968) that efficient transformation occurs if donor and competent recipient cells are grown for a short time in mixed culture. That recombinant production was dependent on *extracellular* DNA, as in the normal transformation procedure, was shown by its sensitivity to DN'ase. Co-transformation frequencies resembled those obtained with DNA prepared from the donor strain by the normal procedure.

In spite of a good deal of searching by a number of groups, no method is so far available for performing tests of dominance and complementation in *B. subtilis*. Such tests are badly needed in studies of the genetic control of sporulation (e.g., Schaeffer, 1969; Balassa, 1969).

Although all the markers of *B. subtilis* have been found to be linked, there is no evidence for linkage of the terminal markers; that is the linkage map remains linear rather than circular. However, an ingenious experiment by Sueoka and Quinn (1968) suggests that markers at *both* ends of the chromosome are associated with membrane-bound DNA, perhaps indicating circularity of the chromosome.

4. Streptomyces coelicolor *and other actinomycetes*

Genetic analysis in this actinomycete has developed along different lines from that in the other bacteria discussed in this Chapter. The basis for genetic transfer is a conjugation system, but transfer probably does not occur in the same orientated fashion found in *E. coli* (Hopwood, 1967).

We have discussed genetic analysis of other bacteria in terms of the one-way transfer of chromosomal fragments from one strain (the donor) to another (the recipient). In the case of sex-factor mediated conjugation, as found in *E. coli* and its relatives and also in *Ps. aeruginosa*, possession of an appropriate sex-factor makes a strain a male and therefore an obligate donor to a strain lacking the sex-factor, whereas in a transformation or transduction system, the experimenter determines the respective roles of the two strains. In the bacterial situation, where all recombinants have normally to be selected, because of their overwhelming outnumbering by asexual progeny, and where complete diploids do not occur, the interpretation of linkage data is very much simplified by knowledge of the donor versus recipient role of each parent. In discussing genetic analysis in *E. coli*, we have ignored its confused state during the period following the discovery of genetic recombination (Lederberg and Tatum, 1945) until the recognition and use of known donor and recipient strains. During that initial period of seven or eight years, linkage data, derived from crosses of F+ strains, were very confusing (e.g., Lederberg *et al.*, 1951) owing to the fact that a cross was a heterogeneous mixture in which transfer of chromosome fragments occurred in both directions between strains: between spontaneous Hfr variants arising in one parent culture and F− variants in the other, and *vice versa*.

When a conjugational system was discovered in *S. coelicolor*, an attempt was therefore made (Hopwood, 1959) to use a system of genetic analysis that should reveal unambiguous linkages in a situation where recombinants were selected from a mating mixture in which the parental strains were playing equal roles; not necessarily in the formation of each zygote, but statistically throughout the population of zygotes. A fairly detailed description of the rationale of this linkage test may be useful here since it should be suitable for the analysis of other conjugational systems that may be discovered: it has already provided the key to the linkage map of a second actinomycete, *Streptomyces rimosus* (Friend and Hopwood, 1971). Its contribution is in the early stages of genetic analysis of a new organism, when the first linkages are to be detected; once the beginnings of a linkage map are known, the addition of further loci to the map is very much simpler.

(a) *Preliminary detection of linkage.* The prerequisite is a pair of parental strains differing in four selectable markers, two in each parent; these may involve such characters as auxotrophy versus prototrophy or sensitivity versus resistance.

In such a four-factor cross, there are 16 possible genotypes of progeny, of which two are parental and the remainder recombinant. Clearly not all classes of recombinant progeny are selectable on media which do not allow growth of the parental classes, but it turns out that nine classes can be recovered, provided that four differently supplemented media are used in parallel, each selecting a different pair of markers (see Table VI). The procedure is to plate the output of the cross (in the case of *S. coelicolor* this consists simply of a mixed culture of the two parental strains grown together for a few days until progeny spores are produced) on the four media and classify samples of the resulting recombinant colonies in respect of the pair of non-selected markers appropriate to each medium. From the total number of colonies on each medium and the frequency of each genotype in the classified samples, an estimate of the observed frequency of each genotype is obtained.

Table VI shows the amount of information yielded by this procedure. We see that the nine recovered genotypes represent at least one member of each complementary pair of recombinant classes, and in two cases both members. In several instances, the frequency of a particular class is estimated on more than one medium; this allows an appraisal of a possible bias in the frequency of recovery of a particular class of progeny under the different selective conditions on different media. The two cases in which both members of a pair of complementary genotypes can be recovered indicate, if members of the pairs have the same frequency, the statistically equal contribution of genetic material by the two parents to the progeny,

TABLE VI

**Analysis of a four-factor cross in _Streptomyces coelicolor_, _his_ + + + ×
+ _met phe str_ (unpublished data of Hopwood)**

Genotypes of selectable progeny[†]	Selective media supplemented with								Average frequency of each pair of complementary genotypes
	Methionine		Histidine Phenylalanine Streptomycin		Phenyl-alanine		Histidine Methionine Streptomycin		
+ + + +	7	16	—	—	72	14	—	—	15
his + + _str_	—	—	41	69	—	—	30	55	62
+ _met_ + _str_	26	57	—	—	—	—	39	72	65
+ + _phe_ _str_	—	—	1	2	13	2	—	—	2
+ + + _str_	3	7	1	2	7	1	3	6	4
+ _met_ + +	60	132	—	—	—	—	—	—	⎫ 108
his + _phe_ _str_	—	—	50	84	—	—	—	—	⎬
+ + _phe_ +	—	—	—	—	0	0	—	—	⎫ 0
his _met_ + _str_	—	—	—	—	—	—	0	0	⎭
Sample size:	96	—	93	—	92	—	72	—	
Total recombinants per plate:	—	212	—	157	—	17	—	133	

Relative recombination frequencies in each interval

		his–met	_his–phe_	_his–str_	_met–phe_	_met–str_	_phe–str_
	⎧	15	15	15	65	62	62
Components:	⎨	2	65	65	2	2	65
	⎪	4	4	108	108	4	4
	⎩	0	108	0	0	108	0
Total:		21	192	188	175	176	131

Segregation of pairs of non-selected alleles

	met^+	_met_		phe^+	_phe_		phe^+	_phe_		met^+	_met_
str^+	7	60	his^+	1	1	str^+	72	0	his^+	3	39
str	3	26	_his_	41	50	_str_	7	13	_his_	30	0

† _his_ = histidine-requiring; _met_ = methionine-requiring; _phe_ = phenylalanine-requiring; _str_ = streptomycin-resistant.

A sample of recombinants from each of the 4 selective media was classified into the 4 possible genotypes: the results are in the left-hand column under each medium. Each value was scaled, using the total recombinant count on that medium, to give the frequency of each genotype per plate: these values are in the right-hand column for each medium, and are averaged over the different media to give the figures in the far right-hand column of the table. These figures, each representing the frequency of one _complementary pair_ of recombinants, are used to derive the relative recombination frequencies in each interval in the middle region of the table.

At the bottom of the table, the numbers of each genotype on each of the 4 selective media are displayed to reveal the independence or otherwise of segregation of alleles at the two non-selected loci: those for the first two media do not deviate significantly from independence, while those for the second two deviate very significantly from independence.

since complementary genotypes are equivalent in respect of the crossing-over patterns required to produce them. Provided these internal checks are satisfied, the various frequencies can then be averaged to give a single estimate for the frequency of each *pair* of complementary recombinant genotypes. This information is enough for the calculation of the *relative* frequencies of recombination between each pair of markers in the cross: six values for the set of four loci (Table VI).

The resulting estimates are relative only since we lack an estimate of the frequency of the pair of progeny classes having the parental arrangement of markers, a value that would appear in the denominator of the ratio of recombinants out of total progeny for each pair of loci if an absolute recombination percentage were calculable. If all six relative recombination frequencies are approximately equal, no linkages are indicated. However in *Streptomyces*, both in *S. coelicolor* (Hopwood, 1959) and in *S. rimosus* (Friend and Hopwood, 1971), the total linkage map is rather short and linkage is likely to be revealed amongst most randomly chosen sets of four markers; this shows itself by one or more of the six relative recombination frequencies being appreciably lower than the others. In the example in Table VI, recombination between *his* and *met* is very much rarer than between the other pairs of markers, while recombination between *phe* and *str* is somewhat less frequent than between the remaining four pairs. Thus *his* and *met* are linked, while the other two loci, if linked to each other and to the first two, do not lie in the short interval between them.

The next step in the analysis (Hopwood, 1969; Friend and Hopwood, 1971) is a consideration of the segregation, on each medium, of each non-selected marker in relation to the other: that is whether they show independence or not. The relationship is best shown by a 2×2 tabulation for each medium (Table VI). In this example we see that the segregation of *his+*/*his* is independent of that of *phe+*/*phe*; and *met+*/*met* is independent of *str+*/*str*. On the other hand, *his+*/*his* is not independent of *str+*/*str*, nor is *phe+*/*phe* independent of *str+*/*str*. These findings indicate that the loci of *phe* and *str* must be linked with the other two loci, and the data are in fact indicative of circular linkage of all four loci (Fig. 17), as the following reasoning will show.

In a case of circular linkage, the total number of cross-overs in a zygote must be even, and if two markers are selected, the number of cross-overs in each of the two arcs defined by the selected markers must be odd. As we see from Fig. 17, when one non-selected marker is in each of these arcs, segregation of the two markers is independent, each of the four genotypes arising by two cross-overs. On the other hand, when both non-selected markers are in the same arc, their segregation is not independent since three gentoypes require two cross-overs and the fourth requires four.

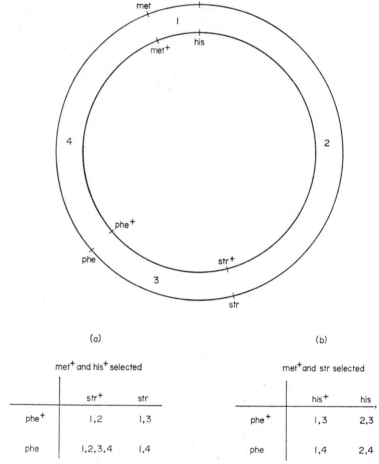

(a)   (b)

met⁺ and his⁺ selected   met⁺ and str selected

	str⁺	str
phe⁺	1,2	1,3
phe	1,2,3,4	1,4

	his⁺	his
phe⁺	1,3	2,3
phe	1,4	2,4

FIG. 17. Criteria for circular linkage in a four-factor cross in *Streptomyces*. Selection is made for alleles at two alternative pairs of loci, either adjacent (a), or non-adjacent (b). The tabulations show the cross-overs required to generate each combination of the non-selected alleles. In (a), the segregation of non-selected alleles is not independent, since one class (*str⁺ phe*) requires a combination of cross-overs in intervals 2, 3 and 4 which are required singly by the other three classes. In (b), non-selected alleles segregate independently because they require homogeneously related pairs of cross-overs (e.g., 1, 3 : 2, 3 versus 1, 4 : 2, 4).

The data under consideration fit the expectations of this model, with a highly significant shortage of the genotype requiring four cross-overs in each case of non-selected markers in the same arc.

All other possible arrangements of the four loci, including linkage on a

TABLE VII

Progeny of the cross depicted in Fig. 18. The data are analysed to choose between two alternative positions for the locus *hom*, either between *str* and *cys*, as illustrated in Fig. 18, or between *str* and *phe* (from Hopwood, 1969)

	hom	arg hom	arg cys hom	arg cys	arg	cys
ura pro	13	49	12	70	4[b]	0
phe ura pro	3	7	6	6[c]	0	0
phe pro	16	68	14	14[c]	0	1[a]
phe	34	28	3	3[c]	0	0
pro	0	0	0	2[a]	0	0
ura	1[b]	0	0	0	0	1[a]
—	0	1[a]	0	0	0	0

All progeny arise by the minimum number of two cross-overs, one in each arc between the selected markers, except for the following:

a These 5 progeny require multiple cross-overs on either hypothesis.

b These 5 progeny require multiple cross-overs if *hom* lies between *str* and *cys*.

c These 23 progeny require multiple cross-overs if *hom* lies between *str* and *phe*.

The order *str–hom–cys* results in a minimum number of multiple cross-over progeny and is therefore chosen.

linear (not circular) map, lead to different predictions in terms of the segregation of the non-selected markers on each of the four loci, and can be rejected. This linkage test appears to be the simplest demonstration of circular linkage of markers by a selective analysis so far described (but see Foss and Stahl's experiment with T4 phage: page 137).

(b) *Sequencing of new markers.* Once a number of intervals on the circular linkage map have been defined by a sequence of markers, further markers can be assigned to a particular interval by a simple selective procedure (Hopwood, 1967) involving plating the products of a cross on a single selective medium. The parents are chosen to differ in several markers, in addition to a new one. Selection is best made for two approximately diametrically situated markers, and the resulting recombinants are classified in respect of the non-selected markers, including the new marker. From the resulting genotype frequencies (Table VII), the position of the new marker, say *hom*, is deduced by a two-stage process.

The first stage consists of considering the percentage frequencies of the two alleles (the "allele ratio") at each locus (Fig. 18). The allele ratios at the selected loci are 356 : 0 and 0 : 356; the allele ratios at non-selected loci take intermediate values, falling on a continuous gradient in each of the two arcs. Thus in Fig. 18 the ratios at successive loci in a clockwise

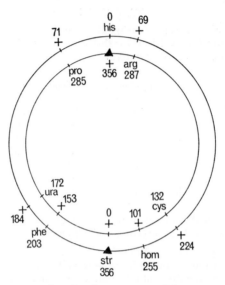

Fig. 18. Marker sequencing in *Streptomyces*. The experiment consisted of selecting recombinants inheriting the two alleles indicated by triangles and classifying them in respect of alleles at the other 6 loci. Numbers indicate the frequency of each allele in the classified sample.

direction, starting with the selected locus at the top of the map, and omitting the locus of *hom*, are as follows: 356 : 0, 287 : 69, 132 : 224, 0 : 356, 153 : 203, 172 : 184, 285 : 71. It follows that any new locus can be assigned two alternative positions by virtue of its allele ratio, one in each arc: for *hom*, with a ratio of *hom*+ : *hom* of 101 : 255, these are between *str* and *cys* or between *str* and *phe*.

The second stage in assignment is a choice between these two alternative map positions. This choice derives from the frequencies of particular genotypes of recombinants, and depends, as in all genetic reasoning (see page 40), on minimizing the total number of cross-overs required to explain the observed data, bearing in mind the particular constraints of circular linkage: that the simplest pattern of recombinant formation requires two cross-overs, one in each arc. The consequences, in terms of cross-overs, of these two alternatives for the *hom* locus are analysed in Table VII, and we see that the position between *str* and *cys* is very much more economical in terms of cross-overs and is therefore chosen.

(c) *Estimation of map distances.* The procedure just outlined is effective for sequencing markers on the *Streptomyces* linkage map, but leads to only very approximate indications of the spacing of loci. This is a consequence

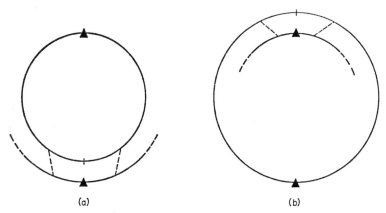

(a)   (b)

FIG. 19. The effect of merodiploidy on linkage estimation in *Streptomyces*.
(a) and (b) represent the two alternative classes of merozygote capable of yielding
progeny inheriting the two alleles indicated by triangles. Note that in (a), crossing-
over will be concentrated in the regions on either side of the bottom locus, whereas
in (b) it will be largely in the intervals flanking the top locus.

of the incomplete nature of the zygotes (which are merodiploids as in other
bacteria), which results in the probability of crossing-over per unit length
of the chromosome being very different in different regions, depending on
the combination of selected markers. In *S. coelicolor* it appears that on
average about one-sixth of the chromosome is duplicated in each zygote
(Hopwood, 1967) and, provided the cross involves identical fertility types
(Vivian and Hopwood, 1970), the duplicated region is random. This means
that, when selection is made for a particular pair of markers, one from each
parent, only those zygotes that contain both selected markers can give
rise to selectable recombinant progeny (Hopwood, 1969) (Fig. 19). In
these zygotes, the duplicated region will, on average, centre on one or
other of the selected markers, so that the probability of duplication, and so of
crossing-over, will be at a maximum near to the selected loci, and will
fall off with distance from them. Thus the gradient of allele frequencies
observed (Fig. 18) is not linear with distance, being much steeper near the
points of selection.

   This problem has been overcome by making use of special heterozygous colonies called heteroclones which arise from merodiploid spores.
The spores arising on one of these colonies are very largely haploid and
the study of segregation amongst the colonies produced by germinating
these spores on non-selective medium allows estimates of recombination in
intervals within the region duplicated in that particular heteroclone
(Hopwood and Sermonti, 1962).

The method of analysing individual heteroclones was inapplicable to the estimation of map intervals over two long segments of the map of *S. coelicolor* devoid of markers. However the lengths of these segments were measured from a population of heteroclones, which were classified simply in respect of which loci were included in the duplicated region of each heteroclone (Hopwood, 1966). By making the assumption of randomness (according to a Poisson process) of duplicated segments amongst heterocolones, it was possible to estimate the lengths of all intervals on the circular map in terms of the probability of the ends of duplicated segments falling within each interval (Edwards, 1966). Unfortunately, the mechanism that gives rise to the chromosome fragments in the heteroclones is still obscure, and may conceivably be non-random, so that we cannot say whether the spacing of loci on the linkage map corresponds to their physical spacing.

(d) *Fine genetic analysis.* Mapping of short regions in *S. coelicolor* has been done by techniques similar to those based on conjugation in *E. coli* (page 105). Short distances have been estimated by the frequency of recombination between pairs of auxotrophic mutations, and the sequence of sites has been determined by following the segregation of one or more outside markers (pages 43–45). A technical difficulty arises from the high frequency of heteroclones developing on the selective plates when selection is between closely linked *complementing* mutations, making the recognition of recombinants difficult (Hopwood, 1971). This is because, according to the theory of heteroclone origin (Hopwood, 1967), haploid recombinants arise when *two* (or a higher even number) cross-overs occur within the duplicated region of a merozygote, provided one is between the selected markers, whereas heterozygous genomes capable of initiating heteroclones arise by a *single* (or a higher odd number) cross-over anywhere within the duplicated segment. Thus heteroclones and haploids have comparable frequencies.

(e) *Dominance and complementation studies.* Functional genetic tests in *S. coelicolor* make use of the heteroclones already discussed. Complementation between auxotrophic mutations is indicated when heteroclones develop on a medium lacking the growth factor required by both parents. The system has also been adapted to the study of complementation between radiation-sensitive mutations (Harold and Hopwood, 1970b).

(f) *Non-selective analysis.* As mentioned earlier (page 94), *S. coelicolor* is almost unique amongst prokaryotes in lending itself to the study of recombination in crosses analysed non-selectively. This method is applicable when the cross involves one parent of the NF and one of the UF fertility type (Hopwood, Harold, Vivian and Ferguson, 1969). In this case up to 100% of the total spore progeny arising from the cross is sexually produced, so that no selection need be applied in order to *identify* sexual progeny. Owing to the asymmetry of the merozygotes in such crosses, sequencing of loci in only the left-hand half of the linkage map is efficient, since merozygotes in which regions in the right-hand half of the map are duplicated are rare. However, selection for an NF (donor) marker immediately proximal to the region to be studied ensures heterozygosity of the critical markers in a high proportion of merozygotes (Hopwood, 1971).

(g) *Genetic analysis in* Thermoactinomyces vulgaris. The scheme of linkage analysis outlined above (page 117) has recently been applied to a thermophilic actinomycete, *Thermoactinomyces vulgaris*. In contrast to the situation in *Streptomyces*, on analysing a four-point cross on the four selective media, it turned out that the great majority of recombinants differed from one or other parent in respect only of a single marker (Hopwood and Ferguson, 1970). This finding would have had limited significance had the cross been analysed on a single selective medium, on which such a result would have had a high probability of occurrence by chance due to the particular linkage relationships of the selected and non-selected markers. The significance of the finding was further increased by the identical result from two further crosses in which the same four markers were involved, but distributed differently between the parents. The conclusion was clear that whatever system of genetic transfer was operating in this organism must have involved the transfer only of short segments of chromosome from one parental strain to the other; indeed this was later shown to occur by transformation (Hopwood and Wright, 1971).

5. Other bacteria

There are a number of other bacteria in which genetic recombination has been reported and in which genetic analysis has started. Studies of such systems, if continued, may be expected to provide pointers to approaches that might be fruitful in other bacteria.

(a) *Rhizobium lupini*. A conjugation system leading to a highly efficient genetic transfer (up to 10% recombinants in a mixed culture) has been reported in *Rhizobium lupini* (Heumann, 1968). Little is yet known about the mechanism of conjugation, but it would appear not to be closely comparable with that in *E. coli* and its relatives. Groups of markers were transferred during the recombination process, and a tentative circular linkage map was deduced from the results of selective recombination analysis of crosses involving 2, 3 or 4 selective auxotrophic markers in addition to a non-selective pigmentation difference. The four-point crosses would appear to lend themselves to analysis by the method used in *S. coelicolor* (Hopwood, 1959: see page 117), but the data were not published in enough detail to test this possibility.

(b) *Serratia marcescens.* Another system of conjugation was described in *Serratia marcescens* by Belser and Bunting (1956). In this case a conjugation system was implicated only by the finding that cellular contact was essential for recombinant production; however, markers were normally transferred singly rather than in groups. As we have seen, this feature was shown particularly clearly in *Thermoactinomyces vulgaris* (see page 125).

(c) *Vibrio cholerae.* In *Vibrio cholerae*, Bhaskaran (1960) studied a sex-factor promoted conjugation system with some resemblances to that in *E. coli* and its relatives. The sex-factor, P, appeared to be bacteriocinogenic, and to confer the property of maleness on strains possessing it (Bhaskaran and Iyer, 1961), transferring partial genomes to recipient (P⁻) strains. Some data on recombinant analysis led to the beginnings of a linkage map (Bhaskaran, 1964).

(d) *Staphylococcus aureus.* Genetic analysis of the chromosome of *Staphylococcus aureus* has employed primarily general transduction by some of the staphylococcal typing phages. Fine-structure maps have been produced for several isolated linkage groups involving, for instance, histidine (Kloos and Pattee, 1965) or isoleucine-valine (Smith and Pattee, 1967) mutants. Mapping has been primarily by two-point selective tests. In the histidine region, Kloos and Pattee (1965) found a striking gradient of recipient ability on the part of *his* mutants in one-factor tests (that is using wild-type donors), related to map position. This gradient, which therefore provided evidence on map order, was probably caused by the non-random ending of transducing fragments resulting in the wild-type allele of mutants near the end of the fragment having reduced chances of integration by crossing-over. The transducing technique, which was fairly conventional, was described by Kloos and Pattee (1965) and also by McClatchy and Rosenblum (1966), who recognized haemolysin-producing transductants arising from crosses of pairs of non-producing mutants by a visual procedure. Complementation tests have been successfully carried out by abortive transduction (Kloos and Pattee, 1965).

A method of gross mapping of the *S. aureus* chromosome was described by Altenbern (1968), based on the same reasoning as that used, apparently independently, by Cerdá-Olmedo, Hanawalt and Guerola (1968) in *E. coli* (see page 128); that nitrosoguanidine causes mutations preferentially at the replication point during DNA synthesis. Since no independent evidence on gross map order exists for *S. aureus*, the map produced could not be confirmed by other means, but the method would appear to have a useful role to play in chromosome mapping in this species.

(e) *Anacystis nidulans.* The blue-green algae have proved refractory to genetic analysis to date. However, in view of their interesting status as complex prokaryotes, it is greatly to be hoped that the technical difficulties involved in their genetical investigation will be overcome. There has been a recent report (Asato and Folsome, 1970) of mapping in *Anacystis nidulans* by the mutation frequency method (see page 128). Convincing doubling of mutation frequency for each of six markers at a particular point in the division cycle of synchronized cultures was observed. However, it was disturbing that the time of doubling for four markers overlapped, and all six were very close, following one another within a period corresponding to only one-fifth of the complete genome replication cycle; it seems strange that six randomly chosen markers should occupy such a short segment of the whole map. Confirmation, by some other method, of the sequence of the six markers based on this approach, or the mapping of further markers, would therefore be very desirable.

C. Specialized mapping procedures independent of recombination

We have already discussed (page 114) in some detail the long-range mapping procedures, marker frequency analysis and density transfer, devised by Yoshikawa and Sueoka (1965a, b) for *B. subtilis*. This treatment seemed appropriate since the methods have justified themselves as valuable mapping techniques (Dubnau *et al.*, 1967). Several other techniques, conceptually related to these procedures to varying degrees, have been described which might provide data on the map location of markers, although they have in fact been used in the situation where the map positions of markers were already known, and information on the origin and direction of chromosomal replication was sought.

Two kinds of study have used P1 transduction in *E. coli*. In one (Berg and Caro, 1967) it was argued, following Yoshikawa and Sueoka (1963a) that, in exponentially growing cultures, and assuming a fixed replication origin, genes near the origin should be twice as frequent as genes near the terminus. Thus, other things being equal, "early" genes should have general transduction frequencies twice as high as "late" genes, with genes of intermediate position having intermediate frequences, and this prediction was fulfilled. In another approach (Abe and Tomizawa, 1967), the DNA replication cycle of an *E. coli* culture was synchronized and growth of the cells was then re-started in the presence of 5-bromouracil (BU) as a density label for the DNA. After varying periods of time, P1 phage was grown on the cells and used to transduce recipients for various markers, after separation of the phage particles in respect of density by gradient centrifugation. The reasoning was that an early marker would be transferred to heavy DNA (half labelled with BU) soon after synchronous growth began; thus this marker could be transduced early by heavy phages. Conversely, a late marker would be transduced by heavy phages only after a longer period of growth of the donor bacteria in BU. In other words, the time of transfer of transducing ability from light to heavy phages was a measure of its distance from the replication origin. A rather more elaborate study of the same kind was made by Wolf, Newman and Glaser (1968).

A different approach was used by Cerdá-Olmedo, Hanawalt and Guerola (1968), dependent on the idea, which was supported by their data, that the mutagen N-methyl-N'-nitro-N-nitrosoguanidine (NTG) is more effective in mutagenesis of DNA at the replication point of the *E. coli* chromosome than in other positions. A culture was synchronized in respect of DNA replication and, after various periods of renewed incubation, treated with NTG and different markers were assayed for mutation. It was found that each marker showed a "burst" of mutation at a characteristic time after the re-initiation of DNA replication, this time being related to map position. This method has also been applied to *Staphylococcus aureus* (see page 126), and to *Neisseria meningitidis* (Jyssum, 1969).

A different approach was made by Vielmetter and Messer (1964) who labelled exponentially growing cultures of *E. coli* with P^{32}, plated samples of the culture, and scored the resulting colonies for mutations at various loci. The mutants were classified as homogeneous, or as heterogeneous (a mixture of mutant and non-mutant cells). The proportion of heterogeneous colonies fell on a gradient related to map distance from the origin of replication, on the assumption that a locus would give rise to a heterogeneous colony when lying in a replicated region of the chromosome and to a homogeneous mutant when lying in an unreplicated region.

VII. GENETIC ANALYSIS IN VIRUSES

A. General considerations

1. *Types of viral genomes*

Cellular organisms are very uniform in the basic organization of the genome: all, as far as we know, carry their inherited genetic information in double-stranded DNA, and this is either organized in a single piece, as in those prokaryotes examined, or else is built into a number of discrete chromosomes, as in eukaryotes. In essentially all systems, recombination between and within linked genes occurs owing to the ubiquity of molecular mechanisms for breaking and repairing DNA molecules.

In contrast, viruses make use of either kind of nucleic acid, RNA or DNA, as genetic material. There are examples of viruses with single-stranded or double-stranded nucleic acid of either kind within the virion, although apparently the nucleic acid always passes through a double-stranded replicative form inside the host cell. Moreover, the total genetic information may be distributed in several different ways. In the case of DNA viruses, the total information is normally in a single molecule (except for certain special cases of "incomplete" viruses dependent on others for some of their functions), but this may be partially duplicated (terminally redundant), as in many bacteriophages, or separable at predetermined positions into more than one piece, as seems to occur in phage T5. In the case of RNA viruses, the total information is often distributed over more than one molecule, which may even be packaged as separate virions, as in many plant viruses. Such fragmented genomes are partially analogous to multi-chromosome genomes of eukaryotes; reassortment of RNA fragments can give rise to recombination of characters in a manner corresponding to that of whole chromosome reassortment in eukaryotes. Conceivably selective pressures for gene reassortment have forced certain RNA viruses to this solution to the recombination problem, in the possible absence of molecular mechanisms for crossing-over RNA molecules, which are still very much in doubt (see pages 144 and 145). DNA viruses, on the other hand, seem to have no difficulties in recombining, sometimes by means of host mechanisms, but often by systems specified by themselves.

The nature of the genomes of particular viruses has an important bearing on their genetic analysis.

2. Special features of virus genetics

(a) *Markers.* Although this Chapter is not concerned with the raw material of genetic studies, mutations, a particular category of markers has become so central to the genetic analysis of viruses that a few words about such "conditional lethal" mutations are warranted. These mutations have largely solved the marker problem for animal viruses and phages, because of their occurrence in nearly all genes. There are two main categories of conditional lethals (Campbell, 1961; Epstein et al., 1963; Fenner, 1969): temperature sensitive (ts) and host-dependent. ts mutations simply render the virus incapable of reproduction at one temperature (the restrictive or non-permissive temperature) at which the wild-type virus can grow, while still able to grow at a lower temperature (the permissive temperature). The conditions for exploiting such mutations, two incubation temperatures, are very easily established, and these markers have been used widely in both phages and animal viruses.

Host-dependent mutations, also called by such names as suppressor-sensitive (*sus*) or amber (*am*), require for their study a pair of host strains, one having and the other lacking a suppressor capable of suppressing the nonsense codon carried by the mutants. The former is the permissive and the latter the restrictive host. Clearly, some preliminary work may be needed before suitable hosts for a new virus are available. Nevertheless such markers have been used in quite a wide range of phages, although barely so far in animal viruses.

(b) *Recombination.* Crosses involving viruses are very different from those of cellular organisms, in that it is currently impossible to study the immediate progeny of pairwise matings between parental viruses. This is because each elementary act of virus reproduction or mating is not followed by maturity; from infection of the host cell by parental virions to maturity of the first complete progeny viruses, several or many rounds of reproduction and possibly of mating have occurred between virus *genomes*. Thus two parental viruses, or their genomes, may enter a host cell, and tens, hundreds, or even thousands of progeny may leave it. A virus cross, as clearly enunciated by the early workers on bacteriophage recombination (Visconti and Delbrück, 1953) is thus a problem in population genetics, in which gene and genotype frequencies are accessible for study only in populations separated in time by several generations. This fact, while complicating the mathematical description of a cross, and in particular the relationship between an observed "recombination frequency" and a genetic distance, has not in practice hindered genetic analysis in many viruses, amongst which certain bacteriophages are extremely well studied from a genetic point of view.

There are analogies here with the mathematical analysis of linkage in eukaryotes (see page 38). Elegant mathematical models, involving assumptions that may be hard to justify, have been devised to relate the frequency of recombination between markers with the distance apart of loci in eukaryotes, but these studies have largely been an end in themselves: most genetic investigations in micro-organisms have proceeded very adequately on the basis of raw recombination frequencies, provided that these were not too large. A similar development has occurred in the study of bacteriophages, particularly T4 and lambda; ingenious mathematical devices have been used to arrive at mapping functions, and these have resulted in some improvement in the consistency of linkage maps of T4, and lambda, although based on assumptions of questionable validity (see pages 137–138). However, the great majority of mapping studies of viruses have used recombination frequencies representing the crude proportion of recombinants, in respect of a pair of markers, amongst the total progeny issuing from a "cross" of the type just outlined. Such figures, while clearly

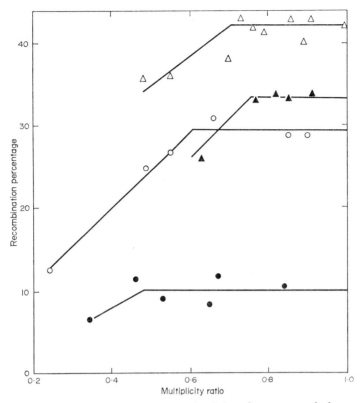

Fig. 20. The influence of the relative multiplicity of two parental phages on the frequency of recombinants. Data from Doermann and Hill (1953). Each curve represents a different pair of markers. The multiplicity ratio represents the multiplicity of the minority parent divided by that of the majority parent.

not equivalent to recombination frequencies in cellular organisms, yield maps which are both consistent and useful, provided that crossing conditions are standardized in certain respects.

The most important variables under the experimenter's immediate control are the multiplicity of infection (m.o.i.: the average number of parental viruses infecting each host cell), and the proportions of the two parental genotypes. It is easy to see that these factors will affect: (1) the proportion of cells infected by both genotypes (and therefore capable of yielding recombinants) compared with those receiving only one or other parental genotype (and therefore incapable of yielding recombinants); and (2) the proportions of the two parental genotypes entering those cells that *are* mixedly infected, which will determine the probability of mating between unlike genomes. Pioneer work on the importance of these variables was done by Doermann and Hill (1953), who established acceptable criteria for crosses with phage T4 which have been adopted in many later studies with this and other viruses. They reasoned that, when the m.o.i. by either parent fell below 3, at least 5% of host bacteria would escape mixed infection; this value has therefore been generally accepted as a lower limit, and most studies have used m.o.i.'s of 3–10 for each parent. Doermann and Hill (1953) investigated experimentally the effect of varying the *relative* multiplicity of the two parents on the recombination percentage between various pairs of loci, with the results shown in Fig. 20. They found a constant, maximum, recombination percentage for a particular pair of loci, as long as the relative multiplicity was above about 0·67. In practice, an equal and adequate m.o.i. by each parent is aimed for, but owing to the fact that a variable proportion of phages may fail to adsorb in a reasonable time, the absolute and relative m.o.i.'s have to be measured in each cross (see page 134), and any crosses in which the criteria are not met are excluded from analysis.

Other factors have an important bearing on observed recombination frequencies, but these are less subject to experimental control. In formulating their mathematical theory of T4 phage crosses, Visconti and Delbrück (1953) had to take into account the possible occurrence of more than one "round of mating" in the lineage of a particular progeny phage. The spread in maturation times of individual phages was also important, since early maturing progeny could have had fewer opportunties to become recombinant than progeny maturing later. Both concepts have firm experimental support. For example the fact that cells mixedly infected with three genotypes (triparental crosses) yield recombinants inheriting markers from all three parents clearly establishes multiple rounds of mating. The fact that cells artificially lysed prematurely, or alternatively delayed in lysis, yield recombination frequencies respectively lower or higher than usual, points to a spread in maturation time, and the gradual drift of recombination frequencies towards 50%.

In fact, as pointed out by Foss and Stahl (1963), the equilibrium recombination frequency must be less than 50%, for two reasons: (1) even with equal average m.o.i., many cells are by chance infected with more viruses of one genotype than of the other, leading to an excess of matings between identical genomes; and (2) the progeny maturing earliest have had few opportunities for mating. These factors reduce the maximum observed recombination frequency to about 40% for phage T4, a system in which recombination appears to be extremely efficient. Some other bacteriophages, for example T7 (Studier, 1969) and P22 (Kolstad and Prell, 1969) give similar values, but in others they are rather lower; for example 20% for lambda (Campbell, 1959) and 6% for P1 (Scott, 1968). In contrast, the maximum values for single-stranded DNA phages such as S13 are very low, in the region of 0·1–0·2% (Tessman, 1965) and the same is true of the double-stranded DNA phage P2 (Lindahl, 1969). The explanation in the former case is probably that the double-stranded replicative forms of the phage, which are presumably those involved in recombination, are confined to particular "attachment sites" in the cell (Sinsheimer, 1968b) and so tend to be spatially separated from one another, conceivably similar spatial separation operates in the case of P2. In animal viruses, recombination frequencies range up to a few per cent, probably again because of spatial separation of different lineages of progeny. In all these cases of low recombination rates, one is clearly including in the calculation of recombination frequencies a large class equivalent to the asexual progeny of cellular organisms.

In view of all these special features of virus crosses it is not surprising that rigorous control of experimental conditions is needed if reproducible linkage values are to be obtained.

(c) *Complementation.* In contrast to the complexities of recombination studies, the study of complementation by virus mutants, at least qualitatively, is rather simple, provided that the mutations concern functions essential for the virus life cycle. The experiment normally consists of mixedly infecting host cells with viruses of two genotypes, neither of which can yield progeny virus under the conditions used. A complementation test with temperature-sensitive mutations would consist of mixed infection followed by incubation at the restrictive temperature and assay of progeny virus at the permissive temperature. The yield would be expressed relative to that of non-mutant virus under the same conditions. Controls of each mutant grown separately at the non-permissive temperature would estimate the yield of virus explicable by leakiness of the mutations and reversion to ts^+.

In the case of many bacteriophages, complementing combinations often give yields around 50% of that of non-mutant phage, making complementation rather easy to diagnose. Non-complementing combinations may give appreciable yields, due to the production of ts^+ *recombinants* during mixed infection, and these are particularly frequent when the mutations concern genes required only late in infection, giving opportunity for considerable recombination before reproduction would be blocked by the ts mutations. Thus cases of weak complementation may be hard to recognize with certainty since they can be confused with recombinant production by such non-complementing mixtures.

Complementation assumes a particularly important role in virus genetics, partly because recombination may be unavailable as an analytical tool, as in (most) RNA viruses, and partly because, at least in the smaller viruses, it becomes a real possibility to recognize and ascribe a function to *every* gene, particularly when there are only three, as in RNA phages, or seven, as in some single-stranded DNA phages. The total number of complementation groups can then readily be determined by the use of conditional lethal mutations. It is becoming possible to determine the order of the genes recognized in this way by means independent of recombination (see page 141), so that the study of complementation, rather than recombination analysis, emerges as the pre-eminent genetical tool.

B. Genetic analysis in particular viruses

1. *Bacteriophages*

(a) *General remarks.* A vast amount of work has been and is being done on the molecular biology of phages. Much of this is directly related to genetic analysis in the sense that we are using the term in this Chapter; especially in view of the development of chemical and physical techniques for the definition of individual genes and of studying the sequences of genes, it becomes hard to delimit what should be included under the heading of genetic analysis. In order to keep the treatment reasonably concise, I shall have to stick, as in the Sections on eukaryotes and prokaryotes, largely to recombination and complementation studies, while realizing that in very many cases analyses of the functions of particular genes are intimately tied up with these studies. A very useful recent review by Calendar (1970) on the regulation of phage development provides access to a great deal of work in these areas.

As far as strict genetic analysis is concerned, T4 and lambda have dominated the field. Quite recently, and directly consequent on the use of conditional lethal mutations (see page 129), genetic analysis of many other phages is going ahead rapidly. These studies rely heavily on the experience gained with T4 and lambda. I shall consider T4 in some detail, and provide references to original papers on the other phages, discussing them only when additional points of analysis emerge.

(b) *Genetic studies of particular phages*

i. *T4 (and T2) coliphage*

Mapping. The rationale and practical details of a T4 cross are concisely described by Edgar (1963; pp. 25–30). The main concern, as we have already seen (page 132), is to standardize and optimize conditions as far as possible so as to achieve reproducible recombination frequencies. Edgar goes into the important features of a phage cross: the choice of host bacteria at a suitable phase of the growth cycle; the achievement of a high enough host cell density so that phage adsorption is rapid; the addition of cyanide during adsorption (necessary for T-even phages), so that the phages adsorbed earliest do not exclude later arrivals and so vitiate attempts to obtain high proportions of mixedly infected cells; dilution by a large factor after allowing a standard period for adsorption, so as to remove the growth-inhibiting effect of the cyanide, and to end the adsorption period; assay of unadsorbed phages, usually by treating a sample with chloroform to eliminate intracellular phages, and plating; removal of unadsorbed phages by antiserum or centrifugation, if they are numerous enough to cause possible confusion with progeny phages; assay of infected bacteria (as "infective centres"), for comparison with the titre of bacteria before infection, to recognize the possible selective loss of multiply infected bacteria by premature lysis, which would result in a much lower frequency of mixed infections than expected; termination of the cross by addition of chloroform; and plating for progeny phages. Armed with information on all these points, one is in a position to reject any unsatisfactory crosses and calculate recombination frequencies from the remainder. As we saw (page 132), crosses are normally rejected if the observed m.o.i. (based on the phages actually adsorbed) of either parent drops below about 3, and if the *relative* m.o.i. of the two parents falls below about 0·67. The latter is best measured on the allele ratios in the *progeny* phage, in case of unequal maturation of genomes carrying different alleles, but this is not always feasable since both alleles may not be unambiguously recognizable; in this case the actual input ratio (the input m.o.i. corrected for any unequal adsorption of the two parental genotypes) is used.

Plating for progeny phages depends on the markers and the experimental design. Some markers are visually recognizable on suitable hosts, and so a non-selective plating can be made, with classification of all progeny classes. Many studies, on the other hand, use selective markers, usually conditional lethals. In a two-factor cross, samples are plated under both restrictive and permissive conditions and the recombination percentage is calculated as:

$2 \times$ (yield of phages assayed under restrictive conditions/yield of phages assayed under permissive conditions) $\times 100$.

The factor 2 allows for the recovery of only one, the wild-type but not the double mutant, of the two possible recombinant classes. As pointed out by Doermann and Parma (1967), short intervals may be overestimated by this procedure, because complementing heterozygotes, common in T4, can mimic wild-type recombinants. The selective approach is, of course, used in fine-structure work (Benzer, 1955, 1961).

In the early studies of linkage in T-even phages, as in the case of cellular organisms, the linkage of certain markers was in doubt. As we saw (page 132), the maximum recombination percentage depends on the efficiency of mating, and varies in different phages, being about 40 in T4; it falls short of 50%, for reasons other than linkage. Thus Doermann and Hill's (1953) finding of three linkage groups, each showing between 35 and 40% recombination with each other, left open the question of whether the groups were linked. Later the question was directly and simply answered in the affirmative by the finding of *ts* (Edgar and Lielausis, 1964) and *amber* (Epstein, *et al.*, 1963) mutations allowing the spanning of all gaps. However, Streisinger and Bruce (1960) had meanwhile devised a sensitive qualitative linkage test. This depended on the following reasoning. When host cells are deliberately infected with unequal multiplicity by two parental phage genotypes, the allele ratio at each locus reflects the input ratio; however, if in a three-factor cross a particular recombinant class in respect of two linked loci is selected, the allele ratio at a third locus among members of this class will not reflect just the multiplicity ratio of the input, if it is linked to the first two loci. They made the cross with multiplicity ratios of 10 : 1 and 1 : 10 and found that the frequency of the allele at the third locus coming from the minority parent was different in the two crosses, indicating linkage of all three loci. An extra refinement in these experiments was premature lysis of the infected bacteria (by treatment with chloroform) to select only the earliest maturing progeny, and so to maximize linkage.

Now that enough phages have been studied to indicate that a single linkage group is the rule (but see T5: page 140), the emphasis in most recombination studies is on gene sequencing rather than on the qualitative evaluation of linkage. As in other systems, two-factor crosses yield a generally consistent sequence of markers, provided that a fairly full matrix of possible crosses is performed, but some anomalies are normally revealed. Sequences established by the use of outside markers in three-factor crosses are generally more consistent than those revealed by two-point data, but when selection is in very short intervals, high negative interference (Chase and Doermann, 1958; Doermann, 1965; Doermann and Parma, 1967) renders sequences ambiguous. Benzer's (1961) use of a series of overlapping deletions to subdivide the r_{II} region of T4 into segments to be used in the qualitative sequencing of a series of point-mutations has already been mentioned (page 46).

The linkage map of T4 is circular, like that of some other phages, but by no means all, and the detection of circularity presents interesting features. Perhaps the simplest demonstration of circularity in T4, although not the first, was the finding by Edgar and Epstein in their studies of large sets of conditional lethals, already referred to, that the map synthesized by the putting together of very many short intervals defined by pairs of mutations was indeed circular. Previously, Streisinger, Edgar and Denhardt (1964) deduced circularity from three-factor crosses involving many fewer markers; each cross revealed a linear sequence of loci, but when the separate sequences were put together they were found to be mutually consistent, provided that a circular arrangement of markers was adopted. Yet another demonstration of circularity was by Foss and Stahl (1963), in a single cross involving four mutually scoreable markers, ingeniously analysed in two ways: by consideration of the pattern of segregation amongst the 16 possible genotypes of progeny (this analysis appears to have some points of contact with the demonstration of circularity in a four-factor cross of *Streptomyces* analysed *selectively*: page 121); and by marker-rescue experiments in which one parent was heavily irradiated with gamma-rays before crossing, when the patterns of inactivation of groups of markers were consistent with a circular arrangement, on the assumption that joint inactivation was a consequence of contiguous arrangement.

I have already introduced the idea of mapping functions in phage recombination (page 130). Now that map intervals throughout the T4 genome have been elegantly analysed in physical terms by Mosig (1968), as we shall see in a moment, an ideal mapping function is perhaps less crucial than before. Nevertheless the development of mathematical ideas on T4 recombination may be instructive. Most of the story can be followed in the papers of Visconti and Delbrück (1953), Hershey (1958), Steinberg and Stahl (1958), Stahl and Steinberg (1964) and Stahl, Edgar and Steinberg (1964). The last of these papers presented a graphical relationship, derived in a partially empirical fashion by computer-choice of variables, between recombination frequency and map distance, which seemed to pass all tests of additivity of intervals with flying colours; however the last word on the subject has probably not been said (see discussion to Doermann and Parma, 1967).

Mosig's (1968) experiment used the inviable phages containing genomes of discrete but fractional length (0·67, 0·77 or 0·90 of normal) that constitute part of the yield of T4 infections. One class of these light phages, usually the two-thirds size, was purified by density gradient centrifugation; it came from a stock carrying an *r* (rapid lysis) marker. Mixed infections were then made, at low multiplicity of the light phage so that hardly any bacteria received more than one, with a normal phage from an r^+ stock carrying one of a series of *am* mutations. Infected bacteria were plated on host bacteria non-permissive for both *am* and *r*. Plaques that developed must have come from infected cells that had received a two-thirds phage carrying the appropriate am^+ allele, which had recombined with the simultaneously infecting viable phage to rescue the am^+ allele. The plaques were simply scored as clear (pure r^+) or mottled (mixed r^+ and r) and the proportion of mottled plaques was calculated.

Arguing from the assumption that two-thirds phages derive their fractional genome from a *random* segment of the whole, the proportion of mottled plaques would be a function of the distance of the r locus from the particular *am* mutation, since it would reflect the probability of a two-thirds genome including the non-selected r as well as the selected am^+. By using three distantly linked r genes as reference points, a highly consistent and additive map emerged, and the assumption of randomness of the two-thirds phages were confirmed.

Another ingenious approach to the measurement of short map intervals in terms of nucleotides was made by Goldberg (1966). The procedure was to add T4 DNA fragments, reduced to particular lengths by nuclease treatment and fractionation, to spheroplasts containing viable phage genomes, and study the inheritance by progeny phages of pairs of markers from the fragmented DNA. The shortest fragments allowing joint marker inheritence defined their distance apart.

Complementation. The study by Edgar, Denhardt and Epstein (1964) of *ts* and *am* mutants is a good example of the methodology of complementation tests in T4, including the advantages and limitations of rapid spot tests for complementation.

As in nearly all systems (see page 48), these complementation tests did not use the *cis* control for comparison with the result of the *trans* test; the yield from the mixed infection by two mutants was compared with that of a wild-type infection. In more critical tests, Stahl, Murray, Nakata and Crasemann (1966) constructed each *cis* double mutant involving pairs of *am* mutations in nearby genes, so that the *cis* and *trans* double infections could be compared. In this way some cases were found in which a fair complementation yield was given by the *trans* test, but this was nevertheless significantly below the yield of the *cis* test, indicating incomplete complementation, presumably due to a polar effect of an *am* mutation on a distal cistron. In this way the direction of transcription of a number of operons on the T4 genome was deduced (Stahl, Crasemann, Yegian, Stahl and Nakata, 1970).

ii. *Lambda.* Genetic analysis in lambda developed in parallel with that in T4, and many of the concepts and techniques are similar. The current state of the art is at least as sophisticated as that in T4 (Dove, 1968). Methods of recombination and complementation analysis, primarily using conditional lethals, are in papers by Campbell (1959, 1961). In the same way as for T4, mapping functions have been devised, based on empirical estimates of parameters such as interference in three-point crosses (Amati and

Meselson, 1965), but these have not gained wide use: for example Parkinson (1968) found raw recombination percentages over the useful range 0·1–3 (they never exceed 10–20%) to be additive (below 0·1%, high negative interference made ordering unreliable). Moreover, physical determinations of map intervals are becoming feasible; Davis and Davidson (1968) used an ingenious method for measuring the distance of deletion mutations from the end of the genome. The procedure was to hybridize DNA from a deletion mutant with wild-type DNA and chart the site of the resulting single-strand loop in electron micrographs.

Being a temperate phage lysogenizing by integration with the host chromosome, the genome of lambda can be mapped in the prophage state as well as in the vegetative state. The problem is then one of *E. coli* genetics, using P1-mediated transduction (Rothman, 1965).

A fascinating study of lambda prophage genetics, which goes far in underlining the basic similarities between genetic analysis in very different biological systems, allowing for the differences in sexual behaviour, is that of Meselson (1967a). He prepared bacteria that were "diploid" for lambda and heterozygous at three of its loci by having one prophage integrated into the bacterial chromosome and the other into an F′ episome (see page 96). Many different sub-cultures of the strain were prepared, each from a single colony, and the free phages liberated by occasional spontaneous lysis of cells in each culture were plated and characterized. The great majority of cultures contained phages of the two parental genotypes, but many contained recombinant genotypes, often a reciprocal pair of recombinants. Detailed analysis of the data led to the conclusion that the system closely resembled that in a diploid eukaryote undergoing mitotic crossing-over, with any recombinational act involving only two of (probably) four strands.

iii. *Other double-stranded DNA phages.* Genetic analyses of nearly all the well-known phages of *E. coli* and *S. typhimurium*, as well as a few other phages, such as those of rhizobia (Orosz and Sik, 1970), have recently been started. Although each study has special interests, the procedures of recombination and complementation tests closely resemble the T4 models, and so I shall merely cite some key papers in which details can be found: T1 (Michalke, 1967; Olligs, 1967); T3 (Hausmann and Gomez, 1967); T5 (Fattig and Lanni, 1965; Lanni, 1969); T7 (Studier, 1969); P1 (Scott, 1968); P2 (Lindahl, 1969); P22 (Gough and Levine, 1968; Kolstad and Prell, 1969). In all these phages, except P2, maximum recombination frequencies are high, although not always as high as in T4 (see page 132), indicating efficient mating. P2 is a special and intriguing case, in showing very low recombination frequencies, reminiscent of those of single-stranded

DNA phages; the possible implication of this finding, which is by no means understood, has been mentioned (page 133). Apart from P2, T5 presents special interest because it seems to inject its DNA in two stages; 8% of the DNA (probably up to the first of several single-strand "nicks") is injected before the rest, and a gene on this segment may be unlinked with genes on the main DNA fraction (Lanni, 1969). This is a case where an unambiguous test of non-linkage would be welcome (see page 176).

iv. *Single-stranded DNA phages.* A review of the genetics of these phages is by Pratt (1969), and of their molecular biology in general by Ray (1968) and Sinsheimer (1968a).

Although the closely related icosahedral phages ØX174 and S13 have both been studied in detail, the following set of papers on S13 possibly gives the clearest idea of the methodology that has been used in complementation and mapping studies. Tessman (1965) described the details of complementation tests involving conditional lethal mutations (the approach was in fact very similar to the classical T4 procedure). Recombination studies were also fairly typical, except for the difficulty, also encountered in P2 (Lindahl, 1969), that recombination frequencies could be inflated by recombination on the selective plates used to assay wild-type recombinants. This problem, in both phages, presumably arises because of the low intrinsic recombination frequencies, leading to very dense background populations of phages on the assay plates if enough recombinant plaques are to arise; it was overcome in both cases by concentrating the host bacteria by centrifugation in order to increase the cell/phage ratio and so decrease the proportion of mixedly infected cells, and by running controls of the two parents mixed immediately before plating.

In a later paper, Baker and Tessman (1967) deduced circularity by a series of three-factor crosses, in which recombinants were selected between a pair of heteroallelic conditional lethal mutations, and following the segregation of a third marker. The latter could be deduced to be an *outside* marker since the other two were alleles. Negative interference was not high enough to cause ambiguity. Interestingly, in this phage, varying the parental input ratio between 10 : 1 and 1 : 10 did not affect the segregation of the non-selected marker. This presumably reflects the low probability of recombination, which means that each recombinant has a single pairwise mating in its lineage.

Genetic analysis of filamentous phages of this group has not progressed so far as that of the icosahedral ones. An interesting complication is the occurrence of double length, diploid particles, which can be heterozygous when they arise from a mixed infection, with frequencies high enough to complicate recombination analysis (Salivar, Henry and Pratt, 1967; Pratt, 1969).

v. *RNA phages.* These viruses are probably the simplest genetic systems, containing so few genes that it has already been possible to determine the sequence of the three genes in R17 by chemical means (Jepperson, Steitz, Gesteland and Spahr, 1970), a fortunate circumstance since recombination is not available as an analytical tool. Complementation studies had already established the existence of three cistrons (e.g., Gussin, 1966). Complementation tests were complicated by the high reversion rates of the *am* mutations, but this difficulty was overcome by the ingenious use of many small samples of the mixedly infected host culture, each containing only about 5 non-permissive cells. The phages produced in each culture were assayed on non-permissive as well as permissive bacteria; plaques on the former indicated the existence of *am*+ phages, presumably due to reverse mutation. Such cultures were ignored, and complementation deduced only if the remaining cultures showed phage yields significantly above control values.

The paper by Jeppesen *et al.* (1970), summarizes much earlier work on the molecular biology of R17, setting the scene for the sequence determination. This depended on the cleavage of the genome by a specific endonuclease, at a point 40% of the length from the 5′ end, into two separable fragments. Another prerequisite was the ability to recognize the nucleotide sequence of the ribosomal binding site for initiation of each of the three proteins: "A protein", replicase and coat protein. The key findings of this complex study were: (1) the binding site of the A protein gene was on the 40% fragment; (2) the binding site of the replicase gene was on the 60% fragment; (3) the binding site of the coat protein gene was mainly on the 40% fragment; but (4) some recognizable nucleotide sequences corresponding to known amino-acid sequences in the interior of the coat protein were on the 60% fragment. The order of genes was thus: 5′ end–A protein–coat protein–replicase–3′ end. This investigation is a fine example of genetic analysis by "molecular mapping" (see page 148).

2. Animal Viruses

(a) *General remarks.* A useful review of animal virus genetics has recently appeared (Fenner, 1970); this includes details of the nucleic acid contained in each class of animal viruses, as well as summaries of complementation and recombination studies. In studies of animal virus genetics it is important to bear in mind that the genome of the virus, if constituted of RNA, may consist of more than one physical unit. This is true both of viruses containing single-stranded RNA (such as influenza virus) and of those containing double-stranded RNA (such as reovirus).

Although recombination was first detected using non-selective characters (in influenza virus in which quite high recombinant frequencies are observed), the genetics of animal viruses is now being explored much more rapidly by the use of temperature-sensitive (*ts*) mutations. Owing to the narrow temperature range tolerated by the animal cells that are hosts for the viruses, the range for experimentation between the restrictive and permissive temperatures is usually quite small, making the work much more exacting than in the case of most bacteriophages.

i. *Complementation.* Complementation using temperature sensitive mutations usually proceeds as follows. Cells are infected with a mixture of the two mutants under study and incubated at the restrictive temperature; if the two mutations are capable of complementation, then progeny virus will be produced from cells mixedly infected by both mutants. At the end of the incubation period, virus is prepared and assayed by plating at the permissive temperature. Controls consisting of the two mutants grown separately at the restrictive temperature and their yield of virus assayed at the permissive temperature are essential, since complementation is often very inefficient; thus yields from the mixed infections are often not much greater than control values. Complementation is usually expressed as the yield from the mixed infection divided by the sum of the two yields when the parents are grown separately. Consistent complementation matrices have been obtained when values as low as 2 for this expression have been accepted as evidence of complementation, for example in Sindbis virus (Burge and Pfefferkorn, 1966).

These authors emphasized the pitfalls of such complementation studies. For example, if parental virus that had adsorbed to the cells but had not initiated infection was not removed by a rigorous washing procedure after a short period of incubation, then the virus harvested at the end of the incubation period could contain significant amounts of parental particles; these could obscure the difference between cases of weak complementation and those of no complementation. Another complication arose from the high multiplicities of infection needed to ensure reasonable levels of complementation. This led to the parental input of virus being of the same order of magnitude as the yield from a complementing mixture. Since the protein components of animal viruses, unlike those of bacteriophages, enter the host cell along with the nucleic acid, the progeny virus might have arisen merely by re-assembly of parental components, with no new synthesis. This possibility was tested by Burge and Pfefferkorn (1967), who found that in complementing combinations of Sindbis virus, three criteria of viral synthesis were satisfied: double-stranded replicative form RNA arose from the single-stranded input; there was new synthesis of viral RNA; and new synthesis of viral proteins occurred.

ii. *Recombination.* A recombination test involving *ts* mutants differs from a complementation test in that a mixedly infected culture is incubated at the *permissive* temperature and recombinant progeny (*ts+*) are assayed at the restrictive temperature. Again, controls are essential, consisting this time of growth of the two parents separately at the permissive temperature and assay of their progeny at the restrictive temperature, for *ts+* reverse mutations. The yield of *ts+* progeny in the cross, after substracting the number of *ts+* reversions in the two controls, is expressed as a percentage of the total yield from the cross, assayed at the permissive temperature to give a "recombination percentage". As in the case of bacteriophages, this value is not equivalent to a recombination percentage in a meiotic analysis, and only has meaning within a closely controlled set of experiments since the products of matings between genetically unlike virus genomes are mixed with an unknown number of progeny produced from matings between identical parental genotypes, as well as the descendents of genomes that have replicated without mating. Little is known of the number of rounds of mating involved in animal virus reproduction, although in at least two studies repeated mating appeared not to be involved since the proportion of recombinant progeny reached a maximum very early in the incubation period (Cooper, 1968, poliovirus; Mackenzie, 1970, influenza virus); however, as discussed below, the interpretation of recombination in both these studies may be in some doubt.

As in the case of complementation in animal viruses, the recognition and measurement of recombination is subject to pitfalls. Dahlberg and Simon (1969) discussed, with reference to Newcastle Disease Virus (NDV), the possibility of a false indication of recombination arising by complementation. They estimated that 11–15% of particles of this virus contain two or more genomes (that is they are diploid or polyploid). Thus when, in a recombination test, the progeny of a mixture of two *ts* mutants grown at the permissive temperature are assayed at the restrictive temperature, some plaques appear; however these arise from heterozygous particles which grow by virtue of complementation, which thus gives the appearance of recombination. Such plaques can even appear to breed true, thus mimicking stable *ts+* recombinants, since a proportion of heterozygous particles is always present in the inoculum. Dahlberg and Simon (1969) concluded that recombination was not, in fact, occurring in NDV.

(b) *Genetic studies of particular animal viruses*

i. *Viruses containing DNA.* Recombination in rabbit-pox virus has been studied for some time using visually recognizable, non-selective markers, but only recently has it been studied selectively using *ts* mutants. Padgett and Tomkins (1968) found that nearly all their mutants complemented one another and all recombined with one another, but the data obtained so far have not led to a linkage map or informative complementation grouping.

The most developed genetic system in a DNA virus so far appears to be that in a human strain of adenovirus, in which complementation and mapping studies appear to be promising (Williams, 1971).

ii. *Viruses containing multi-component RNA*. Recombination has been detected both in viruses containing single stranded RNA (influenza) and in those containing double stranded RNA (reo). In the case of reovirus, the situation appears so far to be relatively simple in that Fields and Joklik (1969) found recombination frequencies between pairs of *ts* mutants to fall into two classes: zero and greater than 2%. The mutants fell clearly into five recombination groups on this criterion, the presumption being that these represented five of the RNA components of the virus, which could reassort in the manner of whole chromosomes. Thus there was no need to invoke crossing-over within RNA molecules. Complementation tests did not give a clear grouping of mutants, probably because the method was not sensitive enough to distinguish complementation clearly from no complementation (see page 142).

In influenza virus, Mackenzie (1970) also obtained two classes of recombination percentages: greater than 1% and less than 0·6%. On this basis, five recombination groups emerged, and these could have corresponded to the five pieces of RNA detected by physical studies (Duesberg, 1968; Pons, Schulze, Hirst and Hauser, 1969). The intra-group recombination values would presumably have had to implicate crossing-over within an RNA molecule; in view of the general interest of this conclusion, more convincing intra-group recombination values than those reported by MacKenzie (0·6 ± 0·4; 0·4 ± 0·3; 0·4 ± 0·2) would be very desirable. Similarly his conclusion that the *groups* could be arranged in a linear order by the finding of additive recombination frequencies between them, implying a regular association between the separate pieces of RNA, also needs corroboration in view of the fact that several values failed to fit his linear arrangement. His argument that, had the linkage groups been independent, inter-group recombination should have been at the level of 25% appears open to question. In the case of animal viruses reproducing inside large animal cells, the progeny of separate virus particles infecting the same cell may have a greater tendency to remain spatially separated, so that free recombination cannot in fact occur, than in the much smaller bacterial cells. As we have just seen, Fields and Joklik (1969) found only 2% recombination between markers presumably on separate RNA molecules in reovirus.

iii. *Viruses containing a single component of RNA.* The studies of Burge and Pfefferkorn (1966, 1967) on Sindbis and of Dahlberg and Simon (1968, 1969) on NDV have already been mentioned. In both cases, complementation tests served to arrange *ts* mutants clearly into complementation groups, but no recombination was observed.

The most problematical study so far is that of Cooper on poliovirus. Complementation in polio is difficult to study (Cooper, 1965), perhaps owing to the fact that the whole genome is translated into a single polypeptide, which only later is cut into unit proteins (Fenner, 1970). In his studies of recombination between pairs of *ts* mutants, Cooper (1968) went to great lengths to standardize what were otherwise very erratic *ts*+ frequencies; he was forced to include a standard cross in each experiment to allow for day-to-day fluctuations in *ts*+ frequencies from the same pair of mutants. His conclusion was, not only that recombination occurred, in which case crossing-over within RNA molecules would have been required, but that the mutations could be ordered in a linear array. However, in order to arrive at a linear arrangement, three-quarters of the *ts* mutants had to be disregarded. It was assumed that these were double mutants, which would therefore not have been expected to yield a simple map, but there was no good independent evidence for this view. In view of the difficulties of standardizing *ts*+ frequencies, which were often not very much greater than those in control parental cultures, further investigation of the system would be very desirable.

Other studies of viruses in this class are on foot-and-mouth disease virus (Pringle, 1968) and vesicular stomatitis virus (Pringle, 1970). In neither case was recombination demonstrated with complete certainty.

3. *Plant viruses*

The general methodology of work with plant viruses is at present very different from that involving bacteriophages and animal viruses. Plant systems lend themselves much less readily to the kinds of genetic tests involving complementation and recombination used in these other groups of viruses. The very low infectivity, per particle, of plant virus preparations is a stumbling block, as is the comparative difficulty of assaying viable virus by the equivalent of plaque or pock counts. Similarly, difficulties of mutant cloning are responsible for the current dearth of good genetic markers for plant viruses. Nevertheless the recent realization that many plant virus genomes are fascinatingly different from those of most other biological systems has been responsible for a spurt of interest in plant virus genomes which will doubtless lead to the development of more refined genetic tests in the near future.

Plant viruses containing DNA have only recently been discovered (e.g.,

7

Shepherd, Bruening and Wakeman, 1970), and it is too early to know whether good experimental systems for studying recombination in these viruses can be developed. For the most part, plant virologists are working with RNA genomes which are usually, although not always (Hull, 1970), single-stranded in the virion. As we discussed earlier, it may be that enzyme systems for recombining parts of RNA molecules have not been evolved. On the other hand it is now clear that plant viruses very often have their total genetic information distributed over more than one RNA molecule, which are often wrapped into more than one nucleoprotein particle. This phenomenon has been called virus "dependence" or "defectiveness", terms which tend to imply shortcomings on the part of the virus; the concept of the virus as consisting of "multiple components" is probably a better one. This topic has been reviewed by Bancroft (1968), and Lister (1969). The general idea is that the pieces of RNA in different "species" of particles differ from each other qualitatively, as well as (often) quantitatively, so that more than one species of particle is required for virus infectivity and reproduction. The subject is too new for the full details to have been worked out for any virus and in some cases differences in interpretation by different workers have not been ironed out. Some of the viruses showing clear evidence of multiple components are tobacco rattle (Lister, 1969), tobacco streak (Fulton, 1970), cowpea mosaic (de Jager and van Kammen, 1970) and alfalfa mosaic (van Vloten-Doting, Dingjan-Versteegh and Jaspars, 1970). Recently, indications that the phenomenon of multiple components may be even more widespread have been obtained. Lane and Kaesberg (1971) found that bromegrass mosaic virus, which had been regarded as a single component virus, has three nucleoprotein components of only very slightly different density, containing genetically different RNAs: two particles each contain one RNA species, while the third contains two, of combined size very nearly the same as the first two. It may even be that tobacco mosaic, the classical simple virus, consists of particles with two kinds of RNA (Fowlkes and Young, 1970)!

The common approach in this work is to try to define the various species of nucleoprotein and RNA in physical terms, and to associate particular viral characters with particular species of RNA or nucleoprotein. Separation of components and remixing in homologous and heterologous (combining components from genetically different virus strains) combinations is a procedure applied to the latter problem. It can be found, for example, that two components A and B, are not infectious singly, but have good infectivity in a mixture. By combining A and B components derived from strains that differ in a well-defined characteristic (which, as we said above are in short supply) such as antigenic properties of the coat protein, and finding whether this characteristic, in the mixed infection, derives from the strain donating the A or B component, it can be deduced on which component the genetic information for the character resides.

The finding of characters associated with the same RNA component would define a group of linked characters. The finding of characters associated with a particular species of nucleoprotein particle would not necessarily do so, owing to the fact that there is not always a unique species of RNA in each particle; this conclusion will probably tend to become more general with the refinement of methods of separating RNA molecules by, for example, acrylamide gel electrophoresis (Loening, 1967).

Assuming that conditional lethal mutants of plant viruses will be isolated (one or two temperature-sensitive mutants have already been obtained: Hariharasubramanian, 1970; J. B. Bancroft, pers. comm.), complementation tests by mixed infection under non-permissive conditions should be feasible in order to define cistrons. On the other hand it is hard to see how meaningful recombination analysis can be performed without very protracted progeny-testing, owing to the problem of cloning out homozygous or haploid virus from the product of a mixed infection by two mutants. Probably the recognition of genes on the same RNA molecule will proceed by the mixing of artificially separated components, as just outlined, while the sequencing of genes on the same molecule will be done by non-genetical methods, as in RNA phages (page 141).

VIII. CONCLUDING REMARKS

This Chapter may turn out to have been written at a time when genetic analysis is undergoing a rather rapid evolution. The classical concepts of segregation, recombination, dominance, complementation and epistasis, with their experimental determination as outlined in preceding Sections of the Chapter, will certainly continue to serve as powerful analytical tools in the study of microbial, as well as more complex, systems. With renewed interest in comparative questions, they will doubtless be applied to an ever increasing range of microbes. This development is particularly obvious in the rapid current extension of genetic analysis to a wide variety of viruses after its almost complete confinement to a couple of bacteriophages for many years, and the same trend is clearly apparent also amongst prokaryotes.

However, what is even more impressive at the present time is the rapid increase in the range of analytical procedures, directed particularly towards the sequencing of genes and sites within genes, which are independent of the classical concept of recombination. In this Chapter we have seen how non-recombinational methods for gene sequencing in prokaryotes, such as density transfer (page 115), marker frequency analysis by transformation (page 114) or transduction (page 127), and mutation frequency analysis (page 128), have been devised in the last few years, and others will no doubt be invented. In small viruses, a battery of molecular biological techniques is being used to analyse the genome (page 141), and these will doubtless be applied to more complex viruses in the near future. However, these lean heavily on complementation, if not on recombination, studies.

Determination of the amino-acid sequences of many proteins is now relatively straightforward. Whereas it might have been that the complete amino-acid sequence of a polypeptide would have defined the order of nucleotides in its cistron, the degeneracy of the genetic code ensures that this is not the case. In order to deduce this order the amino-acid sequences of polypeptides in mutants related in known ways to each other are required (e.g., Brammar, Berger and Yanofsky, 1967); thus recombination is not yet out of business, even in fine-structure studies. However, with the development of chemical techniques of nucleic acid sequencing, the determination of the order of nucleotides in a complete cistron, perhaps from that of its RNA product, is now a possibility, as has already been demonstrated for transfer RNA (Holley, 1968; Smith, Barnett, Brenner and Russell, 1970), and no doubt will shortly be true for ribosomal RNA (Ehresmann, Fellner and Ebel, 1970). Direct sequencing of DNA regions has also begun (e.g., Kelly and Smith, 1970).

It will be particularly fascinating to see the continued development of non-recombinational techniques of "genetic" analysis, which might perhaps be described as "molecular mapping", but it is a reasonable prediction that recombination and complementation will continue to illuminate, and be illuminated by, such studies for a long time to come.

ACKNOWLEDGMENTS

I am grateful to Drs. Keith F. Chater and Anne McVittie for very helpful discussions. It is a pleasure to thank Mrs. Meredyth Limberg for her painstaking preparation of the typescript.

REFERENCES

Abe, M., and Tomizawa, J. (1967). *Proc. Natl. Acad. Sci. U.S.A.*, **58**, 1911–1918.
Adams, M. H. (1959). "Bacteriophages", New York, Interscience.
Ali, A. M. M. (1967). *Canad. J. Genet. Cytol.*, **9**, 473–481.
Allen, S. L. (1964). *Genetics*, **49**, 617–627.
Allen, S. L., and Gibson, I. (1970). *In* "Biology of Tetrahymena" (Ed. A. Elliott). Appleton, Century and Croft, New York.
Altenbern, R. A. (1968). *J. Bact.*, **95**, 1642–1646.
Amati, P., and Meselson, M. (1965). *Genetics*, **51**, 369–379.
Anderson, T. F. (1958). *Cold Spr. Harb. Symp. Quant. Biol.*, **23**, 47–58.
Apirion, D. (1963). *Genet. Res.*, **4**, 276–283.
Arber, W. (1960). *Virology*, **11**, 273–288.

Arditti, R. R., and Strigini, P. (1962). *Sci. Repts. Ist. Super. Sanità*, **2**, 1–8.
Asato, Y., and Folsome, C. E. (1970). *Genetics*, **65**, 407–419.
Bachmann, B. J., and Strickland, W. N. (1965). "Neurospora Bibliography and Index." Yale University Press, New Haven and London.
Bailey, N. T. J. (1961). "Introduction to the Mathematical Theory of Genetic Linkage". Oxford University Press, London.
Baker, R., and Tessman, I. (1967). *Proc. Natl. Acad. Sci. U.S.A.*, **58**, 1438–1445.
Balassa, G. (1969). *Molec. Gen. Genet.*, **104**, 73–103.
Balbinder, E. (1962). *Genetics*, **47**, 545–559.
Bancroft, J. B. (1968). *In* "The Molecular Biology of Viruses" pp. 229–247, *Symp. Soc. Gen. Microbiol.*, **18** (Eds. L. V. Crawford and M. G. P. Stoker). Cambridge University Press.
Baranowska, H. (1970). *Genet. Res.*, **16**, 185–206.
Barben, H. (1966). *Genetica*, **37**, 109–148.
Baron, L. S., Gemski, P., Johnson, E. M., and Wohlhieter, J. A. (1968). *Bact. Rev.*, **32**, 362–369.
Barratt, R. W., Newmeyer, D. Perkins, D. D., and Garnjobst, L. (1954). *Adv. Genet.*, **6**, 1–93.
Bhaskaran, K. (1960). *J. gen. Microbiol.*, **23**, 47–54.
Bhaskaran, K. (1964). *Bull. Wld. Hlth. Org.*, **30**, 845–853.
Bhaskaran, K., and Iyer, S. S. (1961). *Nature, Lond.*, **189**, 1030–1031.
Beale, G. H. (1954). "The Genetics of Paramecium aurelia". Cambridge University Press.
Beisson, J., and Rossignol, M. (1969). *Genet. Res.*, **13**, 85–90.
Belser, W. L., and Bunting, M. I. (1956). *J. Bact.*, **72**, 582–592.
Benzer, S. (1955). *Proc. Natl. Acad. Sci. U.S.A.*, **41**, 344–354.
Benzer, S. (1961). *Proc. Natl. Acad. Sci. U.S.A.*, **47**, 403–415.
Berg, C. M., and Caro, L. G. (1967). *J. Mol. Biol.*, **29**, 419–431.
Boon, T., and Zinder, N. D. (1969). *Proc. Natl. Acad. Sci. U.S.A.*, **64**, 573–577.
Brammar, W. J., Berger, H., and Yanofsky, C. (1967). *Proc. Natl. Acad. Sci.*, **58**, 1499–1506.
Burge, B. W., and Pfefferkorn, E. R. (1966). *Virology*, **30**, 214–223.
Burge, B. W., and Pfefferkorn, E. R. (1967). *J. Virol.*, **1**, 956–962.
Buttin, G. (1963). *J. Mol. Biol.*, **7**, 183–205.
Buttin, G. (1968). *Adv. Enzymol.*, **30**, 81–137.
Buxton, E. W. (1956). *J. gen. Microbiol.*, **15**, 133–139.
Cahn, F. H., and Fox, M. S. (1968). *J. Bact.*, **95**, 867–875.
Calendar, R. (1970). *Ann. Rev. Microbiol.*, **24**, 241–296.
Campbell, A. (1959). *Virology*, **9**, 293–305.
Campbell, A. (1961). *Virology*, **14**, 22–32.
Campbell, A. M. (1962). *Advanc. Genet.*, **11**, 101–145.
Case, M. E., and Giles, N. H. (1962). *Neurospora Newsletter*, **2**, 6–7.
Casselton, L. A. (1965). *Genet. Res.*, **6**, 190–208.
Casselton, L. A., and Lewis, D. (1966). *Genet. Res.*, **8**, 61–72.
Cerdá-Olmedo, E., Hanawalt, P. C., and Guerola, N. (1968). *J. Mol. Biol.*, **33**, 705–719.
Chakrabarty, A. M., and Gunsalus, I. C. (1969a). *Virology*, **38**, 92–104.
Chakrabarty, A. M., and Gunsalus, I. C. (1969b). *Proc. Natl. Acad. Sci. U.S.A.*, **64**, 1217–1223.
Chakrabarty, A. M., and Gunsalus, I. C. (1970). *Bact. Proc.*, p. 35.

Chakrabarty, A. M., Gunsalus, C. F., and Gunsalus, I. C. (1968). *Proc. Natl. Acad. Sci. U.S.A.*, **60**, 168–175.
Chase, M., and Doermann, A. H. (1958). *Genetics*, **43**, 332–353.
Chater, K. F. (1970). *J. gen. Microbiol.*, **63**, 95–109.
Chen, K. C. (1965). *Genetics*, **51**, 509–517.
Chen, K. C., and Olive, L. S. (1965). *Genetics*, **51**, 761–766.
Clark, A. J. (1963). *Genetics*, **48**, 105–120.
Clark, A. J., Maas, W. K., and Low, B. (1969). *Molec. Gen. Genet.*, **105**, 1–15.
Clowes, R. C. (1958). *J. Gen. Microbiol.*, **18**, 154–172.
Clowes, R. C., and Hayes, W. (1968). "Experiments in Microbial Genetics". Blackwell, Oxford.
Clutterbuck, A. J., (1969). *Genetics*, **63**, 317–327.
Cooper, P. D. (1965). *Virology*, **25**, 431–438.
Cooper, P. D. (1968). *Virology*, **35**, 584–596.
Cooper, P. H., Hirshfield, I. N., and Maas, W. K. (1969). *Molec. Gen. Genet.*, **104**, 383–390.
Cove, D. J. (1970). *Proc. Roy. Soc. B.*, **176**, 267–275.
Curtiss, R. (1969). *Ann. Rev. Microbiol.*, **23**, 69–136
Dahlberg, J. E., and Simon, E. H. (1968). *Bact. Proc.*, p. 162.
Dahlberg, J. E., and Simon, E. H. (1969). *Virology*, **38**, 490–493.
Davis, R. W., and Davidson, N. (1968). *Proc. Natl. Acad. Sci. U.S.A.*, **60**, 243–250.
Day, A. W. (1971). *Heredity*, **26**, 346–347.
Day, A. W., and Jones, J. K. (1968). *Genet. Res.*, **11**, 63–81.
Day, A. W., and Jones, J. K. (1969). *Genet. Res.*, **14**, 195–221.
Day, P. R., and Anderson, G. E. (1961). *Genet. Res.*, **2**, 414–423.
Day, P. R., and Roberts, C. F. (1969). *Genetics*, **62**, 265–270.
Dee, J. (1962). *Genet. Res.*, **3**, 11–23.
Dee, J. (1966a). *J. Protozool.*, **13**, 610–616.
Dee, J. (1966b). *Genet. Res.*, **8**, 101–110.
Dee, J., and Poulter, R. T. M. (1970). *Genet. Res.*, **15**, 35–41.
Demerec, M., and Ohta, N. (1964). *Proc. Natl. Acad. Sci. U.S.A.*, **52**, 317–323.
Demerec, M., Goldman, I., and Lahr, E. L. (1958). *Cold Spr. Harb. Symp. Quant. Biol.*, **23**, 59–68.
Doermann, A. H. (1965). *In* "Genetics Today", Vol. 2, pp. 69–80 (Ed. S. J. Gearts). Pergamon, Oxford.
Doermann, A. H., and Hill, M. B. (1953). *Genetics*, **38**, 79–90.
Doermann, A. H., and Parma, D. H. (1967). *J. Cell. Physiol.*, **70**, (Sup. 1), 147–164.
Dorfman, B. (1964). *Genetics*, **50**, 1231–1243.
Dorn, G. L. (1967). *Genetics*, **56**, 619–631.
Dove, W. F. (1968). *Ann. Rev. Genet.*, **2**, 305–340.
Drexler, H. (1970). *Proc. Natl. Acad. Sci.*, **66**, 1083–1088.
Dubnau, D., Goldthwaite, C., Smith, I., and Marmur, J. (1967). *J. Mol. Biol.*, **27**, 163–185.
Duesberg, P. H. (1968). *Proc. Natl. Acad. Sci. U.S.A.*, **59**, 930–937.
Ebersold, W. T. (1967). *Science*, **157**, 447–449.
Ebersold, W. T., Levine, R. P., Levine, E. E., and Olmsted, M. A. (1962). *Genetics*, **47**, 531–543.
Edgar, R. S. (1963). *In* "Methodology in Basic Genetics", pp. 19–36 (Ed. W. J. Burdette). Holden-Day, San Francisco.
Edgar, R. S., and Lielausis, A. (1964). *Genetics*, **49**, 649–662.

Edgar, R. S., Denhardt, G. H., and Epstein, R. H. (1964). *Genetics*, **49**, 635–648.
Edwards, A. W. F. (1966). *Genetics*, **54**, 1185–1187.
Eggertsson, G., and Adelberg, E. A. (1965). *Genetics*, **52**, 319–340.
Ehresmann, C., Fellner, P., and Ebel, J. P. (1970). *Nature, Lond.*, **227**, 1321–1323.
Emeis, C. C. (1966). *Z. Naturforsch*, **21B**, 816–817.
Emerson, S. (1963). *In* "Methodology in Basic Genetics", pp. 167–208 (Ed. W. J. Burdette), Holden-Day, San Francisco.
Emerson, S., and Yu-Sun, C. C. C. (1967). *Genetics*, **55**, 39–47.
Ephrati-Elizur, E. (1968). *Genet. Res.*, **11**, 83–96.
Ephrussi, B. (1953). "Nucleo-cytoplasmic relations in microoganisms". Clarendon, Oxford.
Epstein, R. H., Bolle, A., Steinberg, C. M., Kellenberger, E., Boy de la Tour, E., Chevalley, R., Edgar, R. S., Susman, M., Denhardt, G. H., and Lielausis, A. (1963). *Cold Spr. Harb. Symp. Quant. Biol.*, **28**, 375–394.
Esposito, M. S. (1968). *Genetics*, **58**, 507–527.
Esposito, R. E., and Holliday, R. (1964). *Genetics*, **50**, 1009–1017.
Esser, K., and Straub, J. (1958). *Z. f. Vererbungsl.*, **89**, 729–746.
Fantini, A. A., (1962). *Genetics*, **47**, 161–177.
Fattig, W. D., and Lanni, F. (1965). *Genetics*, **51**, 157–166.
Fenner, F. (1969). *Curr. Topics in Microbiol. and Immunol.*, **48**, 1–28.
Fenner, F. (1970). *Ann. Rev. Microbiol.*, **24**, 297–334.
Fields, B. N., and Joklik, W. K. (1969). *Virology*, **37**, 335–342.
Fincham, J. R. S. (1966). "Genetic Complementation", Benjamin, New York.
Fincham, J. R. S., and Pateman, J. A. (1967). *Genet. Res.*, **9**, 49–62.
Fink, G. R. (1966). *Genetics*, **53**, 445–459.
Fink, G. R., and Roth, J. R. (1968). *J. Mol. Biol.*, **33**, 547–557.
Flores da Cunha, M. (1970). *Genet. Res.*, **16**, 127–144.
Fogel, S., and Hurst, D. D. (1967). *Genetics*, **57**, 455–481.
Foley, J. M., Giles, N. H., and Roberts, C. F. (1965). *Genetics*, **52**, 1247–1263.
Forbes, E. (1959). *Heredity*, **13**, 67–80.
Foss, H. M., and Stahl, F. W. (1963). *Genetics*, **48**, 1659–1672.
Fowlks, E., and Young, R. J. (1970). *Virology*, **42**, 548–550.
Fox, M. S. (1962). *Proc. Natl. Acad. Sci. U.S.A.*, **48**, 1043–1048.
Friend, E. J., and Hopwood, D. A. (1971). *J. Gen. Microbiol. (in press)*.
Frost, L. C. (1961). *Genet. Res.*, **2**, 43–62.
Fuerst, C. R., Jacob, F., and Wollman, E. L. (1956). *C.R. Acad. Sci. Paris*, **243**, 2162–2164.
Fulton, R. W. (1970). *Virology*, **41**, 288–294.
Gans, M., and Masson, M. (1969). *Molec. Gen. Genet.*, **105**, 164–181.
Garen, A. (1960). *Symp. Soc. Gen. Microbiol.*, **10**, 239–247.
Gibson, I. (1970a). *Adv. Morphogenesis*, **8**, 159–208.
Gibson, I. (1970b). *Symp. Soc. Exp. Biol.*, **24**, 379–399.
Gillie, O. J. (1966). *Genet. Res.*, **8**, 9–31.
Glansdorff, N. (1967). *Genetics*, **55**, 49–61.
Glatzer, L., Labrie, D. A., and Armstrong, F. B. (1966). *Genetics*, **54**, 423–432.
Goldberg, E. B. (1966). *Proc. Natl. Acad. Sci. U.S.A.*, **56**, 1457–1463.
Goldschmidt, E. P., and Landman, O. E. (1962). *J. Bact.*, **83**, 690–691.
Goodgal, S. H. (1961). *J. Gen. Physiol.*, **45**, 205–228.
Gough, M., and Levine, M. (1968). *Genetics*, **58**, 161–169.
Gross, S. R. (1962). *Proc. Natl. Acad. Sci. U.S.A.*, **48**, 922–930.

Gross, J., and Englesberg, E. (1959). *Virology*, **9**, 314–331.
Gussin, G. N. (1966). *J. Mol. Biol.*, **21**, 435–453.
Gutz, H. (1966). *J. Bact.*, **92**, 1567–1568.
de Haan, P. G., and Verhoef, C. (1966). *Mut. Res.*, **3**, 111–117.
Hadden, C., and Nester, E. W. (1968). *J. Bact.*, **95**, 876–885.
Haldane, J. B. S. (1919). *J. Genet.*, **8**, 299–309.
Hariharasubramanian, V. (1970). *Virology*, **41**, 389–391.
Harold, R. J., and Hopwood, D. A. (1970a). *Mut. Res.*, **10**, 427–438.
Harold, R. J., and Hopwood, D. A. (1970b). *Mut. Res.*, **10**, 439–448.
Hartman, P. E. (1963). *In* "Methodology in Basic Genetics", pp. 103–128. (Ed. W. J. Burdette). Holden-Day, San Francisco.
Hartman, P. E., Hartman, Z., and Šerman, D. (1960). *J. Gen. Microbiol.*, **22**, 354–368.
Hartman, P. E., Loper, J. C., and Šerman, D. (1960). *J. Gen. Microbiol.*, **22**, 323–353.
Hastie, A. C. (1962). *J. gen. Microbiol.*, **27**, 373–382.
Hastie, A. C. (1964). *Genet. Res.*, **5**, 305–315.
Hastie, A. C. (1967). *Nature, Lond.*, **214**, 249–252.
Hausmann, R., and Gomez, B. (1967). *J. Virol.*, **1**, 779–792.
Hayes, W. (1957). *J. gen. Microbiol.*, **16**, 97–119.
Hayes, W. (1968). "The Genetics of Bacteria and their Viruses", Blackwell, Oxford.
Hayes, W., Jacob, F., and Wollman, E. L. (1963). *In* "Methodology in Basic Genetics", pp. 129–164 (Ed. W. J. Burdette). Holden-Day, San Francisco.
Heagy, F. C., and Roper, J. A. (1952). *Nature, Lond.*, **170**, 713–714.
Hershey, A. D. (1958). *Cold Spr. Harb. Symp. Quant. Biol.*, **23**, 19–46.
Heumann, H. (1968). *Molec. Gen. Genet.*, **102**, 132–144.
Heumann, H. (1970). *Proc. Int. Symp. Genetics of Industrial Microorganisms*, Prague.
Hirota, Y. (1960). *Proc. Natl. Acad. Sci. U.S.A.*, **46**, 57–64.
Holley, R. W. (1968). *Prog. Nucleic Acid Res. and Mol. Biol.*, **8**, 37–47.
Holliday, R. (1961a). *Genet. Res.*, **2**, 204–230.
Holliday, R. (1961b). *Genet. Res.*, **2**, 231–248.
Holliday, R. (1962). *Genet. Res.*, **3**, 472–486.
Holiday, R. (1964a). *Genetics*, **50**, 323–335.
Holliday, R. (1964b). *Genet. Res.*, **5**, 282–304.
Holliday, R. (1965). *Genet. Res.*, **6**, 104–120.
Holliday, R., and Whitehouse, H. L. K. (1970). *Molec. Gen. Genet.*, **107**, 85–93.
Holloway, B. W. (1955). *J. gen. Microbiol.*, **13**, 572–581.
Holloway, B. W. (1969). *Bact. Rev.*, **33**, 419–443.
Holloway, B. W., Monk, M., Hodgins, L., and Fargie, B. (1962). *Virology*, **18**, 89–94.
Hopwood, D. A. (1959). *Ann. N.Y. Acad. Sci.*, **81**, 887–898.
Hopwood, D. A. (1966). *Genetics*, **54**, 1177–1184.
Hopwood, D. A. (1967). *Bact. Rev.*, **31**, 373–403.
Hopwood, D. A. (1969). *Proc. Int. Symp. Genetics and Breeding of Streptomyces*, pp. 5–18. Yugoslav Academy of Sciences and Arts, Zagreb.
Hopwood, D. A. (1970a). *In* "Methods in Microbiology, Vol. 3A, pp. 363–433 (Eds. J. R. Norris and B. W. Ribbons), Academic Press, London.
Hopwood, D. A. (1971). *Proc. 1st. Symp. Genetics of Industrial Microorganisms*, Prague.

Hopwood, D. A., and Ferguson, H. M. (1970). *J. gen. Microbiol.*, **63**, 133–136.
Hopwood, D. A., and Wright, H. M. (1971). *Heredity* (*in* press).
Hopwood, D. A., Harold, R. J., Vivian, A., and Ferguson, H. M. (1969). *Genetics*, **62**, 461–477.
Hopwood, D. A., and Sermonti, G. (1962). *Advanc. Genet.*, **11**, 273–342.
Howard-Flanders, P., Boyce, R. P., and Theriot, L. (1966). *Genetics*, **53**, 1119–1136.
Hull, R. (1970). *In* "The Biology of Large RNA Viruses", pp. 153–164 (Eds. R. D. Barry and B. W. J. Mahy). Academic Press, London.
Ino, I., and Demerec, M. (1968). *Genetics*, **59**, 167–176.
Ishikawa, T. (1962). *Genetics*, **47**, 1755–1770.
Ishikawa, T. (1965). *J. Mol. Biol.*, **13**, 586–591.
Ishitani, C., Ikeda, Y., and Sakaguchi, K. (1965). *J. Gen. Appl. Microbiol.*, **2**, 401–430.
Jackson, S. (1962). *Brit. Med. Bull.*, **18**, 24–26
Jacob, F., and Adelberg, E. A. (1959). *C.R. Acad. Sci. Paris*, **249**, 189–191.
Jacob, F., and Monod, J. (1961). *J. Mol. Biol.*, **3**, 318–356.
Jacob, F., and Wollman, E. L. (1956). *C.R. Acad. Sci. Paris*, **242**, 303–306.
Jacob, F., and Wollman, E. L. (1958). *Symp. Soc. Exp. Biol.*, **12**, 75–92.
Jacob, F., and Wollman, E. L. (1961). "Sexuality and the Genetics of Bacteria". Academic Press, New York.
de Jager, C. P., and van Kammen, A. (1970). *Virology*, **41**, 281–287.
Jeppesen, P. G. N., Steitz, J. A., Gesteland, R. F., and Spahr, P. F. (1970). *Nature, Lond.*, **226**, 230–237.
Jyssum, K. (1969). *J. Bact.*, **99**, 757–763.
Käfer, E. (1958). *Advanc. Genet.*, **9**, 105–145.
Käfer, E. (1965). *Genetics*, **52**, 217–232.
Keitt, G. W., and Langford, M. H. (1941). *Am. J. Bot.*, **28**, 805–820.
Kelly, M. S., and Pritchard, R. H. (1965). *J. Bact.*, **89**, 1314–1321.
Kelly, T. J., and Smith, H. O. (1970). *J. Mol. Biol.*, **51**, 393–409.
Kemp, M. B., and Hegeman, G. D. (1968). *J. Bact.*, **96**, 1488–1499.
Kitani, Y., and Olive, L. S. (1969). *Genetics*, **62**, 23–66.
Kloos, W. E., and Pattee, P. A. (1965). *J. gen. Microbiol.*, **39**, 195–207.
Kolstad, R. A., and Prell, H. H. (1969). *Molec. Gen. Genet.*, **104**, 339–350.
Koltin, Y., and Flexer, A. S. (1969). *J. Cell Sci.*, **4**, 739–749.
Koltin, Y., and Raper, J. R. (1967). *Mol. Gen. Genet.*, **100**, 275–282.
Kuo, T-T., and Stocker, B. A. D. (1969). *J. Bact.*, **98**, 593–598.
Lacks, S. (1966). *Genetics*, **53**, 207–235.
Lane, L. C., and Kaesberg, P. (1971). *Nature New Biology*, **232**, 40–43.
Lanni, Y. T. (1969). *J. Mol. Biol.*, **44**, 173–183.
Lawton, W. D., Morris, B. C., and Burrows, T. W. (1968). *J. gen. Microbiol.*, **52**, 25–34.
Lederberg, J. (1957). *Proc. Natl. Acad. Sci. U.S.A.*, **43**, 1060–1065.
Lederberg, J., and Tatum, E. L. (1946). *Cold Spr. Harb. Symp. Quant. Biol.*, **11**, 113–114.
Lederberg, J., Cavalli, L. L., and Lederberg, E. M. (1952). *Genetics*, **37**, 720–730.
Lederberg, J., Lederberg, E. M., Zinder, N. D., and Lively, E. R. (1951). *Cold Spr. Harb. Symp. Quant. Biol.*, **16**, 413–443.
Lennox, E. S. (1955). *Virology*, **1**, 190–206.
Lerman, L. (1963). *In* "Methodology in Basic Genetics", pp. 83–102. (Ed. W. J. Burdette). Holden-Day, San Francisco.

Leupold, U. (1958). *Cold Spr. Harb. Symp. Quant. Biol.*, **23**, 161–170.

Leupold, U. (1961) *Arch. Julius Klaus-Stift.*, **36**, 89–117.

Leupold, U. (1970). *In* "Methods in Cell Physiology", Vol. 4, pp. 169–177, Academic Press, New York.

Leupold, U., and Gutz, H. (1965). *In* "Genetics Today", Vol. 2, pp. 31–35 (Ed. S. J. Gearts). Pergamon, Oxford.

Levine, R. P., and Ebersold, W. T. (1958). *Cold Spr. Harb. Symp. Quant. Biol.*, **23**, 101–109.

Levine, R. P., and Ebersold, W. T. (1960). *Ann. Rev. Microbiol.*, **14**, 197–216.

Levinthal, M., and Simon, R. D. (1969). *J. Bact.*, **97**, 250–255.

Lewis, C. M., and Fincham, J. R. S. (1970). *Genet. Res.*, **16**, 151–163.

Lewis, D. (1961). *Genet. Res.*, **2**, 141–155.

Lhoas, P. (1967). *Genet. Res.*, **10**, 45–61.

Lindahl, G. (1969). *Virology*, **39**, 839–860.

Lissouba, P., Mousseau, J., Rizet, G., and Rossignol, J. L. (1962). *Advanc. Genet.*, **11**, 343–380.

Lister, R. M. (1969). *Fed. Proc.*, **28**, 1875–1889.

Loening, U. E. (1967). *Biochem. J.*, **102**, 251–257.

Loomis, W. F. (1969). *J. Bact.*, **99**, 65–69.

Loomis, W. F., and Ashworth, J. M. (1968). *J. gen. Microbiol.*, **53**, 181–186.

Loutit, J. S. (1969). *Genet. Res.*, **13**, 91–98.

Loutit, J. S. (1970). *Genet. Res.*, **16**, 179–184.

Loutit, J. S., and Marinus, M. G. (1968). *Genet. Res.*, **12**, 37–44.

Low, B. (1965). *Genet. Res.*, **6**, 469–473.

Low, B. (1967). *J. Bact.*, **93**, 98–106.

Low, B., and Wood, T. H. (1965). *Genet. Res.*, **6**, 300–303.

Luria, S. E., Adams, J. N., and Ting, R. C. (1960). *Virology*, **12**, 348–390.

Luria, S. E., and Delbrück, M. (1943). *Genetics*, **28**, 491–511.

McCully, K. S., and Forbes, E. (1965). *Genet. Res.*, **6**, 352–359.

McFall, E. (1964). *J. Mol. Biol.*, **8**, 746–753.

McFall, E. (1967). *Genetics*, **55**, 91–99.

Mackenzie, J. S. (1970). *J. gen Virol.*, **6**, 63–75.

McClatchy, J. K., and Rosenblum, E. D. (1966). *J. Bact.*, **92**, 580–583.

Manney, T. R. (1964). *Genetics*, **50**, 109–121.

Manney, T. R., and Mortimer, R. K. (1964). *Science*, **143**, 581–582.

Marinus, M. G., and Loutit, J. S. (1969). *Genetics*, **63**, 547–556.

Margolin, P. (1963). *Genetics*, **48**, 441–457.

Martinek, G. W., Ebersold, W. T., and Nakamura, K. (1970). *Genetics*, **64**, Suppl. 41–42.

Mather, K. (1951). "The Measurement of Linkage in Heredity". Methuen, London.

Mee, B. J., and Lee, B. T. O. (1967). *Genetics*, **55**, 709–722.

Meselson, M. (1967a). *J. Cell. Physiol.*, **70** (Sup. 1), 113–118.

Meselson, M. (1967b). *In* "Heritage from Mendel", pp. 81–104. (Ed. R. A. Brink). University of Wisconsin, Madison.

Meynell, E., Meynell, G. G., and Datta, N. (1968). *Bact. Rev.*, **32**, 55–83.

Michalke, W. (1967). *Molec. Gen. Genet.*, **99**, 12–33.

Mills, G. T., and Smith, E. E. B. (1962). *Brit. Med. Bull.*, **18**, 27–30.

Moore, D. (1967). *Genet. Res.*, **9**, 331–342.

Morgan, D. H. (1966). *Genet. Res.*, **7**, 195–206.

Mortimer, R. K., and Hawthorne, D. C. (1969). *In* "The Yeasts", Vol. 1, pp. 385–460 (Eds A. H. Rose and J. S. Harrison). Academic Press, London.
Mosig, G. (1968). *Genetics*, **59**, 137–151.
Murray, N. E. (1963). *Genetics*, **48**, 1163–1183.
Nanney, D. L. (1968). *Ann. Rev. Genet.*, **2**, 121–140.
Nester, E. W., Schafer, M., and Lederberg, J. (1963). *Genetics*, **48**, 529–551.
Newmeyer, D. (1954). *Genetics*, **39**, 604–618.
Norkin, L. C. (1970). *J. Mol. Biol.*, **51**, 633–655.
Novick, R. P. (1969). *Bact. Rev.*, **33**, 210–263.
Oishi, M., and Sueoka, N. (1965). *Proc. Natl. Acad. Sci. U.S.A.*, **54**, 483–491.
Olive, L. S. (1956). *Am. J. Bot.* ,**43**, 97–107.
Oliver, C. P. (1940). *Proc. Natl. Acad. Sci. U.S.A.*, **26**, 452–454.
Olligs, H. (1967). Doctoral Thesis, University of Köln.
Okada, M., Watanabe, T., and Miyake, T. (1968). *J. gen. Microbiol.*, **50**, 241–252.
O'Neil, D. M., Baron, L. S., and Sypherd, P. S. (1967). *J. Bact.*, **99**, 242–247.
Orosz, L., and Sik, T. (1970). *Acta. Microbiol. Acad. Sci. Hung.*, **17**, 185–194.
Ozeki, H. (1956). *Carnegie Inst. Wash. Publ.*, No. 612, 97–106.
Ozeki, H. (1959). *Genetics*, **44**, 457–470.
Ozeki, H., and Ikeda, H. (1968). *Ann. Rev. Genet.*, **2**, 245–278.
Padgett, B. L., and Tomkins, J. K. N. (1968). *Virology*, **36**, 161–167.
Pardee, A. B., Jacob, F., and Monod, J. (1959). *J. Mol. Biol.*, **1**, 165–178.
Parker, J. H., and Sherman, F. (1969). *Genetics*, **62**, 9–22.
Parkinson, J. S. (1968). *Genetics*, **59**, 311–325.
Pearce, U., and Stocker, B. A. D. (1965). *Virology*, **27**, 290–296.
Perkins, D. D. (1949). *Genetics*, **34**, 607–626.
Perkins, D. D. (1959). *Genetics*, **44**, 1185–1208.
Perkins, D. D., Newmeyer, D., Taylor, C. W., and Bennett, D. C. (1969). *Genetica*, **40**, 247–278.
Pittenger, T. H. (1954). *Genetics*, **39**, 326–342.
Pons, M. W., Schulze, I. T., Hirst, G. K., and Hauser, R. (1969). *Virology*, **39**, 250–259.
Pontecorvo, G. (1952). *Adv. Enzymol.*, **13**, 121–149.
Pontecorvo, G. (1959). "Trends in Genetic Analysis", Columbia University Press, New York.
Pontecorvo, G. (1963). *Proc. Roy. Soc. B.*, **158**, 1–23.
Pontecorvo, G., and Käfer, E. (1958). *Advanc. Genet.*, **9**, 71–104.
Pontecorvo, G., and Sermonti, G. (1954). *J. gen. Microbiol.*, **11**, 94–104.
Pontecorvo, G., Roper, J. A., and Forbes, E. (1953). *J. gen. Microbiol.*, **8**, 198–210.
Pontecorvo, G., Roper, J. A., Hemmons, L. M. Macdonald, K. D., and Bufton, A. W. J. (1953). *Advanc. Genet.*, **5**, 141–238.
Poulter, R. T. M., and Dee, J. (1968). *Genet. Res.*, **12**, 71–79.
Pratt, D. (1969). *Ann. Rev. Genet.*, **3**, 343–362.
Prévost, G. (1962). Doctoral Thesis, University of Paris, Series A, No. 3939.
Pringle, C. R. (1968). *J. gen. Virol.*, **2**, 199–202.
Pringle, C. R. (1970). *J. Virol.*, **5**, 559–567.
Pritchard, R. H. (1953). *Caryologia*, Suppl. **6**, 1117.
Pritchard, R. H. (1955). *Heredity*, **9**, 343–371.
Pritchard, R. H. (1960). *Genet. Res.*, **1**, 1–24.
Pritchard, R. H. (1963). *In* "Methodology in Basic Genetics", pp. 228–246 (Ed. W. J. Burdette). Holden-Day, San Francisco.

Prozesky, O. W. (1968). *J. gen. Microbiol.*, **54**, 127–143.
Prud'homme, N. (1970). *Molec. Gen. Genet.*, **107**, 256–271.
Putrament, A. (1964). *Genet. Res.*, **5**, 316–327.
Raper, C. A., and Raper, J. R. (1966). *Genetics*, **54**, 1151–1168.
Raper, J. R., and Miles, P. G. (1958). *Genetics*, **43**, 530–546.
Raper, J. R., and Raudaskoski, M. (1968). *Heredity*, **23**, 109–117.
Ray, D. S. (1968). *In* "Molecular Basis of Virology", pp. 222–254 (Ed. Fraenkel-Conrat). Reinhold, New York.
Richmond, M. H. (1970). *In* "Organization and Control in Prokaryotic and Eukaryotic Cells", pp. 249–277, *Symp. Soc. Gen. Microbiol.*, **20** (Eds. H. P. Charles and B. C. J. G. Knight). Cambridge University Press.
Richter, A. (1961). *Genet. Res.*, **2**, 333–345.
Roberts, C. F. (1967). *Genetics*, **55**, 233–239.
Roper, J. A. (1950). *Nature, Lond.*, **166**, 956–957.
Roper, J. A. (1966). *In* "The Fungi", Vol. 2 pp. 589–617 (Eds. G. C. Ainsworth and A. S. Sussman), Academic Press, New York and London.
Roth, J. R., and Hartman, P. E. (1966). *Virology*, **27**, 297–307.
Roth, J. R., and Sanderson, K. E. (1966). *Genetics*, **53**, 971–979.
Rothman, J. L. (1965). *J. Mol. Biol.*, **12**, 892–912.
Sager, R., and Ramanis, Z. (1967). *Proc. Natl. Acad. Sci. U.S.A.*, **58**, 931–937.
Sager, R., and Ramanis, Z. (1970). *Proc. Natl. Acad. Sci. U.S.A.*, **65**, 593–600.
Salivar, W. O., Henry, T. J., and Pratt, D. (1967). *Virology*, **32**, 41–51.
Sanderson, K. E. (1967). *Bact. Rev.*, **31**, 354–372.
Sanderson, K. E. (1970). *Bact. Rev.*, **34**, 176–193.
Sanderson, K. E., and Demerec, M. (1965). *Genetics*, **51**, 897–913.
Sanderson, K. E., and Hall, C. A. (1970). *Genetics*, **64**, 215–228.
Sarabhai, A. S., Stretton, A. O. W., Brenner, S., and Bolle, A. (1964). *Nature, Lond.*, **201**, 13–17.
Schaeffer, P. (1969). *Bact. Rev.*, **33**, 48–71.
Schneider, H., and Falkow, S. (1964). *J. Bact.*, **88**, 682–689.
Scott, J. R. (1968). *Virology*, **36**, 564–574.
Sermonti, G. (1957). *Genetics*, **42**, 433–443.
Sermonti, G. (1959). *Ann. N.Y. Acad Sci.*, **81**, 950–966.
Sermonti, G. (1969). "Genetics of Antibiotic-producing Microorganisms". Wiley-Interscience, London.
de Serres, F. J. (1962). *Neurospora Newsletter*, **1**, 9–10.
Shepherd, R. J., Bruening, G. E., and Wakeman, R. J. (1970). *Virology*, **41**, 339–347.
Sheppard, D., and Englesberg, E. (1966). *Cold Spr. Harb. Symp. Quant. Biol.*, **31**, 345–347.
Siddiqi, O. H. (1962). *Genet. Res.*, **3**, 69–89.
Simchen, G. (1967). *Genet. Res.*, **9**, 195–210.
Sinha, U., and Ashworth, J. M. (1969). *Proc. Roy. Soc. B.*, **173**, 531–540.
Sinsheimer, R. L. (1968a). *Prog. Nucleic Acid Res. and Mol. Biol.*, **8**, 115–169.
Sinsheimer, R. L. (1968b). *In* "The Molecular Biology of Viruses", pp. 101–123, *Symp. Soc. Gen. Microbiol.*, **18** (Eds L. V. Crawford and M. G. P. Stoker) Cambridge University Press.
Slonimski, P., Petrochilo, E., Netter, P., Dujon, B., Deutsch, J., Coen, D., and Bolotin, M. (1970). *Heredity*, **25**, 492.
Smith, D. A. (1961). *J. gen. Microbiol.*, **24**, 335–353.

Smith, B. R. (1965). *Heredity*, **20**, 257–276.

Smith, C. D., and Pattee, P. A. (1967). *J. Bact.*, **93**, 1832–1838.

Smith, H. O., and Levine, M. (1967). *Virology*, **31**, 207–216.

Smith, J. D., Barnett, L., Brenner, S., and Russell, R. L. (1970). *J. Mol. Biol.*, **54**, 1–14.

Smith, S. M., and Stocker, B. A. D. (1962). *Brit. Med. Bull.*, **18**, 46–51.

Smith-Keary, P. F. (1960). *Heredity*, **14**, 61–71.

Smith-Keary, P. F. (1966). *Genet. Res.*, **8**, 73–82.

Sneath, P. H. A., and Lederberg, J. (1961). *Proc. Natl. Acad. Sci. U.S.A.*, **47**, 86–90.

Snow, R., and Korch, C. T. (1970). *Mol. Gen. Genet.*, **107**, 201–208.

Somers, J. M., and Bevan, E. A. (1969). *Genet. Res.*, **13**, 71–83.

Sonneborn, T. M. (1950). *J. exp. Zool.*, **113**, 87–147.

Sonneborn, T. M. (1970). *Proc. Roy. Soc. B.*, **176**, 347–366.

Stadler, D. R., and Kariya, B. (1969). *Genetics*, **63**, 291–316.

Stadler, D. R., and Towe, A. M. (1963). *Genetics*, **48**, 1323–1344.

Stahl, F. W. (1969). *Genetics*, **61**, Suppl. 1–13.

Stahl, F. W., and Steinberg, C. M. (1964). *Genetics*, **50**, 531–538.

Stahl, F. W., Edgar, R. S., and Steinberg, J. (1964). *Genetics*, **50**, 539–552.

Stahl, F. W., Crasemann, J. M., Yegian, C., Stahl, M. M., and Nakata, A. (1970). *Genetics*, **64**, 157–170.

Stahl, F. W., Murray, N. E., Nakata, A., and Crasemann, J. M. (1966). *Genetics* **54**, 223–232.

Stamberg, J. (1968). *Mol. Gen. Genet.*, **102**, 221–228.

Stanisich, V., and Holloway, B. W. (1969). *Genetics*, **61**, 327–339.

Starling, D. (1969). *Genet. Res.*, **14**, 343–347.

Steinberg, C., and Stahl, F. W. (1958). *Cold Spr. Harb. Symp. Quant. Biol.*, **23**, 42–46.

Stocker, B. A. D., Zinder, N. D., and Lederberg, J. (1953). *J. gen. Microbiol.*, **9**, 410–433.

Streisinger, G., and Bruce, V. (1960). *Genetics*, **45**, 1289–1296.

Streisinger, G., Edgar, R. S., and Denhardt, G. H. (1964). *Proc. Natl. Acad. Sci. U.S.A.*, **51**, 775–779.

Strickland, W. N. (1958a). *Proc. Roy. Soc. B.*, **148**, 533–542.

Strickland, W. N. (1958b). *Proc. Roy. Soc. B.*, **149**, 81–101.

Strickland, W. N. (1960). *J. gen. Microbiol.*, **22**, 583–588.

Strickland, W. N., and Thorpe, D. (1963). *J. gen. Microbiol.*, **33**, 409–412.

Strömnaes, Ö. (1968). *Hereditas*, **59**, 197–220.

Strömnaes, Ö., and Garber, E. D. (1963). *Genetics*, **48**, 653–662.

Studier, F. W. (1969). *Virology*, **39**, 562–574.

Sueoka, N., and Quinn, W. G. (1968). *Cold Spr. Harb. Symp. Quant. Biol.*, **33**, 695–705.

Takahashi, I. (1961). *Biochem. Biophys. Res. Commun.*, **5**, 171–175.

Takahashi, I. (1963). *J. gen. Microbiol.*, **31**, 211–217.

Taylor, A. L. (1970). *Bact. Rev.*, **34**, 155–175.

Taylor, A. L., and Trotter, C. D. (1967). *Bact. Rev.*, **31**, 332–353.

Tessman, E. S. (1965). *Virology*, **25**, 303–321.

Thomas, D. Y., and Wilkie, D. (1968). *Genet. Res.*, **11**, 33–41.

Tinline, R. D. (1962). *Canad. J. Botany*, **40**, 425–437.

Tomasz, A. (1969). *Ann. Rev. Genet.*, **3**, 217–232.

Tuveson, R. W., and Coy, D. O. (1961). *Mycologia*, **53**, 244–253.

Verhoef, C., and de Haan, P. G. (1966). *Mut. Res.*, **3**, 101–110.

Vielmetter, W., and Messer, W. (1964). *Ber. d. Bunsengesellschaft.*, **68**, 742–743.

Visconti, N., and Delbrück, M. (1953). *Genetics*, **38**, 5–33.

Vivian, A., and Hopwood, D. A. (1970). *J. gen. Microbiol*, **64**, 101–117.

Van Vloten-Doting, L., Dingjan-Versteegh, A., and Jaspars, E. M. J. (1970). *Virology*, **40**, 419–430.

Walmsley, R. H. (1969). *Biophys. J.*, **9**, 421–431.

Wang, C-S., and Raper, J. R. (1969). *J. Bact.*, **99**, 291–297.

Wann, M., Mahajan, S. K., and Wood, T. H. (1970). *J. Bact.*, **103**, 601–606.

Warr, J. R., McVittie, A., Randall, J., and Hopkins, J. M. (1966). *Genet. Res.*, **7**, 335–351.

Wheals, A. E. (1970). *Genetics*, **66**, 623–633.

Wheelis, M. L., and Stanier, R. Y. (1970). *Genetics*, **66**, 245–266.

Whitehouse, H. L. K. (1957). *J. Genet.*, **55**, 348–360.

Whitehouse, H. L. K. (1963). *Nature, Lond.*, **199**, 1034–1040.

Whitehouse, H. L. K. (1970). *Biol. Rev.*, **45**, 265–315.

Wilkie, D., Saunders, G., and Linnane, A. W. (1967). *Genet. Res.*, **10**, 199–203.

Williams, J. F. (1971). *Heredity (in press)*.

Williamson, D. H. (1970). *Heredity*, **25**, 491.

Wolf, B., Newman, A., and Glaser, D. A. (1968). *J. Mol. Biol.*, **32**, 611–629.

Wollman, E. L., Jacob, F., and Hayes, W. (1956). *Cold Spr. Harb. Symp. Quant. Biol.*, **21**, 141–162.

Wu, T. T. (1966). *Genetics*. **54**, 405–410.

Yanofsky, C., Carlton, B. C., Guest, J. R., Helinski, D. R., and Henning, U. (1964). *Proc. Natl. Acad. Sci. U.S.A.*, **51**, 266–272.

Yoshikawa, H., and Sueoka, N. (1963a). *Proc. Natl. Acad. Sci. U.S.A.*, **49**, 559–566.

Yoshikawa, H., and Sueoka, N. (1963b). *Proc. Natl. Acad. Sci. U.S.A.*, **49**, 806–813.

Yu-Sun, C. C. C. (1964). *Genetics*, **50**, 987–998.

Zinder, N. D. (1960). *Science*, **131**, 924–926.

Zinder, N. D., and Lederberg, J. (1952). *J. Bact.*, **64**, 679–699.

CHAPTER III

Inoculation Techniques—Effects Due to Quality and Quantity of Inoculum

J. Meyrath and Gerda Suchanek

Department of Applied Microbiology,
Hochschule für Bodenkultur, Wien, Austria

I. THE IDEAL INOCULUM

A. Unicellular vegetative micro-organisms (bacteria and yeasts)

Experience has shown that these two groups of micro-organisms behave very much alike in so far as growth kinetics are concerned. Very often it is desired that the culture develops as soon as possible after inoculation. The problem of inoculum production then is mainly one of obtaining a culture in the exponential phase, as cultures prior to or beyond this stage of development are liable to give rise to lag phases of undefined lengths. (Monod, 1949; Dean and Hinshelwood, 1966). Attempts have been made to define the extent of the lag phase in terms of nutrient supply and other environmental factors as well as size and history of the inoculum culture. But there can be no doubt that such observations will apply only to the particular organism under investigation, as experience has shown that the lag phase is one of the most difficult characters to define.

Since a bacterial culture should ideally develop exponentially immediately after inoculation, the inoculum culture should not only be in the exponential phase of growth but should also be transferred under conditions likely not to lead to any interruption in the regular sequence of events in the production of cellular matter. As is well known, this is best achieved by the use of a large inoculum, and moreover, this inoculum should not be treated in any way, e.g. should not be washed or undergo any change in temperature, as such factors are important in producing a lag phase. In the majority of cases the inoculum size should be in the region of 3–10% of the final population density with heterotrophic bacteria. An inoculum size of this magnitude will allow the development of three to four generations, and it can thus be calculated easily whether this part of the exponential phase and the phase of decreasing rate of multiplication together with the stationary phase can be evaluated in 1 day's work. If there are possibilities for automatic recording of the characteristics under investigation, such considerations are of lesser importance.

From the practical point of view the problem of obtaining an inoculum in the exponential phase at a suitable time of a working day is not as easy as it may sound. Experimentation is required to determine in each case what size of pre-inoculum is required, and in which stage the pre-pre-inoculum should be in order to produce an inoculum in the right phase of growth at 8 or 9 o'clock in the morning. A further consequence of this reasoning is that some kind of regular transfer of the pre-pre-pre (pre$_3$)-inoculum culture will have to be undertaken. While it may be possible to maintain the culture at this stage in the form of a slant, the adjustment of a suitable inoculum size will then present a little more difficulty than will be the case with a "liquid culture" (culture on liquid substrate); in the latter case the inoculum size is easily and adequately obtained by measuring out a suitable volume of the pre$_3$-inoculum culture. It follows that the most ideal case of regular transfer is steady-state culture, a technique which is described elsewhere in this treatise (Evans, Herbert and Tempest, this Series, Vol. 2). While this method of cultivation is undoubtedly the method of choice with regard to maintaining reproducible population density in a particular stage of development over a limited period of time, there are justified objections to it, as this method is conducive to relatively rapid enrichment of mutants; the number of generations to which a culture is subjected can be enormous. Kavanagh (1963) has shown that a mutant needs to be endowed with only a very small advantage with respect to growth rate in order to replace the parent culture in a matter of days. The other extreme of producing an inoculum is a method making use of dormant cells, and reactivating these only a very few generations or transfers prior to actual inoculation. Freeze-dried cells are readily obtained nowadays so that one can consider producing them in large quantities in the form of ampoules for particular investigations, where a high degree of reproducibility is desired or/and when the organism is liable to mutate easily. It should be possible also to keep bacteria at very low temperature (e.g. $-20°C$) over several years as a good starting material for inoculum-build-up, as Carmichael (1962) has shown that even for higher micro-organisms like fungi in the vegetative state very few losses occur over periods as long as 5 years.

Quite often a culture should be examined over as many generations as possible in the exponential phase in batch culture. While it is possible to increase the number of generations by using higher substrate concentrations, there will be limitations caused by excretion of inhibitory substances and/or too high osmotic pressure or other inhibitory effects. In practice the gain is very little; considering that many bacterial substrates are used at about 1% concentration of carbohydrate (or the equivalent of it in the form of other nutrients), a concentration of as much as 8% would only allow for another three generations. This concentration is too high already, for very many cultures, to maintain exponential growth to near exhaustion of substrate in view of the accumulation of excessive quantities of metabolites. A far better way of obtaining a large number of generations in batch culture is to use a very small inoculum. In the extreme case one single viable cell would be sufficient per unit volume of medium. It can easily be calculated and experiment also shows that with one cell/ml (approx. $2 \cdot 10^{-10}$ mg) as inoculum an increase by a factor of 10^{10} representing some 23 generations, can be obtained, a far greater increase in number of generations than is possible by increasing substrate concentration. The problem of lag phase creeps in of course by reducing inoculum size so drastically. Effects of inoculum size in connection with lag phase are discussed later on; at this stage, however, it should be mentioned that inoculum size not only influences length of lag phase but can also affect later stages of development as shown for example for yeasts (Rahn *et al.*, 1951). Admittedly this is infrequent in bacteria or yeasts but if it does occur, it complicates matters considerably; thus all the more attention has to be given to the standardization of inoculum.

There are also instances when a lag phase or a small inoculum *per se* (the two are not necessarily identical) are desired. An example of the first case is represented by *Saccharomyces cerevisiae* with respect to producing a negative Pasteur effect; this property being strictly related to the late lag and early exponential phase (Wikén, 1961). A "large" inoculum, largely abolishing the lag phase, reduces considerably the magnitude of the effect. An example of the necessity of small inocula (with or without lag phase) can be seen in the use of penicillinase-producing bacteria for the assay of penicillin in the form of liquid culture (Irtiza and Meyrath, 1966).

As the study of growth kinetics of microbial cultures has become more universal nowadays, particularly with respect to steady-state culture, the determination of K_s values ("Michaelis" constant) (Tempest, this Series, Vol. 2) is frequently necessary. Very small substrate concentrations are to be established leading to equally small maximal population densities. It is clear that small inocula have to be used for this kind of experiment, in fact so small that a considerable increase (compared to inoculum size) in number of cells is allowed without affecting measurably the concentration of the nutrient for which the K_s value has to be evaluated.

B. Spore-forming bacteria

While with spore-bearing bacteria inocula are usually made in the form of spores, yeasts producing ascospores, ballistospores or even arthrospores are seldom if ever used in the spore form as inocula for metabolic experiments. The use of ascospores for genetic experiments falls outside the scope of this Chapter.

Using vegetative cells of bacteria a high degree of standardization of inoculum culture and pre-inoculum cultures is needed in order to obtain a reproducible lag phase (which can be zero in the extreme case). Resting cells in the form of spores provide a well-defined starting point, and cultures for metabolic experiments can be obtained directly from a suspension of spores. Such spore material can frequently be kept over long periods of time; in the case of *Bacillus subtilis* in distilled water at about 0·5–6 °C for 2–3 months without impairing the quantitative response of the organism in respect of growth kinetics or reaction towards antibiotics (Girolami, 1963). Gardner and Kornberg (1967) kept spores of *Bacillus megaterium* at − 20 °C, for several years. Spored cultures of *Clostridia* have been routinely kept for at least 1 month in sealed containers for the industrial production of butanol and acetone. These cultures give very reproducible fermentations in rate and yield after the first transfer, thus demonstrating the unchanged properties of the spore inoculum.

With vegetative cells the inoculum substrates are preferably of the same composition as those for growing the ensuing cultures; with spore inocula this ideal is not necessarily reached, as good sporulation is not always obtained under conditions of substrate composition suitable for vegetative growth. The choice of a different substrate may then again lead to transfer effects. Inoculum size effects, together with such transfer effects are discussed later on.

A routine procedure commonly used with bacterial spores is to subject them to heat shock prior to inoculation. Experience has shown that the ensuing cultures are a good deal more reproducible and uniform. Furthermore, in most cases the germination phase is shortened and the percentage of germinating spores increased by this sublethal heat treatment. Time and temperature used for this purpose vary widely, e.g. 70°C for 30 min, 80°C for 10 min, or boiling water for 1 min. Optimal results in each case can obviously only be obtained by trial and error. Recently it has been shown that supply of L-leucine in distilled water brings spores of B. megaterium to germination without heat stock. Without L-leucine or heat shock the particular strain used did not germinate (Holdom and Foster, 1967).

While a spore represents a well-defined reproductive particle, standardization in inoculum production is not to be neglected. Of particular importance in a given environment is age of culture. Thus a culture of B. megaterium in the form of a slant kept aerobically at about 20°C changed its sensitivity towards ozone very markedly over a period of 20–80 days (Wuhrmann and Meyrath, 1955). Keeping a Bacillus spore suspension in the cold in distilled water, or a Clostridium culture in a potato broth in sealed containers give reliable inocula over several weeks of experimentation.

C. Filamentous fungi

There are good reasons for treating this group of micro-organisms separately from any of the previous ones, despite the fact that exponential growth over extended ranges of culture development can be obtained. Regarding influence of inoculum size it should be mentioned at this stage that this factor can influence the metabolism of the fungal organism much more profoundly than that of bacteria or yeasts. While in the latter it is above all the very initial phases of growth (lag phase) and only rarely the more advanced phases of growth (late exponential phase and phase of decreasing rate of multiplication or maximal yield of cells) which are affected by inoculum size, the converse is true for filamentous fungi. It is worthy of note that there are a number of references in the literature (see Meyrath, 1963) claiming that lag phase is influenced by inoculum size. There was, however, no evidence for this being so if one defines lag phase according to Lodge and Hinshelwood (1943), the claims referring in fact to a phase (so-called lag phase) of which the reference point is more or less the stage of commencement of *measurable* growth. In these cases a very large part of the exponential phase is taken as lag phase if small inocula are used. From the few purpose-aimed investigations done so far, it is obvious that lag phase is not affected by inoculum size in *Aspergillus oryzae*, neither with conidia nor with mycelium as inoculum. On the other hand quite a few observations with various species and genera did show that advanced stages of growth, including maximum yield of mycelium, were dependent on inoculum size under a variety of cultural conditions (Meyrath and McIntosh, 1963; Ojha and Meyrath, 1967). That these findings call for very thorough standardization of the inoculum is obvious.

One notable difference of fungal (including yeast) cultures from heterotrophic bacteria is the considerably slower growth rate of the former. While bacteria may duplicate in some 15–20 min, filamentous fungi and yeasts will usually duplicate in not less than 3 h. With bacteria many different tests can be done within 1 day, much of the growth cycle on common substrates will be completed within 5–10 h or even less. With fungal cultures growth curves often cover several days. It is also true to say that fungi have not yet as frequently been used as biochemical and physiological tools as have bacteria. As a consequence interest has not centred nearly so much on the exponential phase (despite the fact that fungi can grow exponentially), rather it has been a case of being interested in the metabolites they produce, how they produce them and investigation of factors influencing rate of production and yield. These characters have been studied in comparatively advanced phases of growth, usually in the phase of decreasing rate of multiplication corresponding, to a large extent, to the arithmetically linear phase of growth or respiration or fermentation.

Thus it is of little importance whether the culture exhibits a more or less pronounced lag phase, and hardly anybody tries to obtain an inoculum in the exponential phase of growth for the mere purpose of getting a culture growing exponentially immediately after inoculation. The aims of inoculum standardization will rather be those of obtaining cultures with reproducible properties in the advanced phases of culture development, aims which are much more difficult to achieve in cutures of filamentous fungi than in bacteria or yeasts.

Many important genera of filamentous fungi form asexual spores such as conidia, oidia or sporangiospores, and less abundantly chlamydospores or arthrospores. Similar to the endospores in bacteria, these dormant bodies form well-defined inocula. They too have to be produced under standardized conditions; with regard to substrate composition of inoculum media it should be noted again that there may be differences from the actual test medium, thus rendering the transfer of nutrients together with the inoculum a possibility even if a washed inoculum is used.

An instructive example of the building up of an inoculum of fungi to reach a high goal is given in industrial scale gluconic acid fermentation by

Aspergillus niger, where by trial and error it has been found that a particular medium is required to maintain the culture in the laboratory, another medium is used to produce conidia abundantly, yet another is required for the germination of the conidia which is different again from the actual fermentation medium (Blom *et al.*, 1952).

II. SELECTION AND STANDARDIZATION OF INOCULA ACCORDING TO TYPE OF MICRO-ORGANISMS

A. Bacteria and yeasts

1. Non-spore-bearing bacteria

In the vast majority of cases it will be advisable to use freeze-dried or perhaps deep frozen cultures as starting material. Failing availability of these, solid substrates will provide a suitable means of keeping cultures over longer periods of time. Aerobic cultures can be kept on slopes in screw-capped bottles, closed after incubation, at near-freezing temperatures; anaerobic cultures kept similarly on slopes in test-tubes under pyrogallol (Willis, Hungate, Barnes, this Series, Vol. 3B) or in the form of stabs; the latter usually being used for micro-aerophilic organisms.

Inoculations are best performed with a loop or straight wire and the organism incubated at about optimal temperature until stationary phase is reached. Thereafter the culture is kept at between 0·5 and 6°C, and transferred regularly at weekly or fortnightly intervals.

Tables I and II show a number of examples of how standard inocula are obtained for a variety of bacteria used in the assay of antibiotics, vitamins and amino-acids. From the tables it appears that a sufficient degree of standardization is obtained after two to three transfers, the later two transfers usually being in liquid media. While the actual inoculum suspension can be kept in the refrigerator for 1, sometimes 2 weeks for assay purposes, or for physiological experiments, it has to be taken into account that extension of lag phase is bound to occur following such storage.

If an organism is to be obtained in the exponential phase early in the morning a small inoculum has to be transferred into a fairly large volume of substrate in the previous evening; for most bacteria which are grown at a temperature of about 37°C transfer is made with a straight wire; a minute inoculum taken from a slope diluted in Ringer's solution or broth and transferred immediately into a 500 ml medium will usually produce a culture in the exponential phase after 12–16 h incubation. One has to take into account the fact that an inoculum from a surface culture of about 1 mm^3 may contain about 10^9 cells. As this figure is also near the maximum which can be obtained per ml of substrate, one would in fact have allowed for only a 500-fold increase of cells. In the absence of a lag phase the

TABLE I

Preparation of standardized inocula with various non spore-forming bacteria and yeasts

Organism	Maintenance of stock culture							Inoculum production					Reference
	Cultivation method	Incubation time h	Incubation temp. °C	Storage time days	Storage temp. °C	Transfer amount	Transfer method	Cultivation method	Inoculation technique	Incubation time h	Incubation temp. °C	Suspension in	
Staphylococcus aureus	slant	16–24	32–36	7	4–6	2–3 ml	0.85% NaCl	Roux bottle	glass beads	24	32–36	0.85% NaCl	Dennin (1963)
Staphylococcus epidermidis													
Staph. aureus	slant	16	36	7	4–6	(a) 10 ml (b) loop	beef serum 500 ml broth	Roux bottle; stationary occasional shaking	glass beads; loop	4–6; 12	36; 36	M/20 Phosph. pH 7 broth	Kavanagh and Dennin (1963)
Sarcina lutea	slant	16	30	7	4–6	2–3 ml	0.85% NaCl	Roux bottle	glass beads	16	30	broth	ditto
Bordatella bronchiseptica	slant	16–24	32–35	7	4–6	2–3 ml	0.85% NaCl	Roux bottle	glass beads	24	32–35	0.85% NaCl	ditto
Staph. aureus	lyophil →slant	24	37	7	4–6	2–3 ml		100 ml broth (shaker)		8	37		Girolami (1963)
Corynebacterium	slant	24	37	14	4–6	2–3 ml		100 ml broth (shaker)		18	37	brain heart infusion broth	ditto
Candida albicans	slant	16	37	30	4–6			Sabouraud broth		24–48	37		Gerke et al. (1963)
Candida stellatoidea													ditto
Saccharomyces cerevisiae	slant	16	25	30	4–6					24–48	25		ditto
Candida tropicalis	slant	16	37	30	4–6			Sabouraud (shaker)		(a) 16 transfer 25% to b (b) 3	37; 37	homogenize Waring Blendor	ditto

TABLE II

Preparation of a standardized inoculum of *Staphylococcus aureus* to obtain exponential phase culture at desired time (Kavanagh and Dennin, 1963)

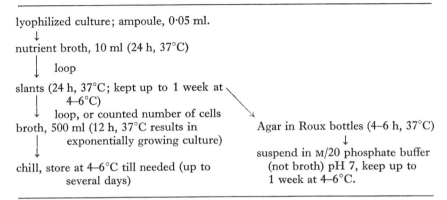

lyophilized culture; ampoule, 0·05 ml.
↓
nutrient broth, 10 ml (24 h, 37°C)
| loop
↓
slants (24 h, 37°C; kept up to 1 week at 4–6°C)
| loop, or counted number of cells
broth, 500 ml (12 h, 37°C results in exponentially growing culture)
|
↓
chill, store at 4–6°C till needed (up to several days)

Agar in Roux bottles (4–6 h, 37°C)
↓
suspend in M/20 phosphate buffer (not broth) pH 7, keep up to 1 week at 4–6°C.

stationary phase would thus already have been reached after 270 min by an organism with as long a generation time as 30 min. If the organism is to be taken in a particular stage of the exponential phase, an inoculum preferably in the exponential phase and of known cell density will have to be used. From a graph or a table either the incubation time or the inoculum size can be taken in order to reach the desired bacterial density at a given time, once the rate of multiplication and the dependence of lag phase on inoculum size are known for the bacterium and the conditions of testing. Dependence of lag phase on size of inoculum is discussed by Dean and Hinshelwood (1966). For many heterotrophic bacteria on rich substrates the lag phase is likely to be short, of the order of 1–2 h, or less if an inoculum from a 16–24 h solid culture is taken, and the inoculum size is in the region of 10^5 cells/ml. Under these conditions a prediction of the stage of growth can be made as near as the equivalent of 1–2 generation times.

For yeasts a procedure essentially similar to that of bacteria can be used. The generation time is considerably longer than for many of the heterotrophic bacteria used for assaying or for other purposes.

2. *Spore-bearing bacteria*

The stepwise preparation of inocula of these organisms is essentially the same as for vegetative cells. Storage and standardization is easier considering the longevity of the spores. An example of the sequence of preparation of inoculum is given in Table III. Stock cultures may be transferred once a month (preferably as pasteurized cultures) if kept in the form of a slope. According to Gardner and Kornberg (1967) a transfer is made once in

TABLE III

Preparations of spored inocula from *Bacillus subtilis* (Girolami, 1963)

| | Maintenance of stock cultures | | | | | Inoculum production | | | | | | | | | | | |
| | | Incubation | | Storage | | 1st stage | | | 2nd stage | | Incubation | | | Pasteurization | | Storage | |
Cultivation method		time h	temp. °C	time days	temp. °C	Cultivation method	Incubation time h	temp. °C	Inoc. size	Cultivation method	time days	temp. °C	Degree of sporulation	time min.	temp. °C	time days	temp. °C
(lyophil) → slant		24	37	30	4–6	shaker 100 ml broth	24	32	5%	shaker 100 ml broth	6–8	28	90%	30	65	60–90	4–6

several years if cultures are stored at $-20°C$. The degree of sporulation of *Bacillus* and *Clostridium* should be 90–95%. A prepared spore inoculum suspension, preferably in water, may be held available easily for 1 month, even up to 2–3 months if kept between 0·5 and 6°C or longer at $-20°C$. The inoculum is heat-shocked before use e.g. heated at 65–68°C for 20 or 30 min or at 80°C for 10 min, or in boiling water for 1 min. In order to minimize transfer effects from the previous culture, and from sporangia in particular, the procedure described by Grecz *et al.* (1962) can be followed. Liberation of clean spores from vegetative sporangia of *Clostridium botulinum* was accomplished by the use of *lytic enzymes* and *ultrasonic irradiation*. Suspensions of crude spores in phosphate buffer (pH 7) were digested with lysozyme (200 µg/ml) and trypsin (100 µg/ml). Rapid lysis of sporangia was induced by ultrasonic irradiation of the reacting mixture at 10 kc for 5 min after 0, 0·5, 1, 2, 4, and 6 h of incubation at 45°C. Intermittent washing of the resulting spore suspension with a solution of lysozyme and trypsin hastened purification of the spore crop. The cleaning procedure was completed by repeated washing of the spores with distilled water. The spores produced by this procedure were clean, as judged by their microscopic appearance, refractility to staining, loss of heat-sensitive toxin, and partition behaviour in a two phase system composed of polyethylene glycol and 3M potassium phosphate buffer (pH 7·1). The cleaning procedure appeared not to affect the viability, resistance to heat and gamma radiation, or the toxic nature of *Cl. botulinum* spores.

Limited predictions, as to the pattern of growth, can be made in a similar way as for non-spore-bearing bacteria; consideration has to be given to an obligate lag phase, i.e. germination phase, of the cultures which under ideal conditions is in the region of 1–2 h. There is also a tendency for *Bacillus subtilis* (and probably other spore-bearers) to grow exponentially only over a limited range of the culture cycle. With our *B. subtilis* strain in a rich substrate exponential growth did not extend over a greater portion than one fifth of the maximum population density. This occurred even if strong agitation under aerobic conditions of the cultures was maintained (Irtiza, 1967). With regards to cell numbers, however, this reduced exponential phase will provide sufficient scope for predictions in the production of large inocula.

B. Filamentous fungi

1. Filamentous fungi with abundant asexual spore production

Many of the commonly used moulds for physiological experiments readily produce asexual spores in the form of conidia. Like bacterial spores this material represents a fairly well-standardized inoculum. Age of the inoculum culture may at times exert a certain influence on the properties of the ensuing cultures. Storage of conidia of *Aspergillus oryzae* derived from cultures on malt wort slopes for some 290 days did not markedly affect amylase production (Meyrath, 1967). The small differences observed with regards to reaching a given stage of fermentation in a given time were obviously due to a reduction in viability of the conidia to about one tenth of the original. Inoculum size does have a marked influence on the behaviour

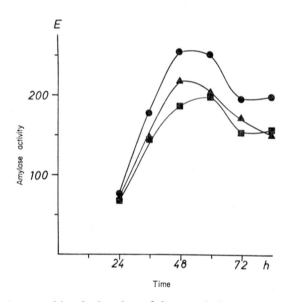

FIG. 1. Influence of incubation time of the sporulating culture (age of conidia) on α-amylase production in *Aspergillus oryzae*, using equal inoculum densities of washed conidia. ●——●, 5-day-old sporulating culture; ▲——▲, 8-day-old sporulating culture; ■——■, 12-day-old sporulating culture.

of fungal cultures, as shown in detail below; but usually, the inoculum size will have to vary quite drastically in order to cause marked effects on growth rate or maximum yield. With the same organism it could be shown (Fig. 1) that conidia from fairly young cultures, i.e. 5 days incubation at optimal temperature on a sporulating substrate, reacted differently from those of older cultures.

Using conidia for inoculum standardization is comparatively easy. There are really only two transfer steps involved: (a) the routine transfer of the stock cultures, (b) inoculation of the mould onto sporulating substrate.

The question of sporulating substrate needs consideration. It will not always be possible, desirable as it might be, to produce the asexual spores on the same substrate which is used for the tests under consideration. Transfer effects of nutrients with washed inocula of conidia are not very frequent, and seem to occur only under extreme conditions (see later).

A good sporulation substrate for a particular organism will often have to be found by trial and error. For some organisms a high osmotic pressure seems to be beneficial for conidiation, e.g. *A. oryzae* on a substrate of high phosphate content or slightly humidified bran (Meyrath, 1957; Meyrath and McIntosh, 1963). Sporangia-producing moulds are frequently cultivated on moistend bread for abundant formation of sporangia (Lockwood *et al.*, 1936). Bartnicki-Garcia and Nickerson (1962) sporulated *Mucor rouxii* on a rich substrate of yeast extract, peptone, and glucose. Sporulation was completed in about 1 week at 28°C. Regarding environmental factors influencing sporulation of the various groups of fungi the compilations and discussions by Cochrane (1958) or the appropriate chapters in "The Fungi" (Ainsworth and Sussman, 1966), may be consulted (see also this Series, Vol. 4).

There are fungi which after some time of cultivation break up into mycelial fragments or some kind of resting cells, e.g. oidia in *Oidium*; in other organisms like *Fusarium* or *Collybia* a similar phenomenon exists; the fragments are known under the terms of conidia or chlamydospores or arthrospores. Most streptomycetes behave in a similar way, and here the second stage of inoculum build-up, i.e. after transfer of the organisms from the stock culture, may be used with advantage for standard inoculum production. In a further stage vegetative cells may be taken from an early growing phase, e.g. the exponential phase. In these cases homogenization is achieved very easily by simply shaking the culture.

2. Fungi without extensive asexual spore production

For this group of organisms one does not have the advantage of resting cells produced in sufficient abundance. Vegetative mycelium will have to be used. The problem of standardization becomes more serious, as the stage of culture development is a further factor influencing the ensuing culture. Brown (1923) and Fries (1938) noted the difficulties of standardization of inocula when they have to be used in the form of small portions of mycelium, and not in the form of spores.

A method, which in principle is still widely used, consists of cutting small portions of a culture developed on the surface of a suitable agar substrate (Fries, 1938; Melin and Lindeberg, 1939).

Schelling (1952) found, in experiments with *Mycelium radicis atrovirens*, that the variation of mycelium development with surface-grown cultures was greatest if the inoculum, consisting of 6 × 6 mm squares of the surface-grown mycelium including the agar substrate, was taken from that portion of a radially developed culture which showed a high proportion of fluffy aerial mycelium. Inoculum from the circumference of larger radii, in particular that from the outermost portion, gave the most reproducible growth curves; those cultures in which the inoculum did not float on the surface soon after inoculation were discarded. With poorly developed aerial mycelium this happened in about 20% of the inoculated samples. A high degree of regularity and reproducibility was obtained if agar blocks of the culture were allowed to grow for a short period (in the above case, with a slow-growing fungus, for 4 days) on fresh agar substrate, whereupon a large proportion of young mycelium was formed. With this method homogeneity of culture development was even maintained if the primary inoculum consisted of mycelium of various ages. Norkrans (1950) used this double transfer method in order to check the suitability of the inoculum. The inoculum had to show a sufficient degree of formation of mycelial tips on the second agar substrate in a given time to be suitable for further experiments. A considerable disadvantage of this method, however, is the transfer of large amounts of substrate. Hence it was necessary to homogenize the inoculum, to allow washing of the cell material and the removal of adhering medium.

Schelling (1952) drew attention to irregular growth of cultures due to the fact that secondary colonies developed when inoculated with (a standardized number of) agar blocks. Their origin was due to individual mycelial fragments breaking loose during or after inoculation.

In experiments with *Trichoderma, Peniophora* and *Fomes* Gundersen (1962) found that the age of the vegetative inoculum plays a considerable role in obtaining growth curves with a low degree of variation. The standard deviation in quantitative growth experiments was considerably larger if the inoculum consisted of mycelium derived from various stages of culture development (i.e. from a radially growing culture, inoculated centrally on the agar surface in a Petri dish), despite the homogenization which followed the recovery of the inoculum. With respect to inoculum size Wikén (1951), Taber (1957) and MacLeod (1959a) found for various fungi that the standard deviation in mycelium production tended to be largest for small inocula.

The method of homogenizing is discussed later, and is certainly an invaluable tool for the treatment of surface-grown mycelium. It has at times also been used with advantage for the treatment of submerged grown fungi, particularly if these appear in the form of pellets (Savage and van der Brook, 1946). Shaken cultures usually and deep cultures very frequently, lead to the formation of mycelial pellets, particularly if inoculum size is small. One method of laboratory deep culture has proven in our hands to be particularly advantageous for the production of uniformly dispersed filamentous growth, i.e. the vibrating stirrer (Meyrath, 1964a). Even very small inocula gave reproducible dispersed filamentous growth. With this material inocula can be taken from a particular stage of culture development. MacLeod (1959a) obtained a "mycelial type" strain when an inoculum from a mature submerged culture consisting of mycelial "discs" was used; however, reproducibility in this case was not very good.

Mycelium in the form of pellets has also been used as inoculum. Karow and Waksman (1947) and Steel *et al.* (1954) used pellet inoculation, produced in two transfer stages, in standardized numbers, to obtain reproducible citric acid fermentation by *Aspergillus niger*. Lockwood and Nelson (1946) did similar experiments on itaconic acid production by *Aspergillus terreus*.

The stages of inoculum production with vegetative mycelium may be few, although occasionally the process may be more complicated than with bacteria: From a non-sporulating stock culture kept on solid substrate a culture can be produced by central inoculation of a suitable solid substrate in a Petri dish, and according to Gundersen (1962) mycelium can be cut out on a circumference from the portion immediately behind (inside) the growing tips of the culture. After homogenization in a suitable suspending medium the material can be used as inoculum. It is obvious that this method of inoculum production, while giving a lesser degree of variation, is rather wasteful in material and it may in fact be difficult to produce enough inoculum for a large set of experiments or for producing large inocula if this were desired. Under these circumstances it will be better to harvest the whole of the surface growth on a slant from the stock culture, homogenize, transfer aliquots to either liquid substrates (incubated as stationary, shaken or deep cultures) or solid substrates and after sufficient mycelium has developed and the culture is in the desired stage of growth, harvest, wash if necessary, homogenize and use as inoculation material.

Regarding the standardization of the density of the homogenized inoculum, it has been customary so far to take a known proportion of a culture from a known stage of development, homogenize for a standard time under given conditions, and use an aliquot of the homogenate as inoculum. Standardization of density can be improved by measuring wet weight of the mycelium before homogenizing, provided the inoculum is derived from a liquid substrate. This is easily done by washing an aliquot of the culture on a fine nylon filter, with repeated pressing by a rubber bung in between the washings in order to remove intermycelial water, and weighing the mycelial pellets directly after squeezing between thumb and forefinger. Taking the wet weight of the homogenized suspension is a little more cumbersome. Filtration of an aliquot of the washed suspension through a No. 4 tared Glass-filter is satisfactory. It will be realized that despite standardization of the suspension with regard to density (weight), one cannot dispense with the other procedures of standardization.

The influence of cold storage of prepared vegetative inoculum does not seem to have received any great attention, an aspect certainly worth consideration for large sets of similar experiments. MacLeod (1959) stored prepared inocula at 12°C for a period not longer than 24 h.

Regarding the suspending medium, water will usually be satisfactory. Saline or buffer may be advantageous at times, in particular if slimy growth is to be dealt with as centrifugation is rendered easier.

Attention may at this stage also be drawn to the effect of heterokaryosis on growth rate. Homokaryons, isolated from natural heterokaryons, exhibited different growth properties between themselves as well as with the parent strain. This has been shown by Jinks (1952) for *Penicillium* and by Taber (1957) for *Isaria*.

3. Inoculum for plant pathogens

Meyers and Simms (1964) described a procedure for obtaining ascospores of *Lindra thalassiae*. Perithecia were allowed to develop (2 weeks maturation) in gas-sterilized (ethylene-oxide) leaves of Thalassia on 1% yeast extract sea-water media. Perithecia were dissected from infected leaves, crushed in sea water or distilled water with a flattened needle in depression slides to release the ascospores. Inoculations were done with Pasteur pipettes on to the plant material.

4. Zoospores as inoculum

Hall (1959) kept cultures of *Phytophthora infestans* on slopes of French bean agar and subcultured at fortnightly intervals. Spore suspensions for inoculation purposes were obtained from cultures grown on agar slopes in 6 oz medicine bottles. The sporangial suspension was either used as inoculum immediately, or kept at a temperature of 10–15°C for about 2 h to allow liberation of zoospores. The latter procedure, according to Hall, was usually only successful if the sporangia came from a culture which had previously been kept at 10–15°C for several days. This procedure was adopted also by Clarke (1966) for the same organism. Separation of the liberated zoospores from the empty sporangia was obtained easily, as the spores concentrated near the surface of the water whereas the sporangia settled out to the bottom of the tube.

In preparing single spore isolates care had to be taken not to cause damage to the zoospores by inoculating devices (Hall, 1959). Drops of the suspension of zoospores, containing about 20 spores per drop, were placed at one side of a Petri dish containing a thin layer of agar, the dish was tilted allowing streaks of spore suspension to run across it and separating the zoospores.

C. Microscopic Algae

The problem of providing a reproducible inoculum has been discussed by Myers (1962), who draws attention to the heterogeneity regarding physiological conditions of the individual cells on an agar surface. No doubt, here, as in the case of other vegetative cells, the actual inoculum used for experimentation is preferably taken from a homogeneous liquid culture where one withdraws the desired inoculum from a suitable stage of culture-development, preferably the exponential phase. Myers kept his cultures, in steady-state continuous culture but the general advisibility of this is open to some doubt. The suitability of continuous culture for inoculum production for various species of *Chlorella*, *Scenedesmus*, *Euglena*, *Anabaena Anacystis* (Myers, 1962) is obviously due to this author's finding that most algae are not notably mutatable.

III. RECOVERY AND PREPARATION OF INOCULUM

A. The sterile room

Very often in a microbiological laboratory work is done which requires the chance of air contamination to be further reduced than is the case in an average laboratory. A sterile room is then useful. Instead of a room, inoculation chambers, or biological hoods can be used. More recently, the usefulness of biological hoods equipped with laminar air flow for microbiological purposes has been recognized (McDade *et al.*, this Series, Vol. 1).

In planning a sterile room several points have to be considered. First of all the location, which is determined by the people using the room, and by the fact that the distance to controlled temperature rooms should be short. The main microbiological laboratory should be close by, a small unit might even be installed within the laboratory. Since natural light is not essential, the facility can be situated in the middle of a building. The size should be kept small, but there should be one or two working areas and some spare table space to deposit a certain amount of equipment (Petri dishes, conical flasks, etc.) prior to inoculation. If more space is needed several subunits should be considered instead of a single larger room. It is necessary that there is an air lock to the sterile room which may be substituted by a double door if there is not enough space. The air lock is equipped with a sink preferably with hot and cold water, disinfectants for general purposes as well as for hand treatments should be provided. An electric hand drier is preferable to paper towels. Lab coats can be changed here, too. If there is no air lock the connecting door should preferably have a window which should be double glazed. Door and window have to be installed with special care in order to avoid any draught in the sterile room.

The room itself as well as the air lock should be easy to clean and inaccessible surfaces and corners should be avoided, since dust may settle there representing a potential contamination source. The walls should be smooth and crack-free and perferably finished with washable paint. Likewise the floor surface should consist of a material allowing rapid cleaning and disinfection if necessary.

The essential services to a sterile room are electricity and gas; hot and cold water, as well as air and steam supply, are useful additions. Not much furniture is needed in a sterile room and conventional laboratory equipment is certainly satisfactory. A dark colour for the surface of the working table is often preferred. Part of the working place might be illuminated from below and thus facilitate colony isolation. For the storage of some sterile glass ware a small bench unit is useful but should be installed in such a way that there is still enough foot space so that one can sit comfortably in front of the bench. Wall cabinets should be avoided to keep the room dust free and to render it easily sterilizable.

To achieve sterility of the room several methods can be applied, which are essentially methods of air sterilization. Since air sterilization is considered in other Chapters in Vol. 1 of this Series only the most commonly used methods are quoted. In practice very often the reduction of air borne micro-organisms in a sterile room is accomplished with ultraviolet irradiation. The wavelength used is 2537 Å, simply because commercially available low pressure mercury vapour type quartz tubes emit 95% of their radiation at this wavelength. A 15 W lamp at a distance of 30 cm has an energy output of the order of 400 microwatts. In order to kill 90% of most organisms 2 to 5 milliwatts/sec/cm^2 are needed (Sykes, 1967). One should keep in mind that u.v. radiation cannot penetrate, that its action is mainly on the surface and that the dose is reduced with the square root of the distance from the source. u.v. lamps are installed near the door. They hang from the ceiling and the switch should be outside the room. The u.v. lamp is turned on prior to the time of use and has to be turned off during the time the room is in use. Better sterilization might be obtained if the air is forced to pass the u.v. tube. If sterile air is available, the room can be ventilated and a slight positive pressure should be maintained inside so that there is no air leakage into the room.

Another way to sterilize the air is by means of aerosols. The chemicals used as bactericidal substances in sprays are propylene glycol, ethylene glycol and triethylene glycol. An older method consists of the use of formaldehyde gas as sterilizing agent. Formaldehyde has to be employed in the proper amount since an excess would form a white film on all surfaces. After a certain reaction time an excess of formaldehyde can be neutralized with ammonia which is vaporized into the air.

Chatigny (1961) describes, in a review article on microbiological laboratory safety, two types of commercially available biological safety hoods. One type has an open front and a filtered exhaust which creates a specific inward air flow whereas the other one is a fully closed hood with air lock entry port, rubber gloves, viewing glass and exhaust system. The latter is recommended for work with highly infectious material or where aerosols are created deliberately (see Darlow, this Series, Vol. 1).

A newer way to create an environment free from micro-organisms is the use of laminar air flow equipment; their value and usefulness is discussed in several papers (Favero and Berquist, 1968; McDade et al., 1968; Barbeito and Taylor, 1968; Staat and Beakley, 1968). This approach to the problem is considered in detail by McDade et al. (this Series, Vol. 1) and will not be discussed here.

B. Inoculation apparatus

1. Metal and glass apparatus, pipettes and syringes

The simplest way to transfer an inoculum is by means of an inoculation wire fixed to a metal holder or glass rod. The wire material is preferably platinum since it is most resistant, is not oxidized and cools off rapidly after flame sterilization. Platinum wire is expensive and therefore is often replaced by nichrome wire, which has the disadvantage of oxidizing rapidly and becoming pitted, so that it needs to be replaced rather frequently. Furthermore nichrome wires do not cool very quickly after flaming so that time must be allowed between sterilization and the actual transfer. The wire can be used as an inoculation needle—the straight wire—or might be bent to form a small hook or a round closed loop.

The needle is used for the transfer of a small inoculum from a culture. Using a loop for this purpose would have the disadvantage of taking too large and undefined an inoculum and the "picking" would be imprecise. Some techniques require a straight wire e.g. stab and streak cultures.

The standard loop has an internal diameter of 3 mm and the wire gauge should be 22–24 (0·5 to 0·6 mm thick). Care should be taken that the loop is perfectly plane and tightly closed otherwise an accurate transfer is not possible. The loop is used to make a liquid–liquid inoculation or to place liquid inocula on to agar surfaces. The solid–liquid or solid–solid transfer is not easily controllable as to the size and quality of inoculum. If a transfer is made from a single colony into a liquid medium the loop should not be introduced directly into the fluid since the deposition of the inoculum might be difficult, the inoculum should be spread on the wall of the container just above the liquid surface then the culture vessel is tilted and the inoculum distributed into the medium.

8

The amount of liquid inoculum on a charged loop depends very much on the way this device is handled. The angle and the speed of withdrawal of the loop from the liquid influences the drop size; increasing the speed and decreasing the angle to the surface increases the volume. The loop is sometimes bent in such a way as to form a right angle to the wire. This loop can be charged by lightly touching the surface in a perfectly parallel manner, thus decreasing the variability between workers. Asheshov and Heagy (1951) used calibrated loops for dilution steps for viable counts of bacteria. They had two sizes: a small loop, 1·5 mm in diameter made of a 0·3 mm platinum or nichrome wire containing 0·001 ml, and a big one 3 mm in diameter of a 0·5 mm wire with a liquid capacity of 0·025 ml. The loops are made by winding the wire around standard steel gauges and are calibrated in the following way: From a tube containing a few ml of water, accurately weighed, a loopful is withdrawn and the difference in weight is determined. With the small loop 10 loopfuls are withdrawn before weighing. The mean of several measurements is taken. It is obvious that the handling of the same loop by different persons is likely to give different results and that from liquid to liquid the drop size will change.

For inoculation of surfaces of solid media Drigalski spatulae or bent rods are used for the even distribution of inocula. They are also used to prepare spread plates if surface colony counts are used to estimate a viable count. The even distribution of the inoculum for surface growth in Roux bottles represents a certain difficulty. Gently tilting the bottle to and fro does not always succeed. Here sterile glass beads, which are introduced together with the liquid inoculum and serve as distributors are usually successful.

Besides the loop the pipette is most frequently used to inoculate culture vessels. Before sterilization the end is plugged with cotton wool and then singed to assure a tight closure of the pipette. Some companies manufacture pipettes with the end already widened for the cotton plug. They are much more easily plugged and can be used with pipette plugging machines. If a very large number of sterile pipettes is used every day the pipettes are placed in square, ovoid or round metal containers or glass tubes for sterilization. For use in small quantities they are preferably wrapped separately in paper. One should be able to see where the tip is (either the tip is marked or is visible because of a special wrapping technique). The normal measuring pipettes for use in total and viable count methods are: 1 ml in 1/100 divisions 2 ml in 1/100 divisions and 5 or 10 ml are used besides special pipettes for particular purposes, e.g. 0·1 ml single mark pipettes or 0·2 ml pipettes with a second mark at 0·1 ml only. Miles and Misra (1938) employed a calibrated dropping pipette which delivered a drop of 0·02 ml. A Pasteur pipette can be calibrated by pushing it gently into a wire gauge of, e.g., 0·036 in (0·09 mm). When the pipette is held firmly a mark is made with a glass file at the junction with the gauge, the pipette is withdrawn and the tip broken off; the new end should be at right angles to the length of the pipette. The drop size is influenced by the dropping rate and the surface tension of the liquid.

Sometimes syringes can be used for the distribution of inocula, especially for small volumes and for test series where rubber-capped flasks or bottles are used. The rubber can be pierced with the syringe whilst sterility is maintained.

To transfer inocula of fungi several workers have used a sterile cork borer to cut cylinders from a Petri dish culture with sporulated mycelium, which are then placed onto new media. Gundersen (1962) constructed a disc cutter which consisted of a stainless steel tube about 2 cm long and 0·3 mm thick of 6 mm internal diameter and was sharpened inside at both ends. A second longer tube, covered with rubber at the end, was welded at right angle on to the cutting tube to enable aseptic handling.

Another unusual inoculation device was used by Wilson (1957). He isolated fungi from soil samples. As an inoculator he used a round plastic scalp massager with all the teeth ending in one plane. After washing in a sterilizing agent and successive rinsing with sterile water and drying with sterile absorbent paper, the massager was inoculated with soil which was sieved into a sterile container and then the teeth were brought into contact with the chosen solid medium. Sterilization of the massager was only necessary when the soil sample was changed. Disposable pipettes have many advantages but are costly. Some low-cost transfer devices have been used in the past e.g. applicator sticks (Giblin 1951; McGuire 1964), sterile tooth picks (Corlett et al., 1965; Shiflett et al., 1966, Lee and Wolfe, 1966) and even soda straws.

No attempt has been made to cover the subject of calibrated loops and dropping pipettes completely. For more details and references see Hartman (1968) on "Loop dilution, dropping pipettes and other means of culture transfer" in his book "Miniaturized Microbial Methods".

Automatic and semi-automatic pipetting devices are manufactured by many companies. In principle the automatic apparatus consists of interchangeable glass syringes to cover the range of volumes required (1 ml up to 500 or even 1000 ml), a fixed or flexible nozzle and the mechanical part which allows the speed of delivery to be varied. Some dispensers have chambers which can be sterilized and sterile solutions can be distributed. As accessories a foot-operated switch or a touch release switch can be obtained (e.g. Filamatic, National Instrument Co., Baltimore, U.S.A.; Dispensor, H. Struers Chemiske Laboratorium, Copenhagen, Denmark). If diluents or media which must be subsequently sterilized are pipetted, slightly more has to be filled into the containers since during autoclaving there is some loss of liquid. The right amount has to be established by repeated trial.

Semi-automatic apparatuses deliver volumes of 1 ml up to 100 ml. The pipetting is done by hand, the filling of the measuring chamber is automatic and operates through a non-return valve. Most of them can be sterilized. In studying the influence of different compounds on the growth rate of micro-organisms many flasks have to be inoculated at the same time. Here, instead of inoculating each flask separately, the whole medium is inoculated and then distributed with such a pipetting device. The reservoir is shaken from time to time, or can be placed on a magnetic stirrer (e.g. Zippette, Jecons, Hertfordshire, England; Struers Chemiske Lab., Copenhagen, Denmark; Chemical Rubber Company, U.S.A.; Quickfit & Quartz Ltd., Stone, Staffordshire, England).

2. Multipoint inculation devices

Biochemical tests on micro-organisms are laborious, time consuming and costly. Several devices are described in the literature to facilitate these procedures. The idea of multipoint inoculators was stimulated by the development of the replica technique used for the isolation of mutants. Whereas the replica plating technique is limited to solid media and does not control the amount of inoculum delivered, multipoint inoculators can be used with liquid, solid and semi-solid media and the size of inoculum is constant. Difficulties in this new method were encountered with diffusion of metabolites as well as spreading of organisms. These were successfully overcome by using a divided, disposable Petri dish developed by Dyos Plastics Ltd., 242 Tohoorth Rise South, Surbiton, Surrey (Sneath and Stevens, 1967). Another technique to avoid these difficulties was proposed by Goodfellow and Gray (1966). They used an aluminium plate fitted into a Petri dish with 25 holes to hold flat-bottomed glass vials. The use of small agar medium discs placed separately in a Petri dish without any further division has also been proposed (Harris, 1963).

The multipoint inoculator by Tarr (1958) is commercially available (J. Biddulph and Co. Ltd., Manchester) and is shown in a schematic side view in Fig. 2. It consists of a middle support (a) which is rotatable and holds a rectangular plate (b) having a multipoint inoculator unit (c) on each side. A unit consists of 25 wire loops, which are fixed to the plate in such a way that a regular square pattern is obtained. At the correct distance from the middle support an inoculator platform (d) is mounted on the base plate (e). On the inoculator platform the master plate (f) a Petri dish containing the inocula is placed and then raised until the loops are charged. The master plate is then lowered again and is changed for a test plate which is inoculated by the same procedure. Between each inoculation step the loops can be flame-sterilized. Fig. 3 shows a surface view of an inoculated plate: (a) Petri dish, (b) droplet of inoculum.

An examination of the usefulness of this multipoint inoculator in combination with Petri dishes having internal divisions was made by

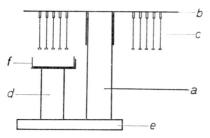

FIG. 2. Schematic representation of multipoint inoculator according to Tarr (1958).

Goodfellow and Gray (1966) and by Sneath and Stevens (1967). Goodfellow and Gray used a normal Petri dish in which 25 glass vials are spaced by means of an aluminium plate drilled with 25 holes. They found that this new technique was very useful for the performance of biochemical tests on bacteria such as acid production, nitrite production, utilization of single

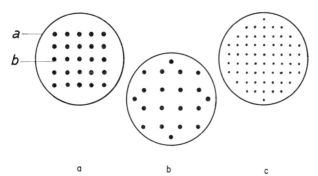

FIG. 3. Various patterns of multipoint inoculators: (a) Tarr (1958) inoculator; (b) Lighthart (1968); (c) Massey and Mattoni (1965).

energy sources and gaseous metabolites, which are identified by chemical methods or trapped as gas bubbles under a cover slip. Almost every test might be adapted for this new method. The slight change in standard conditions must be recognized and represents a limit to the usefulness of these tests for taxonomic classification of new species.

The sterile disposable Petri dish referred to above and used by Sneath and Stevens (1967) is divided into 25 separate square wells holding about

6 ml each. The dish can be closed in two different ways, one position results in a tight fit the other one allows an air space of 2·5 mm. Tests performed successfully are the detection of catalase and nitrite, growth on different media and at different temperatures and hydrolysis of various substrates. In the opinion of the authors testing for gaseous products, e.g. H_2S requires more experience in order to achieve good standardization

An automatic multipoint inoculator was described by Watt et al. (1966). The apparatus consists of a metal casing bearing the mechanical devices. The inoculator arm which is fixed to a vertical shaft at the back of the machine consists of a horizontal plate with 28 inoculating pins arranged in a regular but asymmetrical pattern. Underneath this inoculator unit is a corresponding set of inoculum cups (Oxoid aluminium test-tube caps). A cover of Perspex protects the whole apparatus. A Petri dish is inserted into the right hand side of the unit, thereby closing a micro switch which starts the inoculation cycle. Inoculation pins are lowered into the cups, then lifted and brought into a position above the Petri dish. Then the inoculating pins are lowered, the Petri dish is inoculated and the inoculation arm swings back to its original position. At the end the Petri dish is pushed out, and is covered immediately. This cycle needs about 6–7 sec. The size of inoculum of each pin is 0·006 ml with a standard deviation of 13% as against the 25% of the hand operated unit described by Beech et al. (1955). According to the authors the machine might be adapted to deliver antibiotic impregnated filter paper discs by means of vacuum release from stainless steel tubes. Even streak cultures can be made if the long edge of a slide is used as the inoculating unit.

Beech et al. (1955) constructed a rather heavy and compact inoculator. The apparatus consists of an inoculum tube holder, a multipoint inoculator and a Petri dish holder. All metal parts are of stainless steel. The inoculum tube holder consists of a circular base with 24 equally spaced holes, capable of holding 9 × 50 mm Durham tubes. The tubes are held in position by means of two circular metal discs interleaved with rubber (heat-resistant), all three having the same hole pattern and screwed onto the base plate. The holes in the rubber sheet are slightly smaller, thereby holding the tubes in position. The base plate is further furnished with two vertical guides for the inoculator. The inoculator consists of 24 vertical rods, $\frac{1}{8}$ in. in diameter, the end being truncated to prevent dripping of the inocula. The rods are fixed to the basal surface of an inverted cup in such a way, that their tips correspond with the centre of the tubes which are used to charge the rods. Furthermore the cup has a handle and external grooves to guarantee the right fit on the inoculum tube holder. The Petri dish holder consists of a base plate with two vertical guides, the base is recessed slightly to accomodate the plate. The inoculum delivered is

0·01 ml ± 25%. Several simpler multipoint inoculating devices are described in the literature and some are cited below, a more complete review is given by Hartman (1968). What these apparatuses have in common is their simplicity of construction from cheap and easily available material.

Lighthart (1968) used a $\frac{1}{8}$ in. thick aluminium disc (dia. $5\frac{5}{16}$ in.) with 62 holes holding inoculating spikes ($\frac{1}{8}$ in. dia., 1 in. long). Instead of the spikes, inoculating loops or needles might be fixed to a similar disc. It is essential to have a rack, which holds the tubes in such a way that the centres of the tubes coincide with the inoculator spikes. Since there is no guidance for the inoculator, this condition is not easy to fulfil.

The inoculator of Lovelace and Colwell (1968) consists of a square base with a handle and has 25 stainless steel prongs (1 in. long, $\frac{1}{16}$ in. dia.) which are spaced in such a way that the inoculator can be used with the divided plastic Petri dishes described above.

Seman Jr. (1967) constructed a multipoint inoculator by mounting a base plate with tapped holes on a chrome drawer-pull as support for 20 inoculating units which are fixed into the holes and have aluminium bases (0·64 cm round × 1·27 cm) with a 22 gauge nichrome wire in the centre. The inoculating units are easily removed and render the whole inoculator device very flexible i.e. certain samples of the master plate might be omitted. The same base plate and a retaining disc having smaller holes in the same pattern are spaced with aluminium tubing and kept together with a machine bolt. If Pasteur pipettes are placed into the holes a multipipette inoculator results. A self-contructed test-tube holder alignes the tubes— either filled with inocula or with liquid medium to be tested—in such a way, that they correspond to the inoculator needles. A similar capillary action replicating apparatus as the one mentioned above was used by Massey and Mattoni (1965) to make mass assays of physiological characteristics of algae.

A further modification of such a capillary action multipoint inoculator was made by Hartman and Pateé (1968). Instead of Pasteur pipettes capillary tubes with one end melted to a hook are used. Their replicator could deliver a maximum of 81 inocula at a time. The capillaries were charged from a plastic plate which contained the inocula in wells. Several plastic plates with different numbers of wells are commercially available thus permitting inoculation in a variety of patterns and numbers.

Sirockin (1968) constructed a cheap multiple inoculator of square brass plate into which sewing needles are soldered either with the needle point or needle eye projecting. The plate has three brass legs and a copper tube, which is screwed to the plate and serves as a handle.

Antibiotic disc dispensers are commercially produced (BBL, Division of Bioquest) which space the discs regularly onto an agar plate. But the discs have to be tamped by hand in order to guarantee even diffusion. Bourgeois *et al.* (1968) constructed a new dispenser which can be used together with a standardized method for antibiotic sensitivity testing involving a large Petri dish. This dispenser consists of two main parts: a stand, which is the holder for the dispensing apparatus and centres the Petri dish together with the dispensing assembly. The dispensing apparatus is filled with eight cartridges of antibiotic discs (BBL), a spring operated centre knob starts the delivery mechanism. This apparatus tamps the discs lightly to the surface automatically. The advantage is time saving and simplification.

C. Techniques for recovery of the inoculum

Bacterial, yeast and algal cells grown on solid media either aerobically or anaerobically, can usually be recovered quantitatively by suspending the surface growth in a suitable liquid. If more than a gentle treatment by inoculation needle or loop is required, as in the case of some aerobic spore-formers, addition of a few drops of sterile water to the slants and gradual "homogenization" by a stiff-wire loop through the application of small rotary movements over the surface of the agar, and gradual addition of more water will result in the formation of a homogeneous suspension. Further treatment can then be carried out by pipetting or the use of standardized loops.

Sporing fungi and actinomycetes need somewhat more effort if they are to be used in the form of a homogeneous suspension. Some of these spores (usually asexual) are strongly hydrophobic and addition of dispersing agents such as Tween or polyvinylalcohol (Wetter, 1954) are required. Most of the conidia-forming fungi (including streptomycetes) lend themselves to suspension in liquid without addition of dispersing agents. If conidiation is carried out on the surface of a solid substrate like agar, addition of water, followed by vigorous shaking will result in a suspension. For this treatment it is advantageous to use a medical bottle of suitable size, preferably flat shaped, replace the cotton wool plug used for incubating the culture by a sterile screw cap, add some glass beads if necessary and shake vigorously until a suspension is formed. Suspending will be rendered easier if sporulation is allowed to occur on the surface of liquid substrates, the culture fluid (used whole or in part) acting as dispersing agent. More mycelial fragments will be suspended also by this method and an extra washing step should be included in order to remove culture fluid adhering to the spores.

Mycelial fragments can be removed quite effectively by filtration through a pad of absorbent cotton wool. This can be placed in an ordinary glass funnel fitted with cotton wool in a conical flask. The funnel diameter should be such that it can be covered by a Petri dish or similar lid to prevent contamination during the operation. The whole assembly is wrapped in a sufficiently tough paper for sterilization and stored in this way until used.

Before pouring the conidial suspension on the filter, the bottle containing the inoculum should be allowed to stand just long enough, to allow larger fragments of mycelium or pieces of agar to settle. There is quite a loss of conidia through this filtration process; it can be reduced to some extent by carefully lifting the cotton wool pad with a stiff sterile wire, followed by a rinse with sterile fluid and gentle pressing of the cotton wool.

The yield of recovered inoculum can sometimes be increased by constructing a device which allows the dry-harvesting of the aerial conidia by suction and allowing them to settle in a sufficiently large container (Koch and Dedic, 1957). The dry-harvested spores can then be stored over very

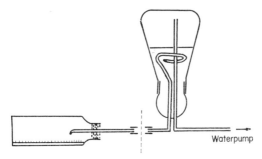

Fig. 4. Apparatus for harvesting surface-grown conidia by air-suction (Witt, 1968).

long periods of time at low temperature. A more convenient way of trapping the conidia is by adapting a container of water to the stream of air-suspended conidia which leads into the trapping fluid in the form of a spiral (Witt, 1968) as shown in Fig. 4. Trapping in liquid, usually water, will also be advantageous from the point of view of counting the conidia. While standardization and use of a dry inoculum (e.g. by weighing) may be possible, though a little difficult, for large inocula (see Allen, 1955 for rust uredospores) it is obviously not possible to standardize inocula accurately in the region of a few thousand spores per sample. Furthermore, a washing procedure will in the vast majority of cases be beneficial by reducing the chance of transfer effects of nutrients adhering to the conidia. The above-described filtration to remove mycelial fragments can be dispensed with if the conidia are recovered by suction, provided there is not too much aerial fluffy mycelium interspersed with the conidia.

Sometimes fungi, particularly conidia-producing moulds, are allowed to spore on substrates such as bran (wheat, rye, rice) or grains (barley in particular); we have also used substrate-impregnated vermiculite with success. The yield of spores on such material can be very high. Bhuyan *et al.* (1961) used barley grains in order to obtain conidia of a more uniform age, as on agar surfaces their strain of *Penicillium chrysogenum* continued to produce new crops of conidia for long periods. On barley grains the vegetative hyphae collapse, as Bhuyan *et al.* point out, after 8–10 days due to the drying of the grains, and the mass of conidia is left as a powdery coating. The method has in fact been used successfully in the penicillin industry for a considerable time. With these substrates recovery of the conidia by suction is not very practical, since the spores are largely inaccessible. Suspending in fluid must be resorted to.

Recovery of sporangiospores of phycomycetes grown on the surface of an agar can usually be achieved by adding a little sterile water and rubbing the surface with a stiff wire or sterile pipette (Bartnicki-Garcia and Nickerson, 1962), or simply by shaking vigorously. Filtration through absorbent cotton wool or similar material will remove mycelial fragments.

Non-sporing mycelium on the surface of agar plates can be recovered by cutting small squares (about 5–6 mm side length) with a sharp small scalpel. Gundersen (1962) constructed a disc cutter of stainless steel for this purpose (see Section B). Discs are usually cut from a culture, inoculated in the centre of a Petri dish, at a chosen radius in order to obtain mycelium of constant and reproducible age. The mycelium portions may then be homogenized by "blending" or other means.

D. Homogenization of inoculum

1. *Results obtained*

Savage and van der Brook (1946) and a little later Dorrell and Page (1947) were probably the first to apply this method as a means of increasing the number of reproductive units in mould mycelium to be used as a vegetative inoculum. The fragmentation of the mycelium of *Penicillium notatum* and *P. chrysogenum*, grown in pellet-like form in a shaker, was carried out in a "Waring Blendor" by Savage and van der Brook. As a result of a 2-min "blending" (a period found to be optimal) the increase in viable mycelial units was such that an inoculum size of 1/50,000 that of unblended seed could be used with at least as high yields as the unblended inoculum. Other dilutions, e.g. 1/4000 that of untreated mycelium gave the same yield of penicillin as the unblended seed in the same incubation period, but the maximum yield of the blended and diluted inoculum was considerably higher. There was no marked difference between single-blended seed and double-blended seed; the latter obtained by blending the contents of one shaker flask after 2 or 3 days and a second time after 4 or 6 days of incubation (blended, mended and blended seed).

After a 2-min blending period the hyphal fragments contained between one and ten cells, with an average of about four cells per fragment. About 20% of the fragments were created by actual fracture of the cell wall. The terminal cells of the rest of the fragments were undamaged, and these fragments appeared to have been formed by fragmentation of the septa, under the physical stress of blending. Longer periods of blending did not bring any advantage with respect to rate of production or maximum yield of penicillin. Cooling of the blendor was necessary after each 2-min period of treatment in order to avoid damage to the inoculum. Blending periods of 10 min or more (with cooling) showed signs of damage to the inoculum.

This method of disintegrating coherent mycelium has been widely used since the appearance of the above paper, although one may have to beware of generalizing the advantages too widely. As Cochrane (1958) points out damage to the mycelium may occur even with the briefest treatments.

Claviceps purpurea, used for the production of ergot alkaloids by Taber and Vining (1958, 1959) had optimal treatment periods of 3–5 sec when the fungus was grown in pellet-like form on a shaker. According to these authors the effect of blending was such that the inoculated mycelial fragments rose much earlier to the surface of the liquid substrate, thus being exposed to better conditions of gaseous exchange. The unblended inoculum developed as a submerged mat, which through folding and buckling eventually reached the surface. The blended inoculum rising soon after inoculation started to form a coherent mat at the surface.

A "low-temperature basidiomycete" was homogenized by Ward and Colotelo (1960) in a "Waring Blendor" (semi micro Monel metal type) for 90 sec in order to obtain optimal disruption. While no detailed kinetic data were taken in these experiments, it appeared that treatment periods as short as 10 sec and as long as 360 sec could be used equally well. These experiments seem to have been carried out at room temperature.

MacLeod (1959a) blended pellet-shape mycelium of *Hirsutella* at 3°C for 1 min, followed by a second treatment after washing and centrifuging.

2. Types of blenders (mixers) used

The "Waring Blendor" seems to have been used very frequently on the American continent. An essential feature of any kind of mixer for this purpose is that it can be run at high speed (e.g. up to 10,000 rpm), with variable settings and that it is sterilizable with no chance of contamination during the treatment. A well-fitting lid is an advantage when the capacity of container used is small in comparison to the volume of culture treated.

3. Other methods of homogenization

If the mycelium is not too strongly coherent, considerable disruption can be obtained by less drastic means than a high-speed mixer. Basidiomycetes grown on the surface of agar have been successfully homogenized, by Wikén et al. (1951) by shaking the mycelium vigorously for about 2 min in a glass-stoppered bottle with about 6 mm glass beads. A further degree of homogenization was advocated by Müller (1957), who after such short and mild treatment filtered the hyphal suspension through a sterile nylon sieve with 0·5 mm pore size.

E. Large scale inoculations

In industrial fermentations, even in the old-established ones like brewing, it was realized very early that quality and quantity of inoculum played a very important role for obtaining either the desired quality or the expected rate of fermentation or final yield. For most fermentations a large inoculum is desired. The main advantages to be gained thereby are a quick start in the development of the cultures and less chance for contaminants taking over.

That a large inoculum is usually beneficial in reducing lag phase is shown elsewhere in this Chapter; on the industrial scale this will also be true in principle, and more particularly one can visualize that a large inoculum will also be conducive to a small lag phase for example in obligately anaerobic fermentations like butanol-acetone production, where no particular efforts are being made to create anaerobic conditions at the start of the fermentation (there are usually small amounts of oxygen present as a result of pressure compensation by air after sterilization). At the other extreme there are strictly aerobic fermentations like vinegar production, where in the submerged process it has become customary to leave something like one third of the previous culture batch in the fermenter, to keep it strongly aerated, and to add fresh alcoholic mash only gradually in order not to disturb the normal metabolism of the acetic acid bacteria. For most fermentations inoculum volume in the form of a relatively young culture is chosen between 3 and 10% of the total culture volume, except when inoculation is carried out with mould conidia. Using such large inoculations one may not always get the advantage of obtaining highest possible yields, but the desirability of shortening lag phase and preserving relative freedom from contaminants are very often more important factors.

Using vegetative cells as inoculum for the main fermenter the build-up of seed material will in principle be the same as for laboratory cultures, except that a few more transfers are required in order to reach such high volumes as are necessary to inoculate fermenters of say 200 m³ capacity.

The technique of inoculation is an important part of the whole process, in particular the stage of transfer of the laboratory culture to the plant is crucial. The problem is one of transferring large volumes of culture aseptically. While in the laboratory inoculation by loop, pipette or syringe can be performed without any severe risk, the transfer of volumes of say 1 litre presents more problems. The pouring over of culture volumes of this size should be done only in sterile rooms or under special hoods. More usual and less risky is the transfer of these cultures under full protection via tubing of glass, rubber or other suitable material. In the simplest arrangement a tubing outlet will then have to be provided for the culture container which can be attached under aseptic conditions to the next culture vessel; the inoculum culture can then be drained. Alternatively the sterile part of the

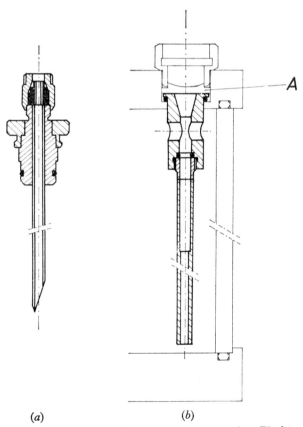

(a)   (b)

FIG. 5. Inoculator device for laboratory fermenter after Fiechter (1965). (a) represents a hollow needle used to pierce rubber A of the inoculating port (b) of the fermenter.

drain pipe of the inoculum vessel can simply be inserted into the orifice of the next culture vessel from which a stream of sterile air is forced. Draining by syphon is simply started by applying positive air pressure inside the inoculum vessel.

The attachment of sterile tubes does of course require some caution. Orifices must never be touched, and while it is inappropriate to use undue haste, the whole procedure should be carried out smoothly with all the apparatus laid out neatly beforehand. The use of glass joints is apt to facilitate the attachment process. A more sophisticated inoculation device has been developed by Fiechter (1965) as shown in Fig. 5. It is used to transfer an inoculum into a sturdily built fermenter.

The transfer of a culture of the order of several litres usually represents inoculation from the laboratory to the plant scale, i.e. inoculation of the seed tank. Suitable arrangements have to be made on both the laboratory fermenter and the seed tank for this transfer. Preferably the attachment is rigid. A scheme of inoculating a seed tank is shown in Fig. 6. The connecting plug A has to be thoroughly steamed before attaching the inoculum line by allowing a small stream of steam under sufficient pressure to pass through it. For this purpose the cap a is not closed absolutely tight. The connecting plug B of the inoculum vessel is kept sterile by fitting tightly cap b. Before attachment of B to A care is taken that sterile air is streaming out of the plug A. Plug C, the air inlet to the inoculum vessel, is tightly closed before attachment to prevent the inoculum from being forced back through the aeration tube into the air filter D. Having carefully removed caps a and b, cap a is put into a strong disinfectant and the attachment of A and B is carried out quickly. Valve E is now closed and an air line is connected on to C thus establishing a pressure inside the inoculum vessel higher than in

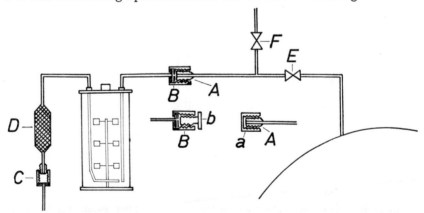

FIG. 6. Inoculation of a seed tank from a laboratory fermentor (for details see text).

the seed tank. At that moment valve E is opened and the inoculum will be forced into the tank. After partly closing valve E again cap a can now be screwed on to plug A. It can be steamed out by closing valve E completely and opening the steam line F. A simplification of the above procedure can be made in the case of anaerobic fermentations. The inoculum vessels not being fitted with aeration devices are connected in a similar way as previously, except that the vessels are turned upside down and connected in a vertical position (Fig. 7). When gas (usually air) pressure is equal in both inoculum vessel and seed tank, the pressure in the latter is dropped slightly and the over-pressure in the inoculum vessel will force out the contents into the tank. Transfer of the contents of the seed tank into larger fermenters must always be made via sterilizable permanent lines.

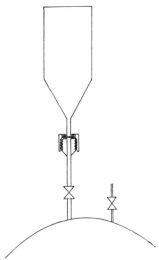

Fig. 7. Inoculation of a seed tank for anaerobic fermentation (for details see text).

F. Automatic inoculation

It is often convenient to arrange to inoculate a culture automatically at a predetermined time to ensure that the cells are at a suitable stage of growth for experimental work during normal working hours. With a bacterium this often involves inoculation during the early hours of the morning.

Automatic inoculation can be achieved by various means. The simplest method is to connect an inoculum vessel to the culture vessel by a flexible tube (Esco rubber is suitable), close the tube with a pinch cock before sterilizing the assembled apparatus and arrange the tubing in a peristaltic pump. The inoculum vessel is incubated under conditions which will allow adequate development of the culture and the contents transferred to the main culture vessel by means of the peristaltic pump activated by a time switch. This method has the advantage of simplicity but the disadvantage that transfer of the inoculum may take an appreciable time to complete.

More rapid transfer of inoculum can be achieved by activating an electric solenoid valve by means of a time switch so admitting sterile filtered air under pressure to the air space above an inoculum culture, and forcing the contents through a connector to the main culture vessel. With an aerobic organism it is convenient to arrange the system so that air passing through the main culture vessel is diverted for a short time to the inoculum vessel by means of a two-way time controlled solenoid valve. The slight positive pressure within the culture vessel prior to inoculation decreases the possibility of contamination and the switching of air prevents back pressure in the culture vessel during inoculation interfering with the transfer. A similar result can be achieved by the use of a time controlled sterile dispensing syringe to transfer inoculum to a culture vessel. With each of these methods it is important that the mechanism used to transfer the inoculum, be it sterile air, peristaltic pump or automatic syringe should be applied only for long enough to transfer the required volume of material. This is easily achieved by the use of two time clocks in circuit.

An alternative method of initiating growth of a culture at a predetermined time has occasionally been used. It consists of circulating cold water through the heating coils of a culture vessel which has previously been inoculated, thereby restricting development of the inoculum until a predetermined time when an electrically operated solenoid valve is activated and water at incubation temperature is switched into the heater system. This has the advantage of simplicity but the disadvantage that control of the condition of the inoculant is difficult to achieve.

The device described by Norris et al. (1970) for automatically monitoring the turbidity of microbial cultures is a useful addition to many inoculation systems since it enables an accurate indication of the early stages in the development of a culture to be recorded for later reference.

IV. EFFECTS OF INOCULUM SIZE AND HISTORY OF INOCULUM

A. Carry-over effects

In bacteria the carry-over of culture fluid together with the inoculum has been known to shorten the lag phase (Lodge and Hinshelwood, 1943). In these cases it seems to be a transfer of essential intermediary metabolites which must be produced by the young culture until a critical concentration of these compounds is reached inside the cells, and in the culture fluid.

Transfer effects seem to have been involved also in some germination tests with fungi, where the compounds in question are already present in the spores (Krishnan *et al.*, 1954; Keitt *et al.*, 1937). This can be concluded from the observation that self-inhibition of germination by crowded spores could be reduced by previous extensive washing of spores harvested without culture fluid.

The possibility of carry-over of trace elements in spores of fungi has been recognized for some time. Thus Shu and Johnson (1947) showed that the delicate citric acid fermentation by *Aspergillus niger* could be upset by the transfer of manganese contained in conidia (washed) when they were harvested from sporulation media containing about 9 ppm manganese. It will be noted that this amount of manganese is rather high and certainly exceeds by far the normal requirement of fungi, which is quoted by Nicholas (1963) to be in the region of 0·005 to 0·01 ppm. Nicholas in fact points out that the chance of transferring trace elements in a washed inoculum of fungal (asexual) spores when grown in the usual laboratory media is very small. For controlled and reproducible experiments it is advisable to prepare synthetic media, containing known amounts of trace elements, where the concentration does not exceed by too much the requirements of fungi since there is, as Shu and Johnson showed, an enrichment in conidia. In the case of bacteria, where the trace element requirement can be very low, transfer of heavy metals with washed cells causing inoculum size effects seem to be more frequent. Nicholas (1963) points out that repeated cultivation of the test organism in media particularly poor in trace elements has to be carried out in order to obtain any response to trace element deficiency in the culture medium.

Regarding the transfer of other nutrients MacLeod (1959a) reports on the nutritious value of mycelium of *Hirsutella*. The observed inoculum size effects, i.e. increase of total yield of mycelium with larger inocula, were thought to be due to sufficient micro nutrients carried over in a large inoculum, an effect similar to the well-known phenomenon reported by Wildiers (1901) where a minimal inoculum size of yeast had to be used to enable its growth, and where a large inoculum has been shown to contain sufficient growth factors to initiate culture development. According to our experience there are, however, also phenomena involved in inoculum size effects with mycelium other than carry-over effects.

B. Genuine effects in various micro-organisms

1. *Bacteria*

Reference has already been made to the well-known effects of inoculum size and history of inoculum in bacterial cultures in so far as influence of the duration of the lag phase is concerned. Further details regarding these phenomena can be found in the literature (Rahn, 1939; Monod, 1942; Lodge and Hinshelwood, 1943; Dean and Hinshelwood, 1966). The reduction of lag phase by large inocula has been thought to be due to the formation of essential intermediate metabolites (Lodge and Hinshelwood, 1943) which have to be produced by the young cultures. A large-inoculum culture would build up a critical concentration of these (diffusable) compounds in the cell (and in the culture fluid) more quickly than a small-inoculum culture. Walker *et al.* (1962) examined the accumulation of a cell division activator in *Bacillus megaterium*. The lag of a small-inoculum culture proved to be much shorter than would be expected if the activator were synthesized at a constant rate by cells at all inoculum concentrations. A small inoculum responds to the stress of dilution by producing greatly increased quantities of activator per cell, and at an increased rate, so that its concentration per cell at the time of the first division is an inverse log/linear function of inoculum density.

Other inoculum size effects are characterized by the requirement of a minimal (threshold) inoculum size, enabling bacterial cultures to grow at all (Grossowicz, 1945; Nakamura and Pitsch, 1961; Fredette and Forget, 1962). The latter have shown that the minimal inoculum size enabling a bacterial culture to develop depends on the composition of the substrate. Out of seven conventionally used substrates tested with 21 different strains of heterotrophic bacteria (representing 14 genera and 21 species, which included aerobes, anaerobes and facultative anaerobes) the best substrate (enabling the use of the highest inoculum dilution) was Trypticase Soy Broth (BBL). Besides suggesting the preference for this particular substrate of many bacteria, these observations show that the lag phase of bacteria can be prolonged *ad infinitum* if the inoculum size is small, despite the presence of viable cells. The above phenomenon is similar to the observations of Wildiers on yeast; however it would appear that in the case of bacteria substances other than vitamins are involved.

This effect has been called an all-or-none-effect by Halmann *et al.* (1967) in their investigations on the growth of *Pasteurella tularensis*. Besides the usual influence of inoculum size on lag phase minimal inoculum densities varying between 2×10^4 and 2×10^7 viable cells per ml, according to strain and substrate composition had to be used to enable the development of *Pasteurella*. In this instance it was a case of heterogeneity of the inoculum, in that within about 2×10^7 cells (for one particular strain) one cell was present which was able to produce a very characteristic "growth-initiating-substance" (GIS). This substance was of a chelating nature (Halmann and Mager, 1967) and quite closely related to the now well-known substances of the sideramine group, compounds of importance in iron-metabolism (see for example Neilands, 1957; Zähner *et al.* 1963). In fact GIS (produced not only by *Pasteurella* itself, but for example also by *Rhizopus*) could be replaced quite effectively, although not fully, by certain sideramines. Blood exhibited a sparing action on GIS, but on its own had little effect. The general observations on the suitability of certain media as against others to support growth of bacterial cultures from small inocula might be due to the presence of small quantities of similar compounds, which have been shown to be of universal importance.

Having quoted these observations on the formation of essential chelating agents and their importance in inoculum size effects of bacteria one should mention some other reports on similar substances, which when added to the culture medium acted as lag phase reducers. Mayer and Traxler (1962) showed that certain compounds which have a potential as metal chelators were stimulatory for *Bacillus subtilis* at low concentrations. This organism also required a critical concentration of manganese for growth initiation. Sergeant *et al.* (1957) observed that in *Bacillus* the lag can be reduced by adding glucose autoclaved in the presence of phosphate. The effective compound was thought to be of a chelating nature. These observations thus provide supporting evidence for the importance of heavy metal ions in connection with lag phase in bacteria (Lodge and Hinshelwood, 1939; Grossowicz, 1945).

Inoculum-size effects can be encountered occasionally when carrying out serial dilutions on agar plates for counting the number of viable cells. While it would be of advantage to establish very high numbers of colonies on agar plates for accurate counting (see Meynell and Meynell, 1965) limits are sometimes observed, for example at colony counts round about 1000 per standard plate (Dean and Hinshelwood, 1966). There is no proportionality any more between the reciprocal of the dilution and the colony count above figures of this magnitude. In this instance it seems that the early developing colonies have inhibited growth of the late developers, which may be a reflection of either the varying sensitivity of bacterial cells to metabolic products with varying culture age or heterogeneity of the cultures with regard to sensitivity towards metabolic products. The effect is of some practical importance as sometimes so-called micro-methods are used for determining the number of viable cells, consisting of the use of large inocula on microscope slide cultures and requiring considerable magnification for enumerating the micro-colonies. Obviously there are justified doubts regarding the universal use of methods involving counting of crowded colonies. One has to conclude, furthermore, that a reduction in lag phase in bacteria by the use of large inocula may not universally be expected.

In fact a case of self-inhibition by large inocula has been reported in the germination of bacterial spores (*Bacillus globigii*) in substrates containing

L-alanine, the latter being necessary for spore germination. The auto-inhibition in this case has been shown to be due to the formation and excretion of a racemase by some early-germinating spores transforming L-alanine into a racemic mixture with D-alanine, the inhibitory substance. With smaller inocula the racemisation effect is so small, due to the low concentration of racemase, that no inhibition of germination occurs (Fey *et al.*, 1964).

The importance of inoculum size in diffusion tests for antibiotics was shown by Wikén (1946) and in more detail by Cooper (1963). Knowing the rate of diffusion of the antibiotic into the agar medium as well as the rate of multiplication and lag phase of the bacteria a fairly accurate prediction can be made regarding the size of the inhibition zone which is to be expected using various inoculum sizes and various antibiotic concentrations. Matters are more complicated if the test organism is able to destroy the antibiotic, e.g. produces penicillinase in a penicillin assay. Inoculum size then plays a still greater role (Barber, 1947). This effect of inoculum size is very much more pronounced in liquid culture with *Bacillus subtilis* (Irtiza and Meyrath, 1966) where it can be observed that no development at all in the presence of 0·5 units of penicillin per ml can take place when inoculum size is too small (below about 4×10^5 spores per ml). On the other hand there is no sensitivity at all to penicillin when inoculum size is large (about $8·5 \times 10^6$ spores per ml). This response is due to the formation of a penicillinase. The establishment of a relatively high concentration of penicillinase at a very early phase in culture development destroys all penicillin present enabling the bacterial culture to grow without any loss of viable cells; with small inocula however, the absolute penicillinase concentration will remain low for a considerable time, enabling penicillin to kill off a proportion of the developing cells until eventually sufficient penicillinase is formed to destroy penicillin, and enable the remainder of the cells to undergo normal, uninhibited development. Further, with very small inocula, the extreme case of no recovery of the inoculum can easily be envisaged, and has been shown to occur. The effect of inoculum size would again be in the delay of establishing a given measurable population density, but not in affecting the late stages of development.

One report on the influence of inoculum size on later stages of culture development comes from Horváth (1967) who shows that competence of *B. subtilis* cells in various stages of growth can vary considerably according to inoculum size. It has not yet been possible to give a full explanation of this phenomenon.

2. *Yeasts*

In yeasts too, inoculum size and history of the inoculum can influence the duration of the lag phase in similar ways to those seen in bacteria

(Morris, 1958). Of interest is the observation described by White and Munns (1953) that the recovery period of a culture after inoculation can be shorter the smaller the inoculum. There are also isolated reports on the influence of inoculum size on later stages of development, e.g. influence on the magnitude of a negative Pasteur effect in *Saccharomyces* (Pfennig and Wikén, 1960) and to some extent on maximal yield of ethanol (Rahn *et al.* 1951).

3. *Moulds and other filamentous fungi*

The importance of inoculum standardization for this group of organisms has been well known in industry where quantitative aspects are of tremendous importance. One example of the influence of the history of a vegetative inoculum on yield, (mycelium) in advanced phases of growth is given

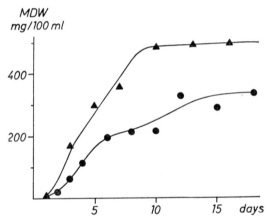

FIG. 8. Influence of the history of inoculum (equal size) on growth in advanced stages of culture development of *Aspergillus oryzae*. MDW = mycelium dry weight. ●——●, inoculum from growing phase; ▲——▲, inoculum from autolytic phase.

in Fig. 8. Filamentous mycelium under submerged conditions of cultivation has been taken from the phase of increase in cell matter as well as from the phase of autolysis. Both inocula had exactly the same mycelial content. Remarkably enough, the culture inoculated with mycelium from the autolytic phase can be seen to have developed faster and to a higher maximum than the culture inoculated with mycelium from the growing phase.

For some time effects have been known to influence the germination of various spores (asexual and sexual) in moulds and higher fungi. Particularly well-known phenomena are to be found in plant pathogenic fungi. The germination of uredospores of *Puccinia graminis* (Allen, 1955) has been shown to be inversely related to the quantity of spores used, the effect being due to an auto-inhibitor produced aerobically, which can be removed by a suitable floatation procedure. The production of the inhibitor is more pronounced in sealed containers and it is fairly heat-stable. It can be removed by glass surfaces. Alternatively, germination of untreated spores is stimulated by dinitrophenol, methylnaphthoquinone and coumarin. Likewise the activity of the inhibitor is overcome by these compounds. This inoculum size effect, while largely unexplained, is at the root of the problem of the erratic and inconsistent germination of rust uredospores and other spores of plant pathogens.

Forsyth (1955) also demonstrated production of self-inhibitory substances during the germination of *P. graminis*. The larger the inoculum size the stronger is the self-inhibition effect. If the test is carried out in Warburg flasks, the effect of the inhibitor can be overcome by placing a solution of silver nitrate into the centre well or by introducing acetone or ammonia into the flask atmosphere. The vapour of trimethylethylene inhibits germination of the uredospores, and the same methods can be used to counteract the inhibition effect by this compound. Since the absorption spectra of the inhibitor and trimethylethylene are identical, there is good evidence that the natural self-inhibitor is in fact trimethylethylene.

Yarwood (1954) working with *Uromyces phaseoli* found, similar to Allen and Forsyth in *P. graminis*, that a self-inhibitor is produced during germination of spores and that the effect is more pronounced in sealed containers. Furthermore (Yarwood 1956a) the germination of *Uromyces* spores largely prevented germination of *Puccinia* and *vice versa*. While self-inhibition of germination of *U. phaseoli* increased with inoculum density, there was stimulation with respect to average length of germ tube (Yarwood, 1956b). Whether the stimulating substance is identical to the inhibiting one is not known.

Besides self-inhibition, self-stimulation of spore germination has been reported, e.g. Padwick (1939), working with *Ophiobolus graminis*, Pine (1955) with *Histoplasma capsulatum*, Hall (1959) with *Phytophthora infestans*. In germination tests with *H. capsulatum* addition of oleic acid enhanced germination of spores in low densities. Mycelium of *Agaricus bisporus* stimulated germination of spores of the same organism (Lösel, 1964). This latter author also found a seven-carbon oléfin as well as *iso*amylalcohol and particularly *iso*valeric acid to stimulate germination (Lösel, 1967). This correlates with the observations by Yarwood (1956b) and Pine (1955) that paraffin oil or oleic acid respectively increase the percentage germination of *Puccinia* and *Histoplasma*. Similarly, Smart (1937) found that a culture filtrate of myxomycetes can stimulate the germination of spores of the same organism.

On the other hand a case of inhibition of germination of *Phycomyces blakesleeanus* by a culture filtrate of the same organism from advanced stages of growth has been reported by Schopfer (1933). This substance appears to be heat labile.

In considering the effects of self-stimulatory and self-inhibitory substances mentioned above, whether produced during germination or in later phases, it has either not been recorded or not been possible to test whether growth at more advanced stages can also be affected differentially. In experiments on inoculum size with *Aspergillus amstelodami* Darling and McArdle (1959) showed that substances excreted by the developing culture stimulated only initial stages of growth from spore inocula, whereas later stages were not affected.

With regards to the influence of inoculum size on later stages of development of fungal cultures one may mention the now well-known effect of the formation of a smaller number of large mycelial pellets with small inocula and the converse with large inocula (Foster, 1939). This phenomenon occurs almost always in shaken cultures, less frequently in deep cultures with forced aeration. While this effect may appear to be the logical thing to happen, it is to be noted that the number of inoculated reproductive particles may exceed by a factor of 100,000 the number of pellets obtained. There is no proportionality between number of reproductive particles and number of pellets. In *Aspergillus oryzae* loose filamentous growth could also be obtained in shaken cultures in media with high phosphate content using both large (10^5 conidia per ml) and small (10 conidia per ml) inocula (Meyrath, 1965). Not only is the size of pellets affected by inoculum size, the shape may be influenced as well. Takahashi *et al.* (1958) reported this for *Aspergillus niger* in shaken cultures.

Of considerable interest is the dependence of the phenomenon of dimorphism on inoculum size. In *Mucor rouxii* Bartnicki-Garcia and Nickerson (1962) found that under aerobic conditions, or under anaerobic conditions in an atmosphere of CO_2, the dissimilarities among growth curves resulting from the different inocula were minor, except for the expected delay in the appearance of measurable growth with small-inoculum cultures. Anaerobically in an atmosphere of N_2 there was decrease in both the rate of growth and maximum yield of cell material with a decrease in inoculum size from $3 \cdot 6 \times 10^5$ to $3 \cdot 6 \times 10^3$ spores per 50 ml. Moreover dimorphism under N_2 was markedly dependent on inoculum size. While yeast-like appearance of the culture was aways observed under an atmosphere of CO_2 it now appeared abundantly under N_2 when large inocula were used. A medium inoculum resulted in the formation of distinctly filamentous growth consisting of short unbranched hyphae. The smallest inoculum produced cultures with long filamentous branched hyphae. This diversity of morphology was maintained during the entire incubation period with only minor variations. It was argued that the formation of yeast-like cells under N_2 from large inocula was due to a rapid utilization of glucose with consequent increased evolution of CO_2 and incomplete removal of this metabolite with the usual flushing with N_2. The authors also realized that this was only part of the explanation as the following observations demonstrate. Using large inocula it could be shown that the increase in weight of cell material during cultivation, which was about a factor of 200, was almost entirely due to the increase in size of the inoculated spores, i.e. from a usual $4.5 \times 5.5 \mu m$ to $18 \mu m$ average diameter. The curious thing is, however, that besides inoculum, size morphogenesis was also influenced by chelating agents, and reversed by heavy metals, in particular zinc. Yeast-like appearance was prevented using "large" (about 10^6 spores per 50 ml) inocula in the presence of EDTA and other strongly chelating agents. Thus, while CO_2 produced by large inocula obviously induced formation of yeast-like cells, one can also argue that the stage of action of CO_2 was important. It would appear that differentiation under the influence of CO_2 occurs only if it acts at a very early stage of culture development, and this probably together with some trace metals

With aerobically grown fungi in shaken cultures the interpretation of inoculum size effects on metabolism may at times be difficult in view of the possible heterogeneity of mycelial growth, i.e. pellets of different size attributable to different inoculum densities. In our own experiments we used for submerged cultivation the vibro-mix method (see also under preparation of inoculum; Meyrath, 1965) which resulted in dispersed filamentous growth even if very small inocula were used. In the example cited below inoculum size effects on metabolism of fungi can be taken as not being due to a secondary phenomenon, such as varied size of mycelial pellets. With respect to influence of inoculum size on growth rate in the exponential phase and in the arithmetically linear phase as well as on maximum yield of cell material, experiments with *Aspergillus oryzae* and other *Aspergillus sp.* gave abundant evidence that such phenomena do in fact exist and that there are no transfer effects involved (Meyrath, 1962, 1963, 1964b; Meyrath and McIntosh, 1963; McIntosh and Meyrath, 1963; McIntosh and Meyrath, 1964). Other organisms reacted similarly, e.g. *Penicillium* (Ojha and Meyrath, 1967), *Claviceps* (Taber, 1957), *Hirsutella* (MacLeod, 1959a, b), *Mucor* (Bartnicki-Garcia and Nickerson, 1962). Such effects have been reported also with tissue cultures of Jerusulem artichokes (Caplin, 1963). As might be expected the yield of metabolites was affected also, e.g. penicillin production in *Penicillium* (Savage and van der Brook, 1946), organic acid production in *Aspergillus* (Meyrath and McIntosh, 1963; Lockwood and Nelson, 1946; Karow and Waksman, 1947), alkaloid production in *Claviceps* (Taber and Vining, 1958). In *A. oryzae* small-inoculum cultures which showed reduced rate of growth and maximum yields of mycelium proved to possess an inefficient metabolism in that more carbohydrate had to be metabolized to produce unit weight of cell matter (Meyrath and McIntosh, 1963). These cultures were also inefficient in assimilating inorganic nitrogen, a large portion of organic nitrogenous compounds being excreted into the medium (McIntosh and

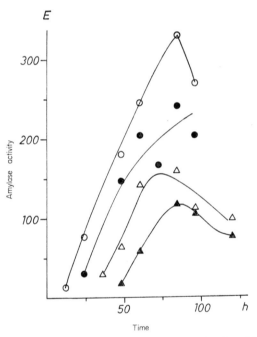

Fig. 9. Influence of inoculum size on α-amylase production (E) on vermiculite substrate in *Aspergillus oryzae*. Conidia served as inoculum. ○——○, 50×10^7 conidia per 10 ml substrate; ●——●, 10^7 conidia; △——△, 10^5 conidia; ▲——▲, 10^3 conidia.

Meyrath, 1963). The latter phenomenon was again tied up with increased amylase production in submerged culture using small inocula (Meyrath, 1965).

Some examples of these effects are represented in Figs, 9–11, Fig 9 shows α-amylase production by *A. oryzae* in the advanced phases of culture development on a rich substrate (in solid form) containing besides starch and inorganic nutrients also yeast extract. That the decrease of maximum yield with inoculum size is by no means the natural result to be expected has been shown repeatedly by the fact that the same organism in deep culture can give a reversal of the above order, as rate of formation in the arithmetically linear phase and maximal yield were considerably larger with small-inoculum cultures than with large ones (Meyrath, 1964b, 1965); with respect to growth, however, the effect was inverse to amylase formation.

One of the better arguments, besides direct evidence, to show that in inoculum size effects of the above nature there is no transfer of nutrients

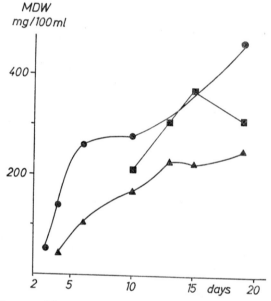

FIG. 10. Influence of inoculum size (conidia) on growth in advanced stages of culture development in *Aspergillus oryzae*. MDW = mycelium dry weight. ●——●, 4×10^6 conidia per 100 ml; ▲——▲, $4 = 10^3$ conidia; ■——■, 40 conidia.

involved, has already been presented elsewhere (Meyrath, 1963). The essence being that with certain substrates "very small" inocula can give higher rates of growth and maximal yields than "small" (100 times larger than "very small") inocula, which in turn resulted in growth rates smaller than with cultures from "large" (1000 times larger than "small") inocula. A similar effect with conidia of *A. oryzae* is shown in Fig. 10, again indicating that there is not a gradual fall-off of growth rate and maximum yield of mycelium with smaller inocula. Other effects of inoculum size are reported in the literature previously mentioned.

It can also be shown *directly* that in the inoculum size effects of *A. oryzae* shown above there is no transfer effect involved, since (a) addition of the ash of conidia of a large inoculum to a small-inoculum culture did not alter growth rate or maximum yield of mycelium (Meyrath, 1963); (b) conidia harvested from various substrates, i.e. partially purified, synthetic substrate, unpurified synthetic substrate and from malt wort, inoculated in large quantities did not show any significant differences in growth habit on a substrate largely freed from trace elements (Table IV after Steiner and Meyrath, unpublished).

TABLE IV

Influence of composition of sporulation substrate on the transfer of trace elements with conidia† of *Aspergillus oryzae*

| Incubation time days | Dry weight mycelium (mg/100 ml)† | | | | | |
| | Growth in unpurified‡ substrate | | | Growth in purified§ substrate | | |
	P‖	U¶	M††	P‖	U¶	M††
3	88	71	66	—	—	—
4	157	154	157	31	35	30
5	223	194	196	56	44	54
6	214	213	239	50	110	83
7·5	235	253	238	74	72	100
8	240	236	238	86	97	84

† Inoculum density = $4 \cdot 10^6$ conidia/100 ml.
‡ Values rounded off or up; average of 3 parallels.
§ Unpurified and purified growth media were the same in composition, consisting of dextrin and inorganic salts with Fe^3 as the only trace element added; the purified substrate was treated by specially prepared cation exchangers, except the salts of Mg and Fe.
‖ P = purified synthetic sporulation substrate.
¶ U = unpurified synthetic sporulation substrate.
†† M = malt wort agar as sporulation substrate.

Nonetheless trace element content in the test substrate plays a very considerable role in the magnitude and the nature of the effects of inoculum size mentioned (McIntosh and Meyrath, 1964). The influence is indirect though, by affecting the production of self-stimulatory and self-inhibitory substances.

That transfer of vitamins or other micro nutrients has not been involved in the above effects may be concluded from all those many observations where the use of small inocula resulted in higher yields of mycelium than large inocula. Furthermore, addition of heat-killed conidia (corresponding in amount to a large inoculum) to a small-inoculum culture had no effect (Meyrath, 1963). That the cause of such effects was in the formation of self-stimulatory and self-inhibitory substances has been shown for *A. oryzae*

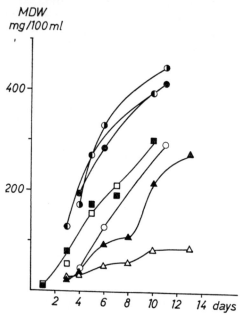

FIG. 11. Demonstration of the formation of self-stimulatory substances in *Aspergillus oryzae*. Growth in synthetic medium as surface cultures. MDW = mycelium dry weight. ○——○, conidia (4×10^6/100 ml) as inoculum; ●——●, conidia plus 5 ml culture filtrate (membrane filtered) from two-day old culture per 100 ml of substrate added at time of inoculation; ◑——◑, as before, but 3-day-old culture filtrate; ◐——◐, as before, but 4-day-old culture filtrate; △——△, washed mycelium (4 mg dry weight per 100 ml) as inoculum from 2-day-old submerged culture; ▲——▲, as before, plus culture filtrate (5 ml/100 ml) from 2-day-old culture; □——□, washed mycelium (4 mg dry weight per 100 ml) as inoculum from 4-day-old submerged culture; ■——■, as before plus culture filtrate from 4-day-old culture.

(Meyrath, 1962; Meyrath and McIntosh, 1965) and for *Penicillium chryso-genum* (Ojha and Meyrath, 1967). Cultures showing inhibitory effects in advanced phases of growth (be it in large- or small-inoculum cultures) proved to possess inhibitory substances in the culture filtrate at early stages of culture development which obviously were not rapidly enough overcome by the stimulatory substances excreted at a later stage. In cultures with comparative stimulation in advanced phases the inhibitory substances, probably excreted to some extent also in very early phases, were soon overcome by stimulatory substances.

This implies, of course, that the stage of action during culture development of these self-produced substances is very important. An example of formation of stimulatory substances is given in Fig. 11, where it can be seen that a culture of *A. oryzae* started with a conidial inoculum is stimulated by addition of small amounts of culture filtrate from the same culture when added at the stage of inoculation. Similar stimulation effects can also be seen in cultures from young mycelial inoculum, but not from old mycelial inoculum, both of which had been derived from the same conidia-inoculated cultures. The fact that the culture inoculated with young mycelium showed a lower growth rate in the linear phase of growth and lower maximal yields may be taken as an indication that the mycelium in very early phases had been affected by inhibitory substances excreted at that stage. The nature of these substances is not yet known, but some evidence has been given that they are of a chelating nature (McIntosh and Meyrath, 1964; Ojha and Meyrath, 1967). Lankford and his school showed the formation of such substances in cultures of *Bacillus* where they had certain regulatory functions on growth and division of the organism concerned. More recently *Pasteurella tularensis* has been shown to form chelating substances, with effects on its own metabolism during the lag phase, the accumulation of the compounds varying according to inoculum size.

Invaluable assistance has been provided by Mrs Ursula Stangl, Mr U. Stahl, Mr H. Prillinger (all from Vienna) in the preparation of this Chapter. Miss Annemarie Mohler (Zürich, Swiss Federal Institute of Technology) and Mr J. H. Shah (Glasgow, Strathclyde University) gave reliable technical assistance in research work hitherto unpublished. Full appreciation is also given to the Science Research Council (G.B.), who under grant No. 1936, supported the investigations on amylase production reported in this article. Thanks are also due to Prof. J. R. Norris (Sittingbourne) for writing the paragraph on automatic inoculation.

REFERENCES

Ainsworth, G. C., and Sussman, A. S. (eds.) (1966). "The Fungi", Vol. II. Academic Press, New York and London.

Allen, P. J. (1955). *Phytopathology*, **45**, 259–266.

Asheshov, J. N., and Heagy, F. C. (1951). *Can. J. Med. Sci.*, **29**, 1–4.

Barbeito, M. S., and Taylor, L. A. (1968). *Appl. Microbiol.*, **16**, 1225–1229.

Barber, M. (1947). *J. Path. Bact.*, **59**, 373–384.

Bartnicki-Garcia, S., and Nickerson, W. J. (1962). *J. Bact.*, **84**, 841–858.

Beech, F. W., Carr, J. G., and Codner, R. C. (1955). *J. gen. Microbiol.*, **13**, 408–410.

Bhuyan, B. K., Ganguli, and Ghosh, D. (1961). *Appl. Microbiol.*, **9** (1), 85–90.

Blom, R. H., Conway, H. F., Crocker, C. K., Farison, R. E., Moyer, A. J. Pfeifer, V. F., and Tranfler, D. H. (1952). *Ind. Engng. Chem. ind. Edn*, **44**, 435–440.

Brown, W. (1923). *Ann. Bot.* XXXVII, 105.

Bourgeois, L. D., Nutt, D. A., and Young, V. M. (1968). *Appl. Microbiol.*, **16**, 1606–1607.

Caplin, S. M. (1963). *Am. J. Bot.*, **50**, 91–94.

Carmichael, J. W. (1962). *Mycologia.* LIV, 432–436.

Chatigny, M. A. (1961). *In* "Advances in Applied Microbiology" (Ed. W. W. Umbreit), Vol. 3, pp. 131–187. Academic Press, New York and London.

Clarke, D. D. (1966). *Trans. Br. mycol. Soc.*, **49** (2), 177–184.

Cochrane, V. W. (1958). "Physiology of Fungi". John Wiley Inc., New York.

Cooper, K. E. (1963). *In* "Analytical Microbiology" (Ed. F. Kavanagh), pp. 85. Academic Press, New York and London.

Corlett, D. A. Jr., Lee, J. S., and Sinnhuber, R. O. (1965). *Appl. Microbiol.*, **13**, 808–817.

Dean, A. C. R., and Hinshelwood, C. (1966). *In* "Growth, Function and Regulation in Bacterial Cells". Clarendon Press, Oxford.

Dennin, L. J. (1963). *In* "Analytical Microbiology" (Ed. F. Kavanagh), pp. 311. Academic Press, New York and London.

Darling, W. M. and McArdle, M. (1959). *Trans. Br. mycol. Soc.*, **42**(2), 235–242.

Dorrell, W. W., and Page, R. M. (1947). *J. Bact.*, **53**, 360–361.

Favero, M. S., and Berquist, R. K. (1968). *Appl. Microbiol.*, **16**, 182–183.

Fey, G., Gould, G. W., and Hitchins, A. D. (1964). *J. gen. Microbiol.*, **35**, 229–236.

Fiechter, A. (1965). *Biotechnol. Bioeng.*, **7**, 101–128.

Forsyth, F. R. (1955). *Can. J. Bot.*, **33**, 363–373.

Foster, J. W. (1939). *Bot. Rev.*, **5**, 207–239.

Fredette, V., and Forget, A. (1962). *Can. J. Microbiol.*, **8**, 315–320.

Fries, N. (1938). *Symb. bot. upsal.*, III : **2**, 13.

Gardner, R., and Kornberg, A. (1967). *J. biol. Chem.*, **242**, 2383–2388.

Gerke, J. R., Levin, J. D., and Pogano, J. F. (1963). *In* "Analytical Microbiology" (Ed. F. Kavanagh), p. 392. Academic Press, New York and London.

Giblin, M. (1951). *Public Health Lab.*, **9**, 82.

Girolami, R. L. (1963). *In* "Analytical Microbiology" (Ed. F. Kavanagh), p. 353.

Goodfellow, M. and Gray, T. R. G. (1966). *In* "Identification Methods for Microbiologists" (Eds. B. M. Gibbs, and F. A. Skinner), Part A, pp. 117–125. Academic Press, New York and London.

Grecz, N., Anellis, A., and Schneider, M. D. (1962). *J. Bact.*, **84**, 552–558.

Grossowicz, N. (1945). *J. Bact.*, **50**, 109–115.

Gundersen, K. (1962). *Fortryk of Friesia.*, VII : 1–9.

Hall, A. (1959). *Trans. Br. mycol. Soc.*, **42**, 15–26.

Halmann, M., Benedict, M., and Mager, J. (1967). *J. gen. Microbiol.*, **49**, 451–460.

Halmann, M., and Mager, J. (1967). *J. gen. Microbiol.*, **49**, 461–468.

Harris, P. J. (1963). *J. appl. Bact.*, **26**, 100.

Hartman, P. A., and Patteé, P. A. (1968). *Appl. Microbiol.*, **16**, 151–153.

Holdom, R. S., and Foster, J. W. (1967). *Antonie van Leeuwenhoek.*, **33**, 413–426.

Horváth, S. (1967). *J. gen. Microbiol.*, **48**, 215–224.

Irtiza, S. N. (1967). "Kinetics of Action of Penicillin and Kinetics of Growth and of Penicillin Production". Ph.D. Thesis, University of Strathclyde, Glasgow (Scotland).

Irtiza, S. N., and Meyrath, J. (1966). *Antonie van Leeuwenhoek.*, **32**, 455.

Jinks, J. L. (1952). *Heredity*, **6**, 77–87.

Karow, E. O. and Waksman, S. A. (1947). *Ind. Engng Chem. ind Edn.*, **39**, 821–825.

Kavanagh, F. (1963). *In* "Analytical Microbiology" (Ed. F. Kavanagh), p. 159. Academic Press, New York and London.

Kavanagh, F. and Dennin, L. J. (1963). *In* "Analytical Microbiology" (Ed. F. Kavanagh), p. 314. Academic Press, New York and London.

Keitt, G. W., Blodgett, E. C., Magie, R. O., and Wilson, E. E. (1937). *Wisconsin Univ. Agr. Exp. Sta. Research Bull.*, **132**, 1–117.

Koch, O. G., and Dedic, G. A. (1957). *Zentbl. Bakt. ParasitKde.*, II, **110**, 178–183.

Krishnan, P. S., Bajaj, V., and Damle, S. P. (1954). *Appl. Microbiol.*, **2**, 303–308.

Lee, J. S., and Wolfe, G. C. (1967). *Fd. Technol.*, **21**, 35–39.

Lighthart, B. (1968). *Appl. Microbiol.*, **16**, 1797–1798.

Lockwood, L. B., and Nelson, G. E. N. (1946). *Archs Biochem.*, **9–10**, 365–374.

Lockwood, L. B., May, O. E., and Ward, G. E. (1936). *J. agric. Res.*, **53**, 849.

Lodge, R. M., and Hinshelwood, C. N. (1939). *J. chem. Soc.*, 1683.

Lodge, R. M., and Hinshelwood, C. N. (1943). *J. chem. Soc.*, 213–219.

Lösel, D. M. (1964). *Ann. Bot.*, **28**, 112.

Lösel, D. M. (1967). *Ann. Bot.*, **31**, 122.

Lovelace, T. E., and Colwell, R. R. (1968). *Appl. Microbiol.*, **16**, 944–945.

McDade, J. J., Akers, R. L., Sabel, F. L., and Walker, R. J. (1968). *Appl. Microbiol.*, **16**, 1086–1092.

McGuire, O. E. (1964). *Public Health Rept.*, **79**, 812–814.

McIntosh, A. F., and Meyrath, J. (1963). *J. gen. Microbiol.*, **33**, 57–66.

McIntosh, A. F., and Meyrath, J. (1964). *Arch. Mikrobiol.*, **48**, 368–380.

MacLeod, D. M. (1959a). *Can. J. Bot.*, **37**, 695–714.

MacLeod, D. M. (1959b). *Can. J. Bot.*, **37**, 819–834.

Massey, R. L., and Mattoni, R. H. T. (1965). *Appl. Microbiol.*, **13**, 798–806.

Mayer, G. D., and Traxler, R. W. (1962). *Bact. Proc.*, G.11, 37.

Melin, E., and Lindeberg, G. (1939). *Bot. Notiser.*, 241–245.

Meyers, S. P., and Sims, J. (1964). *Can. J. Bot.*, **43**, 379–392.

Meynell, G. G., and Meynell, E. (1965). "Theory and Practice in Experimental Bacteriology". Cambridge University Press.

Meyrath, J. (1957). "Über die Bildung von Amylasen durch *Aspergillus flavus-oryzae*". P. G. Keller, Winterthur.

Meyrath, J. (1962). *Experientia*, **18**, 41–42.

Meyrath, J. (1963). *Antonie van Leeuwenhoek*, **29**, 57–78.

Meyrath, J. (1964a). *Experientia*, **20**, 235–236.

Meyrath, J. (1964b). *Experientia*, **20**, 257–258.

Meyrath, J. (1965). *Zentbl. Bakt. ParasitKde.*, II, **119**, 53–73.

Meyrath, J. (1967). *Archives, Inst. Grand-Ducal Luxembourg.*, XXXII, 23–37.

Meyrath, J., and McIntosh, A. F. (1963). *J. gen. Microbiol.*, **33**, 47–56.

Meyrath, J., and McIntosh, A. F. (1965). *Can. J. Microbiol.*, **11**, 67–75.

Miles, A. A., and Misra, S. S. (1938). *J. Hyg., Camb.*, **38**, 732–749.

Monod, J. (1942). "Recherches sur la croissance des cultures bactériennes". Hermann et Cie, Paris.

Monod, J. (1949). *A. Rev. Microbiol.*, **3**, 371–394.

Morris, E. O. (1958). *In* "The Chemistry and Biology of Yeasts" (Ed. A. H. Cook), p. 257. Academic Press, New York and London.

Müller, P. (1957). "Untersuchungen über das Wachstum, die Atmung und den intermediären Stoffwechsel von *Marasmius putillus*" Doctoral thesis, Swiss Federal Institute of Technology. Juris-Verlag, Zürich.

Myers, J. (1962). *In* "Physiology and Biochemistry of Algae" (Ed. R. A. Levin). Academic Press, New York and London.

Nakamura, M., and Pitsch, B. L. (1961). *Can. J. Microbiol.*, **7**, 848–849.

Neilands, J. B. (1957). *Bact. Rev.*, **21**, 101.

Nicholas, D. J. D. (1963). "Plant Physiology" (Ed. F. C. Steward), Vol. III, pp. 363–431. Academic Press, New York and London.

Norkrans, B. (1950). Symb. Bot. upsal., XI : **1**, 1–127.

Norris, J. R., Hewett, A. J. W., Kingham, W. H., and Perry, P. C. B. (1970). *In* "Automation, Mechanization and Data Handling in Microbiology". Society for Applied Bacteriology Technical Series No. 4. Academic Press, New York and London. p. 151–161.

Ojha, M. N., and Meyrath, J. (1967). *Path. Microbiol.*, **30**, 959–965.

Padwick, G. W. (1939). *Ann. appl. Biol.*, **26**, 823–825.

Pfennig, N., and Wikén, T. O. (1960). *Zentbl. Bakt. ParasitKde.*, II, **113**, 491–510.

Pine, L. (1955). *J. Bact.*, **70**, 375–381.

Rahn, O. (1939). "Mathematics in Bacteriology". Burgers Publishing Company, Minneapolis.

Rahn, O., Iske, B., and Zungdis, R. (1951). *Growth*, **15**, 267.

Savage, G. M., and van der Brook, M. J. (1946). *J. Bact.*, **52**, 385–391.

Schelling, C. L. (1952). "Zur Kenntnis der Physiologie von Mycelium Radicis atrovirens Melin mit besonderer Berücksichtigung der Verwertbarkeit verschiedener Kohlenstoffquellen", Doctoral thesis, Swiss Federal Institute of Technology. Juris-Verlag, Zürich.

Schopfer, W. H. (1933). *Compt. rend. soc. phys. et hist. nat. Genève*, **50**, 152–154.

Seman, J. P. Jr. (1967). *Appl. Microbiol.*, **15**, 1514–1516.

Sergeant, T. P., Lankford, C. E., and Traxler, R. W. (1957). *J. Bact.*, **74**, 728–735.

Shiflett, M. A., Lee, J. S., and Sinnhuber, R. O. (1966). *Appl. Microbiol.*, **14**, 411–415.

Shu, P., and Johnson, M. J. (1947). *J. Bact.*, **54**, 161–167.

Sirockin, G. (1968). *Biochem. J.*, **106**, 57P.

Smart, R. F. (1937). *Am. J. Bot.*, **24**, 145–159.

Sneath, P. H. A., and Stevans, M. (1967). *J. appl. Bact.*, **30**, 495–497.

Staat, R. H., and Beakley, J. W. (1968). *Appl. Microbiol.*, **16**, 1478–1482.

Steel, R., Lentz, C. P., and Martin, S. M. (1954). *Can. J. Microbiol.*, **1**, 150–157.

Sykes, G. (1967). "Disinfection and Sterilization", 2nd edn., pp. 171–172. E. and F. N. Spon, Ltd., London.

Taber, W. A. (1957). *Can. J. Microbiol.*, **3**, 803–812.

Taber, W. A., and Vining, L. C. (1959). *Can. J. Microbiol.*, **5**, 418–421.

Takahashi, J., Asai, T., and Yamada, K. (1958). *J. agric. chem. Soc. Japan.*, **32**, 501–506.

Tarr, H. A. (1958). *Mon. Bull. Minist. Hlth.*, **17**, 64.

Walker, J. R., Lankford, C. E., and Reeves, J. B. (1962). *Bact. Proc.*, **62**, 37.

Ward, E. W. B., and Colotelo, N. (1960). *Can. J. Microbiol.*, **6**, 545–556.

Watt, P. R., Jeffries, L., and Price, S. A. (1966). *In* "Identification Methods for Microbiologists" (Eds. Gibbs, B. M. and Skinner, F. A.), Part A, pp. 125–131. Academic Press, New York and London.

Wetter, C. (1954). *Arch. Mikrobiol.*, **20**, 216–272.

White, J., and Munns, D. J. (1953). *J. Inst. Brew.*, **59**, 405.

Wikén, T. O. (1946). *Ark. for Bot.* **33A** (No. 3), 1.

Wikén, T. O. (1961). *Sci. Repts. Ist. Super. Sanità.*, 309–325.

Wikén, T. O., Keller, H. G., Schelling, C. L., and Stöckli, A. (1951). *Experientia*, **7**, 237.

Wildiers, E. (1901). *La Cellule*, XVIII, 311–333.

Wilson, H. A. (1957). *Soil Sci.*, **21**, (2), 239.

Witt, H. (1968). Final year thesis, University of Agriculture, Vienna, (Austria).

Wuhrmann, K., and Meyrath, J. (1955). *Path. Microbiol.*, **18**, 1060–1069.

Yarwood, C. E. (1954). *Proc. natn. Acad. Sci. U.S.A.*, **40**, 374–377.

Yarwood, C. E. (1956). *Phtopathology*, **46**, 540–544.

Yarwood, C. E. (1956). *Mycologia*, **48**, 20–24.

Zähner, H., Deir, A., Hess-Leisinger, K., Hütter, R., and Keller-Schierlein, W. (1963). *Arch. Mikrobiol.*, **45**, 119.

CHAPTER IV

Laboratory Assessment of Antibacterial Activity

D. F. Spooner and G. Sykes

The Boots Company Ltd., Nottingham, England

I. GENERAL SURVEY

The purpose of this Chapter is to describe and examine the methods used for the *in vitro* and *in vivo* assessment of antibacterial agents. It is not the intention to include the microbiological assay of antibiotics, although the methods employed are similar to some of those used in the wider field of antibacterial activity described below.

At the outset it is important to understand clearly the terminology used. Antibacterial substances and preparations are classified as disinfectants, antiseptics or chemotherapeutic agents. A *disinfectant* is used to eliminate or destroy infection, and so must be capable of killing a wide range of bacteria, but not usually bacterial spores. The term applies primarily to agents used for treating inanimate objects and materials, although there is no valid reason why it should be so confined.

An *antiseptic* is also used to control or eliminate bacterial infection (the literal interpretation of the term being "against putrefaction") and as such it should have antibacterial properties similar to those of a disinfectant. The term is confined, however, to those agents used on the skin and other living tissues. Because of this, the idea is commonly accepted that the range of activity need not be as great as that of a disinfectant, concern being only with organisms associated with infections of the skin. In some quarters it is believed that in an antiseptic bacteriostatic activity alone is adequate, its function being simply to contain an infection until the natural body resistances can take over.

A *chemotherapeutic agent* is an antibacterial substance administered systemically for the treatment of an infection. It may be either bacteriostatic or bactericidal in its action, its main function, like that of an antiseptic, being to prevent the multiplication of the infective organisms and so allow the body functions to deal more effectively with the infection.

Without exception, disinfectants and antiseptics are formulated with solubilizing, emulsifying or suspending agents, according to the intended purpose, and these can affect radically the apparent activity of the original substance: rarely is the activity of a finished product related directly to that of the constituent antibacterial substance(s) *per se*. It is essential, then, to assess each formulation, and even each variation within a formulation, on its own merits. In chemotherapy, however, the situation is somewhat different because nearly always the agent is administered in simple aqueous solution or suspension, and only occasionally is a slow release or other type of base used.

Besides the formulation itself, other parameters which influence activity are the final concentration of the active agent or preparation, the type of organism involved (including adapted mutants or variants of an original strain), the period of exposure or treatment and the presence of organic or suspended materials. The level of infection, i.e. the number of cells to be dealt with, the presence of stimulating or antagonizing substances, temperature, oxygen tension and pH value are also significant. All must be taken into account in considering the performance requirements of an antibacterial agent, as well as in devising tests, laboratory or otherwise, to assess performance. Each type or group of compounds, and sometimes each compound within a group, has its own characteristics within these parameters, so that they can only be discussed in general terms. Temperature coefficient, concentration exponent or dilution coefficient, the effect of organic matter and so on are entirely individual properties.

Because of these many variables no single laboratory test can be expected to give a complete evaluation. Several tests are needed under different conditions and with a selection of organisms, and even then only an approximation to actual field performance can be obtained. One aspect is certain—

the nearer the laboratory test conditions approach those occurring in practice the more are the results likely to be directly applicable.

Whatever the type of test, be it bacteriostatic or bactericidal and carried out either *in vitro* or *in vivo*, the principle is the same, i.e. a given bacterial population is submitted to treatment by the chosen agent and survival or growth determined at given intervals. The conditions used for making these determinations and the methods for assessing growth or survival may vary considerably, but none departs from the basic principle. Each of the numerous factors which come into play can affect significantly the result obtained. Their influences are discussed in the following pages.

II. BACTERICIDAL AND BACTERIOSTATIC ACTIVITY

Most chemical antibacterial agents possess both bactericidal and bacteriostatic properties, the action depending largely on the concentration of the agent employed, but some few, for example the acridines, are acknowledged to be mainly bacteriostatic and only slowly bactericidal whereas others, for example the chlorine releasing compounds, are exclusively bactericidal.

A. Bacteriostatic activity

Bacteriostasis is a theoretical concept. It is applied to the action of a chemical antagonist on an organism under conditions where growth can normally occur. It means, in effect, that each cell in a bacterial population is prevented from growing. But the individual cells in such a population are far from uniform: each is in a different state of growth and development and so each will respond differently to an adverse situation. In practice, then, the state of bacteriostasis means that the most resistant cells are inhibited, and so, by inference, the action on the less resistant ones is more than that of simple inhibitions; in fact, they succumb progressively more easily and most actually die. This action is a time dependent one, so that if the state of bacteriostasis is maintained for a prolonged period all of the cells ultimately cease to be viable through their inability to complete the metabolic cycle.

This knife edge situation between gradual loss of viability and ability to recover cannot be defined within any given parameter because clearly even the smallest change in the conditions might tip the balance one way or the other so that the cells either succumb more readily or, much less desirably in the present context, overcome the adverse conditions and begin to grow. Thus, it is not unknown for a diluted disinfectant containing organic matter to kill virtually all of a bacterial inoculum within a relatively short time, but for the few least sensitive survivors to continue to grow and so give rise to a large viable population within a few days (see below).

In spite of these restrictions bacteriostasis has a place in assessing antibacterial activity. The minimum inhibitory concentration (MIC) is often determined to ensure that in a bactericidal test the carryover is not sufficient to prevent the growth of surviving cells and so nullify the test, and in chemotherapy it is used to indicate whether an adequate concentration may be attained at the site of infection.

B. Bactericidal activity

Stated in the simplest terms, when a bacterial cell is placed in a nutrient environment and is unable to multiply beyond more than a few generations it is acknowledged to be dead. The mechanism of the lethal process may be protein denaturation, enzyme inactivation, damage to a membrane or the blocking of an essential metabolic path, but the result is the same. Thus, when an organism has been subjected to the action of a chemical antibacterial agent and is unable to recover, a bactericidal action has taken place.

But the matter is not quite so straightforward, for what is meant by the ability to recover? Normal, untreated cultures each have their own lag phase, generation time, optimum temperature for growth and nutritional requirements, but a treated, or "damaged", organism may differ in these respects. Thus it is known that after some treatments a temperature below the normal optimum gives the best recoveries or that an essential bridging metabolite may allow a moribund cell to survive: and there is the important factor of carryover of an inhibitory amount of the bactericide, a point particularly significant where an adsorbed surface-active substance is involved.

In testing for bactericidal activity it is not simply a matter of trying to recover organisms under normal cultural conditions; some thought is needed to determine the optimum conditions. This was not adequately appreciated until relatively recently and was stimulated by the appearance of the surface-active quaternary ammonium compounds whose activities at first were greatly over assessed.

The process of death is not instantaneous in a bacterial population: it is a progressive one, each cell succumbing according to its individual resistance. The net result is the typical, but not universal, logarithmic form of the time–survivor curve, although not infrequently a tailing occurs towards the end of the process due to the presence presumably of a few cells of high resistance.

The results obtained can be subjected to statistical examination. A good example of this is the detailed analysis of the Rideal–Walker and similar tests by Dodd (1969).

III. *IN VITRO* TESTING

A. Assessment of bacteriostatic activity

Bacteriostatic activity can be determined on solid or in liquid media. Each depends on assessing by some means the extent of inhibition of growth. As has already been indicated bacteriostasis is a dynamic situation and is dependent not only on the agent concerned but also on a number of other factors, particularly time of contact, temperature and the nutritional environment in which the organisms find themselves, as well as the type and number of cells involved.

The significance of time of contact, or period of exposure of the organisms to the bacteriostatic influence, already mentioned briefly above (p. 213), is that in some conditions an organism might be inhibited from growing during a certain period of time but after a longer period may recover and be able to multiply regularly. It is important, then, in making an assessment to standardize and record the period of incubation of the test or to use a sufficiently long incubation time to meet this eventuality.

1. Tests using solid media

Two types of test can be applied using solid media. In the one, the medium is pre-inoculated with the test organism and the drug or other bacteriostatic agent allowed to diffuse into the agar and so produce a zone of inhibition of growth: in the other, the drug is incorporated into the agar and the ability of an organism to grow on the surface of such a medium assessed. The former gives mainly a qualitative response, but it can be used semiquantitatively and, with careful attention to detail, can be made fully quantitative, as in the assay of antibiotics or some of the vitamins of the B group.

All tests of this type depend on the compatibility of the test substance with agar and with other constituents of the medium, and on its stability under the test conditions in relation to pH value and other factors: the type of medium itself can also influence significantly the response obtained. Many studies of these points have been reported during the past decades, but most of the publications have dealt with a particular detailed aspect as it affects one compound or type of compound. For this reason, it is inappropriate to discuss them in this Chapter: suffice it to say that in any study the points referred to all merit consideration.

It is appropriate here to make some observations on the preparation of samples for testing. A compound can only exert an antibacterial action when it is in aqueous solution or sometimes as a finely dispersed emulsion: a suspension *per se* is inert apart from that proportion which is in solution. Many antibacterial compounds dissolve very slowly in water, and often a solution aid is required. In some instances this can be done by means of a water-miscible organic solvent such as ethanol or acetone. The compound is first dissolved to a relatively high concentration in the solvent which is then added with agitation to a large volume of water or buffer solution. This has the effect of producing a temporary, but well dispersed, emulsion or suspension, thus affording the compound the maximum opportunity to dissolve in the water phase. Care must be taken to ensure that the final solvent concentration in the water does not itself exert any action; in this respect acetone is much more suitable than is ethanol, which is slowly lethal to some bacteria and moulds even at 5% concentration (Sykes, 1965).

Other ways, most common in the disinfectants and antiseptics fields, are by means of soap solubilizers or emulsifying agents. Both are used mainly with phenolic compounds and their derivatives. The former depends on micelle formation to keep the phenol in solution—usually at a much higher concentration than is normally achieved in water; the latter uses gelatin and casein to give a finely dispersed and reasonably stable emulsion, thus affording the phenolic substance the maximum opportunity to encounter and act on the bacterial cells.

Drugs used in chemotherapy often have low solubilities in water and so frequently are administered as a suspension. To obtain a uniform suspension for experimental purposes a vehicle such as 5% of gum acacia or 1% of carboxymethyl cellulose in water is used.

(a) *Diffusion methods.* These can be considered from two aspects: those which allow a single compound or preparation to be assessed against a number of organisms simultaneously and those which enable several preparations to be compared against a single test organism. In the first, two types of test are used, the agar cup method and the "ditch plate" method. In the second, the agar cup method can also be used, but a variation on this, the surface diffusion method, is the more common. In either instance the medium can be a simple nutrient agar but if the preparation to be assessed is intended for use as an antiseptic it is usual to include serum (inactivated ox or horse serum) at a concentration of 5 or 10%. Incubation is normally for 24 h.

The agar cup method is applicable to semisolid and liquid samples. A plate of sterile agar, poured to a depth of ca. 4 mm, is allowed to set and a single cup, 15 mm diam, cut from the centre of the plate with a sterile cork borer. After allowing the agar surface to dry for an hour or so in the incubator a loopful of a suitably diluted broth culture of each of 6 or 8 test organisms is streaked radially from the cup to the edge of the plate, after which the cup is filled with the test preparation and the plate incubated at 37°C. The extent of inhibition of growth from the edge of the cup gives a measure of the activity of the preparation, so that a comparison of its relative activities against the range of test organisms is readily obtained. By using several plates simultaneously the activities of several preparations can be compared; but only approximately (Fig. 1).

Fig. 1. Assessment of bacteriostatic activity using solid medium. Inhibition of organisms streaked from cup containing formulated antibacterial compound.

In *the ditch plate test* the cup is replaced by a ditch cut in the agar along the diameter of the plate. The ditch is filled with the test material and the selected cultures streaked across the agar surface at right angles to the ditch. The procedure is then as with the cup test.

This test is more commonly used to assess the activity of a bacteriostatic agent itself rather than that of its preparations. To do this the ditch is filled with a dilution of the agent prepared in nutrient agar (this gives greater ease of handling the plate when being transferred to and from the incubator). After incubation note is made of the growth, if any, on the ditch agar itself and of the extent of inhibition beyond the ditch.

To assess the relative activities of several preparations a variation of the agar cup method can be employed, as already indicated. In this, instead of streak inoculating the surface of the agar, the inoculum is introduced into the molten agar before it is poured, the organisms thereby being distributed evenly through the depth of the agar, or it can be flooded on the surface of the agar (and the excess culture liquid drained off), thus giving a uniform spread of organisms over the surface of the plate. In either instance only one test organism can be used to each plate. The test samples are then filled into the appropriate number of 15 mm cups cut in the agar: up to 5 cups can be used in each plate. After incubation a comparison of the zones of inhibition round each cup gives a measure of the relative activities of the samples tested (Fig. 2).

The alternative *surface diffusion tests* also use either of the two methods of inoculation just described but they dispense with the cut out cups. Instead the test sample is placed on the surface of the inoculated agar and the zones of inhibition around each sample compared. Various means are

Fig. 2. Assessment of bacteriostatic activity using solid medium. Zone of inhibition in agar inoculated with *Staphylococcus aureus* around cup containing formulated antibacterial compound.

employed to contain the sample within a small area of the agar surface. In one, small sterile fish-spine insulator beads (obtainable from electrical engineers) are used, the beads being first touched on the test solution so that they pick up a fixed volume and then placed on the surface of the agar. In another, filter paper discs or pads 6 mm diam (Antibiotic Assay Discs, Whatman Ltd., London) are either dipped in the solution, drained for a fixed short period and placed on the agar, or a small measured volume is put on by pipette and then the disc placed on the agar. In yet a third method, small porcelain or steel cylinders ("penicylinders") are placed on the agar and a fixed volume of the test solution, 2 or 3 drops, put into each cylinder.

If the preparation to be tested is a solid or semisolid which does not soften and flow at 37°C the sample can be placed directly on the surface of the agar within a small area of ca. 1 cm diam. This is a useful way for determining the release of an antibacterial agent from an ointment base or a powder.

A novel adaptation of this method (Sykes, 1965), which is more sensitive and gives more clear-cut results, uses a serum agar containing 0·5% of glucose and 0·5% of sterile calcium carbonate in even suspension. A single loopful of a chosen culture, covering an area of ca. 4 mm diam, is placed on the agar surface and the area covered by the test preparation. Provided an organism is used which will ferment glucose, e.g. a *Staphylococcus aureus*, the result after incubation will be a clear plate (where the staphylococci have grown and caused the calcium carbonate to dissolve) with small areas of white underneath the test sample area (where the organisms have been inhibited and so failed to dissolve the carbonate).

(b) *The gradient plate test.* This is, in effect, an extension of the diffusion methods just described. It is mainly of use in chemotherapy for determining the resistance of an organism to an antibiotic or other agent and for determining cross resistance to two or more antibiotics. It can also be used in the wider sphere to determine the approximate inhibitory concentration of any substance which will diffuse into agar, and particularly for observing the combined effects of different concentrations of two substances.

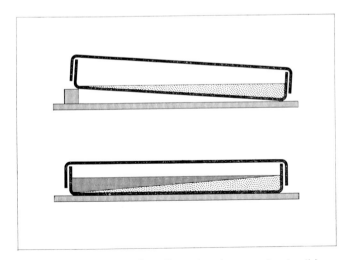

FIG. 3. Preparation of gradient plate (see text for details).

The test, as described by Szybalski (1952), is carried out in Petri plates, each plate being poured with two layers of agar (Fig. 3). The first layer is made by pouring 10 ml of molten agar into the plate which is pitched at such an angle that the agar just covers the entire bottom of the plate. After this has set the plate is put in the normal horizontal position and a second 10 ml of agar, this time containing the antibacterial agent, added. The agar surface is then streaked in the same line as the slope of the agar and incubated. During incubation the agent diffuses vertically through the plain agar thus giving a gradient concentration, resulting [*sic*] in the organism growing, or being inhibited, up to a certain point on the streak.

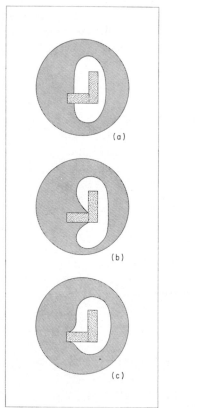
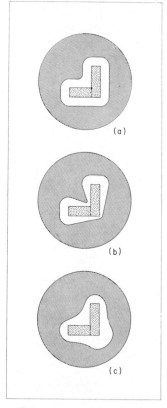

FIG. 4. Paper strip test for combined action. Diagrams on right show tests with pairs of antibacterial compounds; only one of the pair is antibacterial in tests illustrated on left. (a) Indifference, (b) antagonism, (c) "synergism". Clear areas represent inhibition of growth.

By incorporating a different substance in each of the two layers the effects of various combinations of the substances can be studied and the possibility of synergism or antagonism determined.

(c) *Tests for combined action.* Synergistic or antagonistic effects of two anti-bacterial agents, or the potentiating effect of otherwise inactive substances, can be most easily observed with solid media. This can be demonstrated most simply by applying two test solutions at predetermined points on the surface of the nutrient agar, inoculated with the test organism, so that their zones of inhibition overlap. During incubation the test compounds diffuse together and the extent and shape of the inhibition areas within the area of overlap indicates whether stimulation or antagonism has taken place.

Results are sometimes more clearly interpretable when the test compounds are applied on paper strips placed on the agar surface at right angles to each other (King, Knox and Woodroffe, 1953). The size and shape of the zone of inhibition near the area of intersection of the strips compared with those at the opposite ends indicate antagonism or synergism (see Fig. 4).

A paper strip can also be used in combination with the gradient plate test (see (b) above) to study interaction between two antibacterial agents. The bottom layer of agar in the plate contains no antibacterial compound and the upper layer contains one of the agents at the desired upper limit of concentration. A suspension of the test organism is streaked or flooded over the surface of the agar. A sterile strip of filter paper, dipped and drained from an aqueous solution of the second antibiotic, is placed centrally across the plate along the axis of the concentration gradient of the agent in the agar. As a result of diffusion the area around the strip contains all the possible combination ratios of the two agents and the organisms respond accordingly (Streitfield and Suslaw, 1954) (see Fig. 5).

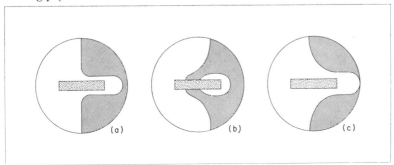

Fig. 5. Paper strip and gradient plate test for combined action. (a) Antibacterial compound in agar is inert with respect to that on strip, (b) higher concentrations of compound in agar antagonize that on strip, (c) compounds show additive or synergistic activity.

(d) *Dilution tests in agar.* This group of tests, in which the test substance is incorporated in the agar and the growth of organisms on the surface observed, is complementary in technique to the diffusion methods. The virtue of the test is that it allows the activities of different concentrations of the substance to be assessed against a wide range of organisms with the minimum of equipment.

First it is necessary to prepare plates of a suitable series of known dilutions of the substance in nutrient agar by adding different volumes of a concentrated solution to molten agar cooled to ca. 45°C and then poured in the usual way. The dilutions chosen should span as near as possible the expected inhibitory concentration of the substance. Each plate is then inoculated with the chosen test cultures and after incubation the density of growth observed. By using the multipoint inoculator of Hale and Inkley (1965) up to 27 cultures can be tested on each plate. The method is described in detail by Croshaw, Hale and Spooner (1969) (Figs 6 and 7).

It is important in this test to use the right density, within limits, of inoculum of each of the test organisms, otherwise the endpoint may be indeterminate: too few cells do not provide an adequate challenge to the

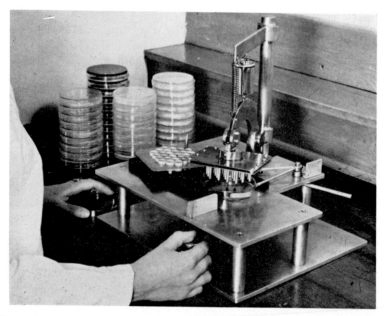

Fig. 6. Assessment of bacteriostatic activity using solid medium. Simultaneous inoculation of an agar plate with a range of test cultures by means of a multi-point inoculator.

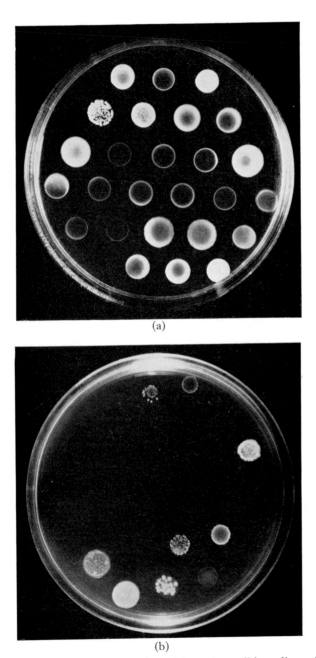

(a)

(b)

FIG. 7. Assessment of bacteriostatic activity using solid medium. (a) Control plate, (b) test compound in agar has inhibited most of the test organisms.

bacteriostatic activity, and too dense an inoculum might give an endpoint, as represented by a relatively few survivors, spread over several dilutions. Equally important is to ensure that each organism, including the most nutritionally sensitive ones, has an equal opportunity to grow. Thus, whilst staphylococci and Gram-negative bacteria will grow readily on ordinary nutrient agar, haemolytic streptococci may only grow with difficulty. This can most easily be met by adding a small amount (5%) of whole blood or of serum to the liquid inoculum culture immediately before use, so that each drop carried over by the inoculator also provides a thin layer of blood or serum to encourage growth.

When a large number of different bacteria are used in agar dilution tests much information can rapidly accumulate, particularly if many compounds are tested. Methods of recording results directly on to punch cards and of processing the information with a card sorter and computer have been described (Cobb *et al.*, 1970) (Fig. 8).

2. *Tests in liquid media*

(a) *Methods.* Tests in liquid media are more direct than those employing agar because factors such as diffusion and reaction with agar are avoided, but they are generally more time consuming to carry out. The inhibition of growth can be observed microscopically or it can be estimated quantitatively. In the serial dilution type of test, organisms are contacted with graded concentrations of the test substance in a nutrient medium and the minimum concentration preventing detectable growth taken as a measure of bacteriostatic activity. Alternatively, a single concentration of test compound can be added to the cells and their rate of growth then compared with that of control cells growing normally (Fig. 9).

Bacteriostatic activity may be usefully assessed microscopically by a microcultural technique (see Postgate, this Series, Vol. 1, and Quastel, 1966) simply by adding a test compound to the environment of the growing bacteria. By this means antibacterial activity can be observed even on organisms located intracellularly, as Showacre *et al.* (1961) found when they worked with *Salmonella typhi* in tissue culture cells. Subjective comparison of the growth rate of treated cells with that of control cells in the absence of compound can be made or a more permanent record obtained by the elegant technique of phase-contrast cinemicrography (Pulvertaft, 1952). Similarly a rapid assessment of antibiotic sensitivity can be made by observing the inhibition of microcolony formation on the surface of agar containing graded concentrations of the antibiotic. This can be best determined with the aid of a low power microscope which gives a result in < 4 h (Mahouny and Chadwick, 1965).

There are innumerable minor modifications of the well known serial

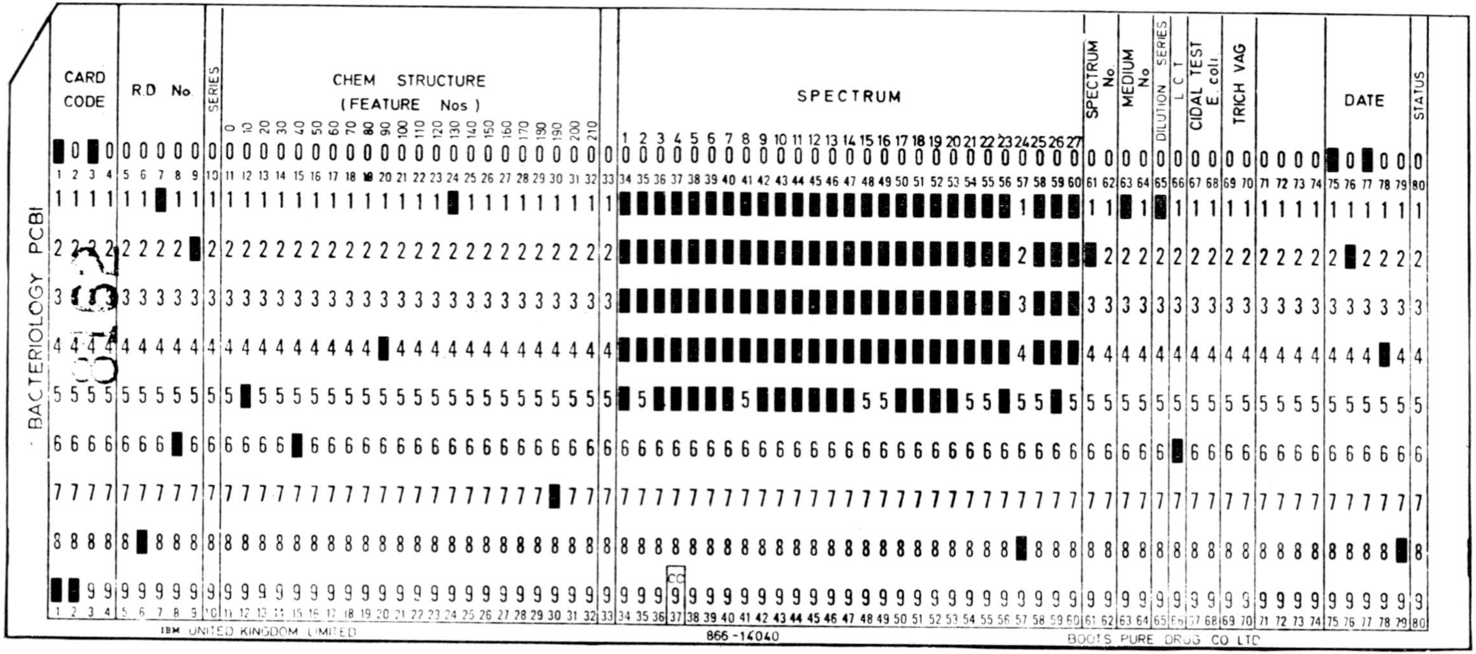

FIG. 8. Punch card record of bacteriostatic activity of a compound. Details of chemical structure, spectrum of test organisms, medium and dilution series are included.

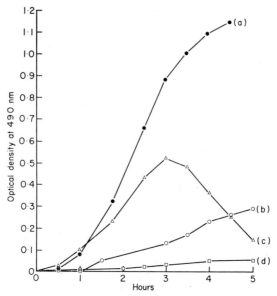

FIG. 9. Turbidimetric assessment of the qualitative effect of 1 μg/ml of some antibiotics on the growth of *E. coli*. (a) Control, (b) ampicillin, (c) chloramphenicol, (d) aureomycin. Antibiotics added to each culture at zero time.

dilution test but the principle is very simple and does not require elaborate apparatus. Dilutions of the antibacterial substance in a nutrient medium are inoculated with the test organism and incubated at a suitable temperature. The lowest concentration of the substance which causes complete inhibition of growth of the bacteria after a set time is taken as the inhibitory concentration (Fig. 10). In practice dilution tests are time consuming because, unlike agar diffusion tests, the concentration gradients are discontinuous and preparation of the separate dilutions is tedious. The test is also subject to many errors.

When more accurate assessments of antibacterial activity are desired measurements of the response of growing bacteria to added test substances can be made in an absorptiometer or nephelometer. Accuracy can be still further increased by measuring the degree of inhibition produced by each of several concentrations of a test compound. Dose-response curves can then be derived and compounds compared quantitatively in terms of the amount of each required to produce a fixed degree of growth inhibition. Measurements of this type have proved valuable in studies of chemical structure in relation to bacteriostatic activity as small changes in activity due to minor changes in structure may go unnoticed if complete suppression of growth is used as a criterion. If growth measurements are plotted against log concentration of test compound the concentration of drug that allows growth to a population density which is 50% of that obtained in the absence of drug in the same experiment can be obtained. Log-probability paper is used for this purpose (see Kavanagh, 1963, and Boyce and Roberts, this Series, Vol. 7A). It is usual, and convenient, to express concentrations of compounds on a weight basis, but comparisons are sometimes facilitated when activities are expressed in molar ratios.

(a)

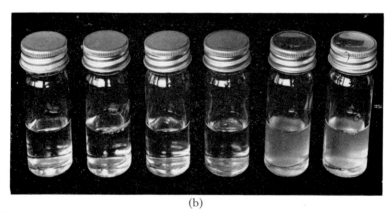

(b)

Fig. 10. Assessment of bacteriostatic activity in liquid medium. Inhibition of *Mycobacterium tuberculosis* by a compound serially diluted in (a) Long's medium, (b) Kirchner medium which contains Tween 80.

Bacteriostatic activity in liquid media, however assessed, is affected by a large number of factors. It follows that comparisons between different laboratories cannot be considered meaningful unless standard reference compounds are included.

(b) *Test preliminaries. Glassware* must be carefully cleaned and rinsed, particularly when sensitive organisms like *Mycobacterium tuberculosis* are used in synthetic media. Tubes should be scratch-free so that the absence of growth in dilution tests is easily apparent. Particularly when the growth medium contains little or no protein it is important that residues of anti-bacterial compound from previous tests are not left on glassware and that the detergent used for cleaning is also completely removed (Elliott and Georgala, this Series, Vol. 1). Cotton wool plugs are best avoided but if they are used glassware should not be sterilized with dry heat, because of the production of toxic volatile materials from cotton wool on heating.

Pipettes, which should be changed at each dilution stage to avoid carry-over, must also be thoroughly cleaned. When working with pathogens pipettes should be immersed immediately after use in a hypochlorite solution before being sterilized in the autoclave. Phenolic disinfectants are not suitable as they leave oily, inhibitory residues.

Solutions of compounds must be prepared aseptically for tests in liquid media, in contrast to those on agar, to avoid interference by chance contaminants. Nevertheless it has been our experience that, even in highly nutrient media, organic compounds are seldom the source of contaminating organisms, so that in preliminary tests sterilization is unnecessary. When sterility is considered vital, membrane filtration of a stock solution is the technique of choice unless it is certain that the compound can withstand heat sterilization. Insoluble heat-labile compounds pose a problem which can be solved by irradiation (see Sykes, this Series, Vol. 1) or the membrane filtration of a saturated solution after the removal of the insoluble residue by centrifugation.

Solubility is frequently aided if the particle size is first diminished by grinding or fluid energy milling (Temperly and Blythe, 1968).

Several sorts of *dilution series* are used. In the **geometric series,** including the ubiquitous 2-fold serial dilution, the dilutions are separated by a constant proportion. A 2-fold series carries a built-in error of $+100$ or -50% in a single test and no significance can be attached to less than a 4-fold difference in activity ratios. As the dilutions are prepared by transference of liquid from one tube to the next, errors in measuring volumes of liquid can be cumulative, one error being passed on and added to the next. Thus in a 10-tube geometric series a small pipetting error is raised to the tenth power and becomes a serious one (see Kavanagh, 1963). A geometric series is used when a test compound may show activity within wide limits and when accuracy is not paramount. **Arithmetic series** take longer to prepare and are only applicable when limited ranges of concentrations are required. Such a series is prepared by pipetting volumes of diluent diminishing by a constant amount. These are then made up to a standard volume by the addition of increasing volumes of reactant. Fluid is not carried from one tube to another so that errors are not transmitted.

A **harmonic series** of dilutions is prepared by pipetting increasing volumes of diluent into successive tubes and adding a unit volume of test compound solution to each. The interval between dilutions becomes smaller as the series progresses. Pipetting errors are not transmitted but this method is wasteful of material.

Ingram (1962) has described another dilution series, the **"centred"** **series**. This combines different geometric series and changes to a smaller dilution-interval over a portion of the total dilution range. The smallest interval is thus "centred" on the range of maximum information. To help in the selection of suitable dilution series Ingram (1962) gives a useful table of concentrations, expressed as percentages, which can be obtained in 12 successive tubes by eight geometric dilution series.

In recent years several pieces of equipment have been designed for the semi-automatic or automatic preparation of dilutions (see Cobb *et al.*, 1970).

(c) *Culture conditions. Medium* composition can have a profound effect. Where *in vitro* studies of the interaction between compound and organism are undertaken a completely synthetic medium is obviously advantageous because the chances of reaction between test compound and components of the medium are diminished. However, if compounds for chemotherapeutic use are being sought, complex media, perhaps containing serum or other body fluid, may have their place. In any case convenience often demands that an undefined medium is used when a large number of compounds are screened *in vitro*. As with tests in agar the compatibility of the test compound and medium has to be considered. Precipitation must be avoided not only because of loss of compound but because of occlusion of the endpoint. Most tests utilize incubation at 37°C so that the stability of the compound in the medium at this temperature should be checked. An unstable compound is not necessarily contra-indicated for chemotherapeutic use as therapeutic levels in the body can often be maintained effectively.

Components of a medium affecting the antibacterial action of a compound drastically have frequently provided clues in unravelling the mode of action. For instance, *Escherichia coli* has shown a 10^3-fold difference in its response to certain sulphonamides in different media and the finding that this is largely due to the content of *p*-aminobenzoic acid in the medium led to our understanding of the mechanism of this group of drugs (Woods, 1962). Similarly, the observation that the antibacterial activity of crude notatin samples depended on the glucose content of the test medium led to the identification of the active constituent as the peroxide-forming enzyme, glucose oxidase (Coulthard *et al.*, 1945).

Serum proteins frequently bind organic compounds and therefore lower their bacteriostatic activities. The degree of binding may be important in the assessment of the potential of a new agent and this can be measured in a preliminary way by adding serum to the medium before more sophisticated tests such as equilibrium dialysis are used. Serum proteins from different bacterial species bind drugs to differing degrees, and the degree of binding for human serum cannot necessarily be predicted from observations of bacteriostatic activity in the presence of animal serum. However, it must be remembered that the pH value of human serum rapidly rises to > 8.0 on standing so that a high concentration might render a culture medium inimical to growth. On rare occasions serum has been observed to enhance, rather than diminish, the bacteriostatic activities of a substance, e.g. phenoxypropylamine (Youmans, 1945).

The inclusion of a dispersing agent in a test medium, e.g. Tween 80 in media for *M. tuberculosis*, has often been shown to enhance the activity of compounds. On the other hand albumin is included with Tween 80 to detoxify the minute amounts of fatty acids liberated by the dispersing agent and this protein can reduce the activity of test compounds. These components obviously complicate assessments with *M. tuberculosis* as may others necessary for the growth of demanding organisms *in vitro*.

The pH value of liquid medium affects the ionization of many compounds and therefore their bacteriostatic activity, since a major factor in the toxicity of many organic molecules is the concentration of the undissociated base in the environment of the cell. Similarly, the organism itself may be rendered more or less sensitive by changes in pH value. Pathogenic bacteria normally grow best at pH values near neutrality, but a chemotherapeutic agent may be required to act under conditions outside these limits in certain loci in the body, e.g. in the urinary tract. Thus, a knowledge of the effect of pH on bacteriostatic activity of a compound is important. However, buffer systems have to be chosen with caution as some are toxic to bacteria and others may be metabolized during growth. The choice of buffers for bacteriological work is dealt with in detail by Munro, this Series, Vol. 2.

Changes in the pH value of the test medium occurring during growth may affect the antibacterial activity of some compounds. If glucose is present, for instance, cultures of *S. aureus* or *E. coli* may fall as low as pH 4·9. On the other hand the aerobic metabolism of amino-acids may liberate unassimulated ammonia and result in a marked rise in the pH value of the medium. When highly reproducible conditions are required, automatic pH control can be used (see Munro, this Series, Vol. 2).

The degree of aeration of a culture is sometimes accelerated during determinations of antibacterial activity in liquid media in order to speed growth and thus to shorten the experimental period. This can affect the result by oxidizing labile test compounds, thereby diminishing their activity, or by altering the sensitivity of the test organism (Kavanagh, 1963). On the other hand some compounds like streptomycin, for instance, show greatly reduced antibacterial activity under anaerobic conditions.

Size and age of inoculum affect the apparent activity of a compound. If too small an inoculum is used a test may be prolonged to an impracticable degree before detectable growth occurs. On the other hand, when large inocula are employed the progeny of even a very small number of viable cells can cause turbidity, making interpretation difficult. Similarly if individual cells can destroy the agent, as with penicillin and penicillinase-producing strains of bacteria, the endpoint is a balance between the rate of destruction of penicillin by the cells and the rate of kill by the antibiotic. Thus it is always desirable for dilution tests to be made at different inoculum levels.

The age of the cells can also affect the result of dilution tests, but little information has yet accumulated concerning the effect of antibacterial agents on different stages of the growth cycle as may be obtained from studies with synchronous cultures (see Helmstetter, this Series, Vol. 1).

Incubation time and temperature must be carefully standardized so that results are reproducible. The incubation temperature affects the generation time of the test organism, the stability of the test compound and its reaction with the bacteria. Changes of temperature alter the activity of each compound differently.

The optimum temperature for some metabolic activities is not always identical with that for growth and this phenomenon may affect the sensitivity of a test organism. For instance, the sensitivity of *Bacillus cereus* to penicillin is greatly enhanced when the incubation temperature is increased to 41°C, a temperature that suppresses the synthesis of the inducible enzyme penicillinase (Knox and Collard, 1952).

Incubation time normally depends on the intrinsic growth rate of the test organism, and in most tests tubes are incubated for 24–48 h. The earlier a dilution test is read the more active may a compound appear to be because the minority of insensitive cells will not have had time to multiply sufficiently. Also, if a bacteriostatic compound is unstable it may be inactivated during incubation, thus allowing inhibited cells to multiply and the endpoint to change.

In tests not involving macroscopic assessment of growth, incubation times are often limited to a few hours during which the growth rate is assessed. Under these conditions stability of the compound and the risk of missing a low level of bacteriostatic activity due to overgrowth by an insensitive minority of cells are less important.

(d) *Growth assessment*. In the conventional dilution test turbidity is used as the criterion of growth, although it should be remembered that tubes showing no visible growth may contain up to 10^7 cells/ml. If there is an interfering turbidity, due perhaps to an insoluble compound, the end-point can be confirmed by microscopic examination, by subculture on to agar, or by the addition of a pH indicator or a reducible dye such as tri-phenyl tetrazolium chloride (see Jacob, this Series, Vol. 2).

Other methods of measuring the degree of growth include turbidimetry, nephelometry, dry weight, chemical content (Mallette, this Series, Vol. 1), total counts (Postgate, this Series, Vol. 1), viable counts (Postgate, this Series, Vol. 1) and a variety of metabolic changes involving pH (Munro, this Series, Vol. 2), redox potential (Jacob, this Series, Vol. 2) and gaseous exchange.

The use of photoelectric instruments to assess the degree of growth is relatively easy, rapid, accurate and objective, but their limitations must be clearly understood. Automatic equipment has been designed so that a large number of tubes can be read or multiple observations made on a single culture after the addition of sublethal amounts of an antibacterial compound at different times in the growth cycle (see Cobb *et al.*, 1970).

Change in the apparent rate of growth of bacteria in the presence of an antibacterial agent may be due to the death of some of the cells or a reduction in the growth rate of all the cells, or a combination of both phenomena. This can best be investigated by simultaneous total and viable counts.

Measurements of metabolic activity of a culture as a method of assessing bacteriostatic activity have to be interpreted with care. The test compound may inhibit the metabolic activity to a greater extent than it inhibits growth or its presence may interfere with the measurements. For instance, the effect of compounds on the gas exchange of growing cells, measured by a manometric method, provides accurate and quantitative information. However, reductions of gas exchange in the presence of a compound do not necessarily mean that the ability of the cells to grow and divide is impaired. The added compounds may themselves take up oxygen, react with carbon dioxide or inhibit the respiratory processes of the bacteria rather than their growth. The use of suitable controls can elucidate the true course of the inhibition but these techniques are best used to study the mode of action of an antibacterial compound rather than to assess its activity.

(e) *Tests for combined action*. The serial dilution test is frequently used to investigate the activity of a combination of antibacterial drugs quantitatively. Tubes are arranged in a "chess-board" pattern, the two agents being serially diluted in rows at right angles to each other so that all possible combinations attainable by the dilution series are obtained. After inoculation and incubation, bacteriostatic endpoints are noted and tubes without growth are subcultured on to agar to determine whether the combined action of the compounds has killed the cells or merely inhibited them. The minimum inhibitory concentrations and minimum bactericidal concentrations of the combinations are then compared with those of the agents alone to determine whether the combinations are synergistic or antagonistic. Because of the errors in the serial dilution test only pronounced effects can be detected with any certainty and the preparation of the large number of dilutions is time consuming. Similar information can be obtained more easily and in greater detail if response curves are compared by measuring the degree of growth of the test organism in an absorptiometer or nephelometer after incubating cultures with several concentrations of the individual compounds or their mixtures.

B. Assessment of bactericidal activity

1. *Phenol coefficients*

(a) *Historical.* In recent years changes have taken place in the approach to the assessment of disinfectants. Hitherto an arbitrary test, such as a phenol coefficient evaluation, has been applied and from this an assumed use-dilution calculated. Nowadays the approach is more realistic and the procedure is to some extent reversed. The desire is for the laboratory conditions to match as closely as possible those found in practice, and even to adjust these conditions further, so that the result obtained is as near as possible to that found to be effective in practice.

As early as 1750 Sir John Pringle attempted to obtain "coefficients" for a number of compounds in terms of their abilities to prevent the putrefaction of raw meat, and this can be taken as the precursor of all of the present day phenol coefficient tests, and of other methods for assessing disinfectants. Years later Koch in 1881 described his silk thread technique in which the threads, impregnated with anthrax spores, are immersed for a given period in a disinfectant solution and then transferred to a nutrient medium, or sometimes planted in test animals, to assess the point at which the spores are just killed. The method was later adapted for testing the activities of disinfectants against nonsporing bacteria such as *Escherichia coli* and *Salm. typhi*. In 1897 Kronig and Paul published a modification of the Koch method in which organisms were dried on garnets instead of being soaked into silk threads. Their findings from these tests led Kronig and Paul to the important discovery, fundamental to all work in disinfection, that bacterial death is not an instantaneous act but is an orderly and measurable process.

The first real attempt at devising a standard, reproducible test was by Rideal and Walker (1903). Here the performance of a disinfectant in relation to that of phenol was assessed, and from it a phenol ratio, or phenol coefficient, was derived. Later the test was made somewhat more realistic by Chick and Martin (1908) by including dried human faeces (later altered to dried yeast cells). Later still, several other techniques were published, amongst which were the U.S.F.D.A. (Food and Drug Administration) (Ruehle and Brewer, 1931) and the U.S.A.O.A.C. (Association of Official Agricultural Chemists, 1955), phenol coefficient techniques, all of which were only variations on a basic theme. The A.O.A.C. method, with additions, is now the official method for assessing disinfectants in the United States. Each of the tests employs a different strain of *Salm. typhi* as the test organism.

Over the years much effort has been devoted to attempts to improve the precision and reproducibility of these tests and the most up to date modifications to the Rideal–Walker and Chick–Martin methods are enshrined in British Standards 541 (British Standards Institution, 1934, with addenda) and 808 (British Standards Institution, 1938, with addenda), respectively. An extension of the Rideal–Walker test was introduced in 1961 (British Standards Institution, 1961) which gives a phenol coefficient with *S. aureus* instead of with *Salm. typhi*. Since the publication of the A.O.A.C. phenol coefficient method the Use-Dilution Confirmation test was added (A.O.A.C., 1960), a test first proposed by Stuart, Ortenzio and Friedl (1953) to confirm that the use-dilution derived from the phenol coefficient test is one that works in practice, at least in some conditions.

All of these tests were designed in the first place to assess the phenolic, or coal tar, disinfectants, but misguidedly they became commonly used for assessing all types of disinfectants, often with grossly misleading results. The outstanding example of this was the very high phenol coefficients attributed initially to the quaternary ammonium compounds (q.a.c.) which arose because of the carryover of inhibitory amounts of the compounds into the recovery medium, thus preventing the surviving cells from growing and giving falsely negative responses. To overcome this deficiency several workers, from 1941 onwards, suggested incorporating an inactivating, or quenching, agent either in the final recovery medium or at one of the preceding dilution stages. Inactivators are also used in other types of lethal tests. They are discussed in more detail in Section 1 (e) below.

In recent years it has been increasingly appreciated that the phenol coefficient type of test has many disadvantages (see Sykes, 1965) not the least of which is that it is too unrealistic to give meaningful results and that it requires too large a correction factor to give reliable use-dilutions. Sykes (1958) drew attention to these and proposed (1961) the outline of a test scheme in which a disinfectant is assessed on its own performance, instead of comparatively, against the more resistant organisms, *S. aureus* and *Pseudomonas aeruginosa*. About the same time the British Standard for the Laboratory Evaluation of Quaternary Ammonium Compounds was published (British Standards Institution, 1960). This followed the same lines of thought, and, although addressed to one group of compounds, is equally applicable to all types of liquid disinfectants. Further examination of these two structures resulted in the development by Kelsey and Sykes (1969) of their scheme for the evaluation of disinfectants for hospital use, and this, with certain modifications, is now being considered as the basic performance test for this purpose.

(b) *The Rideal–Walker test*. The rigorous specifications for carrying out this test are laid down in detail in BS 541 : 1934, already referred to, to which several addenda have been added from time to time. (A new British Standard, incorporating these and other modifications, is at present in preparation.) The test is essentially a comparative one in which chosen dilutions of a disinfectant are matched against standard dilutions of phenol in their action against a *Salm. typhi*, and from this the phenol coefficient for the disinfectant calculated.

The organism used is *Salm. typhi* NCTC 786. This originally was a deliberate mixture of the Hopkins and Rawlins strains, but obviously with the frequent subculturing it has undergone over the years this mixture no longer obtains; nevertheless, the present culture, now preserved in the freeze dried state, retains its standard resistance to phenol. In use the culture is maintained by weekly transfers on an agar medium containing 25 g of Oxoid Nutrient Broth No. 2 and 12 g of agar in 1 litre of distilled water, pH 7·4–7·6 after sterilization, and sterilized in 5 ml quantities in tubes at 121°C for 15 min (only one sterilization is permitted). (Until recently the medium was made with Lab-lemco and Allen and Hanbury's Eupeptone, but the latter is not now manufactured, and the Oxoid medium gives comparable results.) For test purposes, a subculture is made from the agar into a 5 ml tube of Oxoid Nutrient Broth No. 2 in which it is incubated at 37°C subcultures being made at 24 h intervals. Only cultures between the 3rd and 14th generation may be used. Each test requires four dilutions of the disinfectant and one of phenol, although other combinations may be used. The disinfectant dilutions are made in distilled water from an initial 1% dilution of the product. The phenol dilutions are also made in distilled water.

The dilutions of the disinfectant are graded in an arithmetic series, with spacings generally in units of 25 in the range 1 : 75 to 1 : 200 and in units of 50 in the range 1 : 200 to 1 : 400, after which they are in units of 100. The phenol dilutions specified are 1 : 95, 1 : 100 and so on to 1 : 115. Five ml of each dilution are required, and the test is carried out in a water bath at 17–18°C.

To perform the test, shake the 18–24 h culture of *Salm. typhi* and allow it to settle and cool in the water bath at 17–18°C. At zero time, add 0·2 ml to the first 5 ml tube of disinfectant and shake gently. Thirty seconds later add 0·2 ml of the culture to the second tube of disinfectant, and repeat until each of the 5 tubes in the test has been inoculated. Thirty seconds later, i.e. $2\frac{1}{2}$ min after being inoculated, subculture 1×4 mm standard loopful from the first tube into a 5 ml tube of liquid medium. Repeat this subculture routine until all 5 tubes have been subcultured, and repeat further until each tube has been subcultured after $2\frac{1}{2}$, 5, $7\frac{1}{2}$ and 10 min. Incubate all of the subculture tubes at 37°C for 48 h and record the results as growth (+) or no growth (−).

To calculate the Rideal–Walker coefficient, divide the dilution of the disinfectant which gives survivors at $2\frac{1}{2}$ and 5 min, but not at $7\frac{1}{2}$ and 10 min, by that dilution of phenol which gives the same response: interpolation, but not extrapolation, is permitted. Thus, in a test giving

		Time (min) culture exposed to disinfectant			
Disinfectant Dilution		$2\frac{1}{2}$	5	$7\frac{1}{2}$	10
A	1 in 250	−	−	−	−
A	1 in 300	+	−	−	−
A	1 in 350	+	+	−	−
A	1 in 400	+	+	+	−
Phenol	1 in 95	+	+	−	−

the Rideal–Walker coefficient = 350/95 = 3·7 (approx.).

It used to be assumed that a dilution 20 times that of the coefficient (in the example quoted above, $20 \times 3·7 = 1 : 74$) gives an effective use-dilution.

(c) *The Rideal–Walker* Staphylococcus aureus *test.* The procedure followed (British Standards Institution, 1961) is the same as that of the standard Rideal–Walker test except that: (a) the test organism is *S. aureus* NCTC 3750; (b) the medium contains (g/l): Oxoid peptone, 10; Lab-lemco, 5; sodium chloride, 5; pH, ca. 7·0; (c) the phenol dilutions to be used are 1 : 85; 1 : 90 and so on up to 1 : 105. The calculation of the coefficient is the same as in the standard test.

(d) *The Chick–Martin test.* This procedure gives a phenol coefficient for a disinfectant in conditions more realistic than those of the Rideal–Walker test in that organic matter, in the form of killed yeast cells, is present, thus causing the disinfectant to act in the presence of "organic soiling material", as might be found in a normal disinfecting situation.

Although until quite recently the media used differed from those given in the Rideal–Walker technique they are now the same. The culture is the "S" strain of *Salm. typhi* (NCTC 3390) and its maintenance and propagation is the same as with the Rideal–Walker strain.

The yeast suspension is made from an ordinary baker's yeast creamed with water into a concentrated (40%) suspension, sterilized at 121°C for 15 min and finally diluted with water to give an accurate 5% dry wt suspension.

The test dilutions of the disinfectant are made from an initial 2% dilution in water, but here they are in a geometric series dilutions differing from one another by steps of 10%. The phenol dilutions specified are 2·0, 1·8, 1·62 and 1·46%. The test is made at 20°C.

The procedure is as follows. Mix 2 ml of a 24 h culture of the *Salm. typhi* S strain with 48 ml of the yeast suspension and place in the water bath at 20°C along with the tubes each containing 2·5 ml of the disinfectant and phenol dilutions.

To the first tube of 2·5 ml of disinfectant add 2·5 ml of the yeast-culture suspension and shake gently. At 30 sec intervals inoculate the remainder of the tubes similarly. Exactly 30 min later subculture 1×4 mm loopful of the mixtures each into 2×5 ml tubes of the liquid recovery medium and incubate at 37°C for 48 h.

The Chick–Martin coefficient is obtained "by dividing the mean of the highest concentration of phenol permitting growth in both cultures and the lowest concentration showing absence of growth in both cultures by the corresponding mean concentration of the disinfectant". Thus, a test with results

$$
\begin{array}{ll}
\text{Phenol \%} & \text{Disinfectant \%} \\
\left.\begin{array}{l} 2\cdot0 \;\; -- \\ 1\cdot8 \;\; -- \\ 1\cdot62 ++ \\ 1\cdot46 ++ \end{array}\right\} \begin{array}{l}\text{mean}\\ = 1\cdot71\end{array} &
\left.\begin{array}{l} 0\cdot457 -- \\ 0\cdot411 -- \\ 0\cdot370 ++ \\ 0\cdot333 ++ \end{array}\right\} \begin{array}{l}\text{mean}\\ = 0\cdot395\end{array}
\end{array}
$$

gives a coefficient of $1\cdot71/0\cdot390 = 4\cdot4$.

Until recently the main use of this coefficient was for approving the use-dilution of a disinfectant under the Diseases of Animals Act, 1950. This has now been superseded by other more direct tests (see 2(a)).

(e) *The United States Association of Official Agricultural Chemists (A.O.A.C.) tests.* These are described in detail in the Official Methods of the Association (A.O.A.C., 1960). The first phase of these tests, the phenol coefficient assessment, resembles closely the U.S.F.D.A. method, which appears to be but little used these days. The test organisms are yet another strain of *Salm. typhi*, the Hopkins strain 26 (ATCC 6539), and another *S. aureus* (FDA 209, ATCC 6538). Each is maintained on a nutrient agar slope and for use in the test grown for at least 4 daily transfers at 37°C in a liquid medium containing (g/l): Difco beef extract, 5; Armour peptone, 10; sodium chloride, 5; pH, 6·8.

The disinfectant dilutions, prepared from a master 1% dilution, "should cover killing limits of disinfectant in 5–15 min and should at the same time be close enough for accuracy". The phenol dilutions specified are 1 : 90 and 1 : 100 with the *Salm. typhi* and 1 : 60 and 1 : 70 with the *S. aureus*. Five ml amounts of each are required in the test which is made at 20°C.

The procedure is as follows. To each 5 ml tube of disinfectant or phenol dilution add 0·5 ml of the test culture. At the end of 5, 10 and 15 min disinfection time subculture 1 × 4 mm loopful to 10 ml of the chosen recovery medium and read the results after 48 h at 37°C. The recovery media to be used are: (a) for phenolic-type disinfectants, that described above; (b) for those containing mercury and other heavy metals, one containing (g/l): L-cystein, 0·75; agar, 0·75; casein digest (pancreatic), 15; yeast extract, 5; sodium chloride, 2·5; sodium thioglycollate, 0·5; pH 7·0; (c) for those containing cationic surface-active substances, the same as (a), but with 0·7 g of lecithin (Azolectin) and 5 g of Tween 80 added.

A test is valid only if the phenol responses obtained are

		5 min	10 min	15 min
For *Salm. typhi*	1 : 90	+ or 0	+ or 0	0
	1 : 100	+	+	+ or 0
For *S. aureus*	1 : 60	+	0	0
	1 : 70	+	+	+

From these and the responses of the disinfectant dilutions the phenol coefficient is calculated in the same way as in the Rideal–Walker test, and from this the presumed use-dilution is estimated, by multiplying the coefficient by 20.

To confirm this value, the second phase of the test, the use-dilution test, is applied. In this, cleaned and polished stainless steel cylinders, 8 mm diam × 10 mm long, called in the test "carriers", are immersed in groups of 10 in a 0·1% asparagine solution and sterilized. Twenty such carriers are then placed in 20 ml of the test culture (either a 48–54 h culture of *Salmonella choleraesuis* ATCC 10708 or a 24 h culture of *S. aureus* ATCC 6538) and then transferred and stood vertically on a filter paper in a Petri dish to drain. The dish is then put in the incubator to allow the carriers to dry for not more than 60 min, after which one carrier is placed in each of 10 tubes containing 10 ml of the disinfectant use-dilution. After exactly 10 min each cylinder is transferred to a 10 ml tube of the appropriate recovery broth and incubated at 37°C for 48 h. Phenol dilutions, as in the first phase of the test, are used to check the level of resistance of the cultures.

The presumed use-dilution of the disinfectant is satisfactory if no growth occurs from any of the 10 carriers. Growth from any one carrier means that the presumed use-dilution must be adjusted: this involves a recheck. Comparisons in the U.S.A. over a number of years of these results with those from the straight phenol coefficient values show the former to be much more reliable, especially in hospital use for the control of staphylococcal cross infections.

(f) *Defects in phenol coefficient tests.* The first requirement in any comparative biological assessment is that "like" shall be tested against "like", in other words, only phenolic preparations should be considered when phenol is the standard. This was the original intention of the Rideal–Walker test but, as already indicated, it has been extended, wrongly, over the years to embrace every type of disinfectant and antiseptic. The coefficient concept is, therefore, wrong: any antibacterial preparation should be judged only on its own merits and performance. This does not rule out the phenol coefficient type of test entirely. It is convenient for routine batch-to-batch or process control of a standard disinfectant formulation.

A second defect is that the endpoint of the test depends on the transfer of one (or more) viable cell in the loopful and on its subsequent growth in the recovery medium. Towards the end of a disinfection process when the viable count has been reduced from an initial value of the order of 10^6–10^7/ml to 10^2 or less/ml, the tailing portion of the killing curve has been reached, and it becomes a matter of statistical chance whether a viable cell is picked up by transfer of ca. 0.02 $(1/50)$ ml. On this transfer depends the endpoint of the test and consequently the coefficient of the disinfectant fluid.

The statistics of the case have been studied by a number of workers including Rahn (1945), who concluded that a variation of 360% is possible (but unlikely), by Stuart, Ortenzio and Friedl (1958) who stated that "a $\pm 12\%$ tolerance would have to be acknowledged". More recently Dodd (1969) examined the problem in detail.

A further defect is the choice of *Salm. typhi* as the test organism. In its day, this was a pathogen of considerable public health importance, and in this context it is still important, but in recent years other organisms, notably *S. aureus*, *Ps. aeruginosa* and *Proteus* spp., have become dominant, especially in the hospital environment, and they are also generally more resistant. The use of a salmonella as the standard has meant, therefore, the introduction of a large correction factor—20 is the common value—to obtain a use-dilution effective against these more resistant types. This again is wrong in principle because of the widely differing concentration coefficients, or dilution factors, of the different types of disinfectant.

Other comments which can be made concern the choice of recovery conditions, especially the incubation temperature, because it is now known that organisms after adverse treatment, "damaged" organisms, tend to recover better at temperatures below their normal optimum. Finally, there is the question of the disinfection period. Most of the tests allow only a few minutes, but it is rare in practice for such a short period to be needed. There are occasions when a rapid disinfection is required but in most situations a period of several hours, even overnight, is adequate. The short period kill, of course, provides a margin of safety, and a test spanning several hours might be more difficult to manipulate in the laboratory. Some compromise period, therefore, between the extremes merits consideration.

2. *Other standard methods*

(a) *Tests on disinfectants for agricultural use.* Up to 1970 disinfectants for all agricultural uses were approved on the basis of their performance in the Chick–Martin test, the use-dilution award being the Chick–Martin value plus 10% of this, the sum being rounded off to the nearest 0·5 and then multiplied by 20. Late in 1970 a new Order was made (Statutory Instrument 1970 No. 1372) in which disinfectants are approved according to their intended use as follows:

(i) for use specifically against anthrax, brucellosis, contagious bovine pleuro-pneumonia and glanders, and for purposes other than (ii) (iii) and (iv) below [this is the general group];

(ii) for use against tuberculosis;

(iii) for use against foot-and-mouth disease;

(iv) for use against fowl pest (Newcastle disease, fowl plague).

Certain other physical and labelling requirements are also specified.

The general purpose disinfectants, under (i) are required, at the use-dilution recommended, to reduce the viable count of a *Salm. choleraesuis* culture in a yeast suspension by a factor of at least 10^4 in 30 min at 4°C. The test is made as follows.

The culture of *Salm. choleraesuis* NCTC 5364, is kept in the freeze dried state and maintained for current use on a nutrient agar slope. For use in the test inoculate and incubate it in Oxoid No. 2 Nutrient Broth at 37°C for 24 h, so that it contains at least 10^8 viable cells/ml. Add 4 ml of this to 96 ml of a 5% yeast suspension prepared after the manner of that used in the Chick–Martin test (1(d)). Dispense 2·5 ml amounts into sterile test-tubes (15 × 150 mm) and cool in a water bath at 4 ± 0.5°C. Prepare the disinfectant dilutions in sterile water of hardness 340 ppm (WHO formula) at the concentration recommended for use and at a concentration, say, 20% weaker and one 20% stronger. Add 2·5 ml of the disinfectant dilution, also at 4°C, and hold the mixture at this temperature, shaking the tubes every 10 min. At the end of 30 min transfer 0·1 ml (5 drops from a 50 dropper pipette) of the mixture to 10 ml of nutrient broth containing 5% of horse serum. Transfer 1 ml of this to 5 × 9 ml tubes of nutrient broth and incubate at 37°C, reading the result after 48 h. If there is no growth in two or more of the five tubes, the disinfectant is approved for use at that dilution.

For disinfectants intended for use in category (ii) the procedure is similar to that just described, but the test organism is *Mycobacterium fortuitum* NCTC 8573. The growth and recovery medium is Oxoid No. 2 Nutrient Broth containing 0·05% of Tween 80 and the disinfecting period is extended to 60 min. The inoculum is grown for 7 days at 37°C and the recovery incubation period is also 7 days.

No tests can be made against foot-and-mouth or Newcastle disease virus in the United Kingdom without authority (and this authority is exclusive to the Animal Virus Research Unit, Pirbright, England). For the latter, however, it has been suggested that the response with influenza A virus gives a good indication of the likely use-dilution.

(b) *The Kelsey–Sykes test.* Amongst the biggest users of disinfectants are the hospitals, and the uses to which they are put are many. They therefore merit special consideration. Sometimes a relatively quick action of a few minutes is required, but more often the disinfection can take place over several hours, frequently overnight. Likewise the type of infection to be dealt with can vary considerably, but the organisms causing most trouble, being the most resistant, are strains of *S. aureus* and *Ps. aeruginosa*, and less frequently *Proteus* and *Klebsiella* spp. together with *E. coli* and others of the Enterobacteriaceae.

The outline of a suitable test for these purposes is described in BS 3286: 1960, a Method for the Laboratory Evaluation of Quaternary Ammonium Compounds. Here a broad structure is described for assessing, in effect, any disinfectant, allowing for variations in the test culture(s), disinfection period, the presence of organic soiling matter and other parameters. Using this as a basis Sykes (unpublished) devised a technique applicable to most hospital uses. This at least had a number of advantages over existing methods and simulated much more closely the actual conditions under which disinfectants are used. It employs three test organisms, *S. aureus* NCTC 4163, *Ps. aeruginosa* NCTC 6749 and *Proteus vulgaris* NCTC 4635: others such as *E. coli* can be added as desired. On similar lines, Kelsey, Beeby and Whitehouse (1965) produced a capacity use-dilution test, its main feature being that instead of making one addition of a large inoculum of the test organism, with or without organic matter, additions are made in six increments, thus giving some measure of the capacity of the disinfectant dilution to cope with successive bacterial invasions.

In the light of experience with these two tests Kelsey and Sykes (1969) produced a modified version which incorporates the principle of the capacity test but which is less cumbersome and is equally informative.

The test organisms are the same as those prescribed in the Sykes method. Disinfectant dilutions are made in water of 300 p/m hardness (17·5 ml of a 10% (w/v) solution of $CaCl_2.6H_2O$ and 5 ml of a 10% solution of $MgSO_4.7H_2O$ in 3300 ml of distilled water, sterilized in the autoclave). Any number of dilutions may be used, but the critical ones are (i) that recommended by the manufacturer, (ii) one, say, 20 or 25% weaker than this and (iii) one, say, 20 or 25% stronger. The yeast suspension is prepared as a 5% (dry wt) suspension in water, but in the test 2 ml are added to 3 ml of water to give a final 2% suspension. The cultures are maintained in the freeze dried state and for current use are kept on agar slopes: for use in the test, each is grown in "a good nutrient broth" for 24 h. Alternatively, they may be grown on an agar slope and the cells washed off with a small volume of broth.

The test is made to simulate "clean" and "dirty" conditions, the former by using a simple broth and the latter by using the yeast suspension (20% of inactivated horse serum can also be used) in which to suspend the organisms, the final concentration of the bacterial cells being ca. $10^9/ml$. All incubations are at 30–32°C, not at 37°C.

The procedure is as follows. To 3 ml of the disinfectant dilution add 1 ml of the bacterial suspension in broth, yeast or serum and shake gently. After 8 min remove a sample of the mixture with a 50 dropper pipette and transfer 1 drop to each of 5 tubes of a liquid recovery medium. Alternatively, 5 drops may be placed on a nutrient agar plate. Two minutes later, i.e. 10 min after the first inoculation, inoculate the tube with a further 1 ml of the bacterial suspension and 8 min later subculture as before. A further 2 min later, i.e. 20 min after the first inoculation, repeat the process again, thus giving three incremental additions of bacteria (and yeast or serum). After 48 h at 30–32°C record the numbers of tubes showing growth, or the numbers of colonies from each drop on the surface plate culture.

A disinfectant is satisfactory for use at the initial dilution if there is no growth in 2 or more of the 5 tubes subcultured after the second incremental addition or if there are not more than 5 colonies from the 5 drops on the agar plate. It is recommended that the tubes of disinfectant be kept for about 5 days after which further cultures or counts be again made, without further inoculation, to determine whether the occasional and undetected surviving cell may have developed—this can occur if the disinfectant dilution is on the borderline of lethal activity.

This procedure at first gave reasonably reproducible results in collaborative trials in different laboratories but later divergencies appeared due to variations in the resistance of the test organism, particularly with *Ps. aeruginosa*. This, in turn, was found to be due to variations in the media used for growing the initial cultures. This aspect is under investigation, but it appears that a synthetic medium based on that of Wright and Mundy may solve the problem.

To confirm the reliability of the use-dilution obtained from the laboratory assessment, Kelsey and Maurer (1966) proposed an essentially practical in-use test.

To quote "Samples are taken from: bucket contents at the end of a cleaning session, liquid wrung from mops, . . . liquid in which urine bottles are rinsed or stored, laboratory discard jars, linen steepage tanks . . . and so on according to circumstances". Transfer 1 ml of this liquid to 9 ml of $\frac{1}{4}$ strength Ringer's solution (containing an inactivator if necessary) and place 10 drops from a 50 dropper pipette on the surface of an agar plate. The disinfectant is considered to be adequate if not more than five of the 10 drops show growth after incubation for 48 h at, preferably, 30–32°C.

(c) *Other methods.* Apart from those described in the preceding paragraphs several other methods have been published, most of which are concerned with different ways of assessing the endpoint. Every disinfection process involves a numerical reduction in the numbers of viable cells, and every test so far described employs this parameter, even though, as in the phenol coefficient tests, it may be hidden. But even here, although an apparently complete kill is required, the initial inoculum is defined, and so the "kill" is, in effect, a reduction in the viable population by a factor of the order of 10^4–10^5, and this is the order now acceptable by most workers for assessments of this type.

Amongst the other endpoint methods suggested are a nephelometric one, an extinction-time technique, the use of enzyme inhibition, increase in cell volume and even affinity for wool. All these are mentioned by Sykes (1965). Other special and specific tests merit a reference.

Sporicidal tests are included in the A.O.A.C. (1960) and the German Society for Hygiene (1959) schedules. The A.O.A.C. method suggests using the spores of *Bacillus subtilis* and *Clostridium sporogenes* as typical, but others may be used as preferred. It is strictly a surface disinfection test. The German method specifies a saline suspension of spores of *B. subtilis* or *Bacillus mesentericus* which have been heated at 60°C for 60 min.

Tuberculocidal activity is important in some spheres, particularly in hospital work, but activity against other bacteria does not necessarily mean efficacy against *M. tuberculosis* or any of the other acid-fast organisms. Because of this the A.O.A.C. (1960) includes a specific test which is a simple adaptation of the phenol coefficient method, although phenol is not included. The strain used is H37Rv grown for 3 weeks at 35°C in a liquid medium containing asparagine, glycerol and inorganic salts. The contact time specified is 10 min and the subculture volume is 2 loopfuls into the above medium to which 1% of serum is added: incubation is for 3 weeks.

Although most of the *chlorine-releasing compounds* are assessed on the amount of chlorine available, the answer so obtained is not strictly correct because the rate of release of chlorine differs with different compounds. Because of this, the A.O.A.C. (1960) included a rather unusual test which is similar in some respects to Kelsey's original capacity test. *Salm. typhi* and *S. aureus* are the test organisms. First 0·5 ml of a broth culture is added to 10 ml of a series of solutions of the preparation diluted so that they contain 200, 100 and 50 ppm of available chlorine. One minute later a loopful is subcultured into a liquid recovery medium. Thirty seconds later the inoculation and subculture processes are repeated and this is continued until 10 increments have been added. The compound or preparation is considered to be satisfactory if it gives at least as good a response as do the equivalent dilutions of a sodium hypochlorite solution. This is normally: no growth up to the fifth increment at 200 ppm; none up to the third increment at 100 ppm and none up to the second increment at 50 ppm.

3. *Tests on solid media*

Solid media are used far less in making bactericidal assessments than in bacteriostatic ones. One of the problems is that zones of no growth on an agar plate give no indication whether the bacteria within the zones have been killed or only inhibited. Only direct cultures from the zones can give this information, and it is most conveniently done by taking replica samples from the surface of the agar (Fig. 11). The technique of Lederberg and Lederberg (1952), devised in the first place for selecting resistant mutants, uses pads of velvet for this purpose. The pads are first pressed on the test area and then on a sterile nutrient agar, thus transferring viable cells, if any, from one agar surface to the other and allowing them to grow. The velvet, however, only transfers ca. 1% of the organisms on a surface, thus allowing appreciable numbers of viable cells to go undetected. To meet this, Parry (1961) modified the technique by making agar imprint transfers instead of velvet pad ones. The agar is filled into bottle screwtops so that the flat upper agar surface just stands proud of the rim of the cap. The agar surface is first pressed on the test area and then either it can be incubated directly (assuming that it does not pick up enough of the anti-

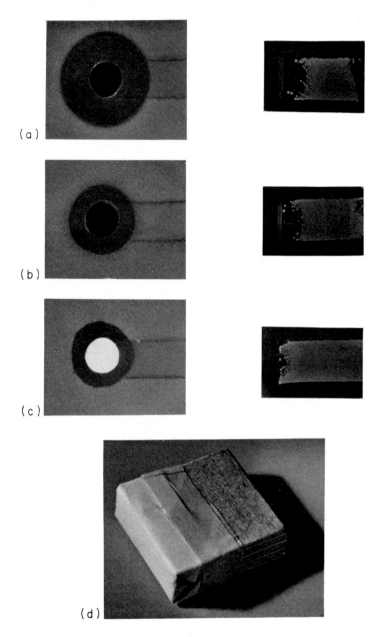

FIG. 11. On left zones of inhibition of growth and, on right, corresponding replicas. Solutions in cups (a) and (b) show some bactericidal activity but when the compound is formulated (c) this is greatly reduced; (d) replication pad.

bacterial agent to inhibit growth) or it can be pressed on the surface of a nutrient agar plate, thus leaving an "imprint" of the first surface (Fig. 12).

The method is not only applicable to agar surfaces but to any surface from which an adequate imprint can be obtained. As such it can be used in the assessment of surface disinfection properties (see 5 below).

Fig. 12. Assessment of bactericidal activity using agar. (a) Bottle screwtop filled with agar, (b) imprint showing growth, (c) imprint showing absence of growth in centre.

As a complement to this, Chabbert (1960) overcame the inhibiting carry-over problem by using thin cellophane tambours which just fit conveniently into an agar plate. A sterile agar plate is poured and on it placed paper strips previously dipped into a solution of the test antibiotic or other antibacterial agent. The tambour is then laid in contact with the agar surface and inoculated with a sensitive organism. During incubation the agent diffuses from the paper and through the cellophane membrane. At the same time nutrients from the agar also diffuse through the cellophane.

After 6–18 h, when the "reaction" is thought to have been completed, the tambour is removed to a fresh sterile agar plate and the incubation continued. During this second incubation period the surviving cells are encouraged to grow thus yielding detectable colonies. Care must be taken that the antibacterial agent is not absorbed by the cellophane sufficiently to inhibit growth: this should be checked with control cultures and plates.

4. Surface disinfection

(a) *Historical.* In this Section only the treatment of inert surfaces is discussed: skin disinfection is dealt with in Section V.E.

Many situations in food catering, and in industry generally, as well as in hospitals and the home, call for organisms on surfaces to be disinfected in contrast to those in aqueous suspension. The mechanisms are not the same because, in the first place, organisms on surfaces are almost invariably protected by soiling materials such as fats, serum, pus or other organic material and, secondly, the organisms may be dried on the surface. A disinfectant, therefore, does not have such ready access to the contaminants as it does when they are in suspension. And the problem is further complicated by the fact that some surfaces are porous and so may further protect the organisms. Because of these differences, tests with organisms in suspension do not necessarily give results which can be translated immediately to surface treatments and most workers agree that a degree of detergency is essential in a disinfectant formulation.

Koch and Kronig and Paul were ahead of their time when they put forward their silk thread and garnet tests before the end of the last century; many years elapsed before further developments took place. Several methods were then proposed involving organisms dried on cover slips, glass cylinders (the forerunner of the A.O.A.C. Use-dilution Confirmation Test) and small squares of materials such as glass, metal, rubber or linoleum, and various recovery techniques were proposed (see Sykes, 1965).

One of the problems in laboratory testing is to persuade organisms to remain viable during the drying period, where the death rate can be such that as little as 0.1% of the initial population, and sometimes much less, survives. To meet this, and to simulate natural conditions, a variety of soiling materials has been suggested, ranging from simple nutrient broth to "soups" and a mixture of raw egg, milk, butter and other fats. Some of these present formidable obstacles and are, in fact, excessive, even for laboratory work. No disinfectant should be expected to treat effectively a dirty surface: it should be a cardinal rule that a surface should be reasonably physically clean before disinfection is attempted.

(b) *Dairy disinfection. The Hoy can test.* Glass slides, rubber strips and metal strips infected with suitable organisms dried on with whole or skim milk have formed the basis of earlier tests in assessing disinfectants for dairy purposes, but the major tests now used in this country are the Hoy can test and the concomitant Lisboa tube test. The former has been the official method for some years and the latter is becoming acknowledged for the same purpose. The feature of all dairy disinfection is rapidity of action, and all tests follow this principle.

Following earlier work with metal trays Hoy and Clegg (1953) developed the method now known as the Hoy can test which uses 10 gallon milk churns. The churns are first soiled with a microbiologically poor quality milk over the whole of the inside surface, allowed to drain and left inverted overnight. Next morning the cans are treated with the disinfectant solution in a prescribed manner and after a fixed short period the action stopped by adding a neutralizing solution, e.g. sodium thiosulphate for chlorine releasing compounds. After draining off the residual solution the insides of the cans are rinsed and squeegeed with $\frac{1}{4}$ strength Ringers solution. Suitable portions of this rinsing are then plated to determine the surviving bacteria.

The Lisboa tube test. Because of the awkwardness in handling large milk churns, Lisboa (1958) developed a method using short lengths of stainless steel tubing, and results from this compare well with those from the Hoy can test. Each length of steel tubing, 33 cm × 3·25 cm internal diam and closed at both ends with removable rubber caps, is soiled with raw milk in the same way as in the can test, sealed and incubated at 30°C for 4 h. Fifty ml of disinfectant solution is added and the tube recapped and rolled horizontally so that the inside is treated uniformly. After exactly 1 min the disinfectant is quickly drained, 25 ml of neutralizing solution added, the residual milk film removed with a close fitting plunger squeegee and the resultant solution plated for surviving organisms.

(c) *Other tests.* Sundry other tests applicable to situations in the food and catering industries have been proposed from time to time. Some use, for example, empty beer glasses, soiled cups and plates and cutlery, whilst others concentrate on the detergent aspect of the problem.

There is still the need for a suitable test for assessing surface disinfection properties in general and the Lisboa tube type of test seems to be the most promising. Short disinfection periods of up to 5 min seem to be the most appropriate, but the problem of a suitable, and not over-protective, soiling material remains unsolved.

(d) *Spray disinfection.* Although not looked on favourably by many workers, spray disinfection, or "fogging", for treating surfaces has found a place in some quarters and the A.O.A.C. (1960) produced a tentative procedure. It covers the application of hand and pressurized spray methods in general, and these are certainly being used more and more extensively. Briefly, the method consists of inoculating 1 in. squares of glass or metal with either a *S. aureus* or *Salm. choleraesuis*, the same strains as used in other A.O.A.C. tests, allowing them to dry in the air and then spraying on the disinfectant as instructed. After 10 min contact the squares are cultured in 20 ml of medium containing, as necessary, an appropriate inactivating agent.

(e) *Textiles disinfection*. Because of the special position of textiles in their importance as bacterial carriers in hospital cross infection, and because of the susceptibilities of cotton and wool to particular forms of microbial spoilage (mainly fungal), techniques have been devised using such materials as test surfaces. Various techniques have been used in the past to recover organisms from sheets and blankets by, for instance, beating them in the horizontal position with agar plates a few inches below, but this is awkward and highly empirical. In recent years it has been improved somewhat and a modification of a later procedure was used by Dickinson, Wagg and Litchfield (1970) in their detailed investigation of the bactericidal action of formaldehyde on blankets. In this a piece of blanket was tied over a squat metal cylinder 10 cm in diam and a weight of 120 g dropped 13 cm on the blanket. This released organisms and fibres, all of which were allowed to settle on an agar plate in the bottom of the cylinder. The counts gave a measure of the organisms likely to be shed from the blanket before and after treatment, and so a measure of the efficacy of the treatment.

To get a total, rather than a release, count the same authors used a method in which a 1 ft² piece of blanket was stitched to the centre of an ordinary blanket. To assess the bacterial count and thus the value of a treatment the square was removed after an appropriate handling and from it a $\frac{1}{4}$ in² section cut out. This was shaken in a 1 oz vial containing 5 ml of broth with 0·1% of Tween 80 plus a small amount of egg yolk (1 yolk in 500 ml of medium) on an homogenizer and then plated. The counts so obtained were up to 10³-fold greater than by the percussion method, and more consistent.

The German Society for Hygiene (1959) employs pieces of linen each ca. 1 cm². About 100 pieces are placed in 15 ml of a broth culture of *S. aureus*, *E. coli*, *Ps. aeruginosa* or *M. tuberculosis*. After 15 min they are transferred to a dish and covered with the disinfectant solution. At suitable time intervals several of the squares are rinsed in an inactivating solution, then transferred to a tube of liquid culture medium and incubated.

5. *Oral disinfectant testing*

Of all the assessments of *in vivo* applications of disinfectants oral treatments are the most intractable. This arises because of the wide variation in the bacterial count in the mouth, even from minute to minute, the variable flow rate of saliva and the extreme fluctuations in its pH value. Several workers have described rinse tests for assessing mouthwashes, gargles and antiseptic sweets and lozenges, and have claimed reductions in the bacterial content of the mouth, albeit temporary ones for an hour or so, and their reported findings are difficult to repeat. Buccal swabs and scrapings have also been tried in place of mouth rinsings, but with no more success.

Because of these difficulties and inconsistencies resort is most frequently made to *in vitro* assessments, but using saliva as the vehicle and treatments at 37°C, to simulate more closely natural conditions. Two types of test have been employed; one applicable to mouthwashes and gargles, where the contact time is short, the other to antiseptic sweets and lozenges, where the treatment time is longer. The saliva can be natural, freshly collected material or a synthetic one made as follows: triturate 0·1% g of gastric mucin in a few drops of a solution containing 0·12% of sodium bicarbonate and 0·18% of potassium dihydrogen phosphate, continue adding more solution with mixing until the volume is 100 ml, sterilize in the autoclave. In either instance an inoculum is added, usually of a mixture of cultures of *S. aureus* and *Streptococcus pyogenes* as representing typical mouth infecting organisms— Gram-negative bacteria are of no real significance in this context. In the test for the more prolonged antiseptic activity one lozenge or sweet is dissolved in 5 ml of water at 37°C and 0·5 ml of saliva-organisms mixture added. At intervals between 1 and 10 min one loopful is subcultured into broth or plated with serum agar—the former giving an "all-or-nothing" response and the latter a quantitative assessment.

In the more rapid test suitable dilutions of the mouthwash or gargle are used and the contact time reduced to 10, 20 or 30 sec, thus simulating the short period treatment associated with a gargle.

6. *Air disinfection*

The treatment of airborne organisms presents an entirely different problem from that associated with organisms in liquid suspension or on surfaces. The chemistry and biology of the lethal process is the same but the mechanism of how the disinfectant gains access is quite different and the agents effective are, in the most part, quite different. Organisms released into the air from a dry surface are rarely airborne alone; they are usually attached to a larger particle of dust or a fibre. Similarly organisms discharged or sprayed from a liquid are rarely sprayed singly; they occur more often in clumps, and when discharged dry rapidly, in a fraction of a second, with the dried residue of the suspending fluid around them. Both act in a protective role to the organism and reduce the efficacy of the bactericide.

To exert its lethal action a bactericidal aerosol must act either by impingement, that is, a droplet of the bactericide must come in contact with the bacterial particle, or, if volatile, must dissolve in the residual moisture surrounding the particle, and it is an interesting fact that most of the effective aerosols are volatile or have a relatively high vapour pressure. They are also active at much lower concentrations in the air than in normal aqueous solution. Thus some workers have reported that one part of sodium hypochlorite in 40×10^6 parts (w/v) of air, and others 2·1 ml of a 1% solution in 1000 ft³ of air, will give substantial kills of airborne bacteria in a few minutes; similar activity has been reported for 0·1–0·4 mg of propylene glycol/ft³ of air (see Sykes, 1965).

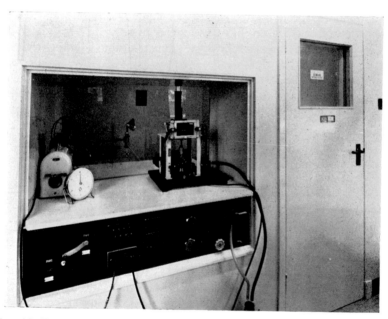

Fig. 13. Evaluation of bactericidal aerosols. Test room fitted with slit sampler and containing fan and Collison inhaler for dispersal of test organism.

Because of the particle dissolution phenomenon, humidity is another controlling factor, high RH values in general being more predisposing to activity than low ones: 60–70% RH is the accepted optimum for most compounds, notable exceptions being the glycols, ethylene or propylene, which, because they are only active at high concentrations in water, are more effective at lower RH values around 30%.

Other aspects on which much work has been done are the methods of dispersal of organisms from aqueous suspension or from dusts, the stabilities of the resultant bacterial aerosol and methods of sampling. Several books have been published and symposia held in recent years on the general subject of aerobiology.

For some years there has been a British Standard for evaluating bactericidal aerosols (British Standards Institution, 1956). In this the te.t is required to be made in a closed room of approximately cubic dimensions and 1000 ft³ capacity. Fans ensure the uniform mixing of the dispersed bacteria and bactericide. The test organism is a *Staphylococcus albus*, chosen because it is a nonclumping strain. It is sprayed into the air by means of a Collison inhaler which gives droplets such that the resultant aerosol consists principally of single cell particles, and its survival, in terms of the total number of viable cells/unit volume of air, determined at preselected intervals by means of a slit sampler (Fig. 13).

The procedure is best described in tabular form as follows:

Period (min)	Slit sampling or other procedure	
0	Disperse the bacterial aerosol	
3–3½	Take a first control plate (30 sec)	(i)
4½–5	Take a second control plate (30 sec)	(ii)
5½–6	Disperse the bactericide	
7½–8	Take the first test plate (30 sec)	(iii)
9½–11½	Take the second test plate (2 min)	(iv)
12½–13	Take a check plate (30 sec)	(v)

Counts (i) and (ii) should be similar if the aerosol is stable. Using the mean of these two, the % survivals at 1½–2 min and 4–6 min exposure can be calculated, and so the efficacy of the bactericide determined.

A bactericide is considered to be satisfactory if it effects a kill of not less than 85% in the 4–6 min period when the RH is 55–65% and the temperature 20°C. A standard reference germicide, cyclopetanol-1-carboxylic acid, is recommended as a means of checking the test conditions.

This test is made under static atmospheric conditions which are never met in practice. In more natural conditions there is always some degree of ventilation (thus reducing the aerial concentration of the germicide) and the personnel present are constantly shedding organisms (thus giving a continuous source of contamination). Both of these significantly reduce the apparent efficacy of a germicide. To overcome this a continuous flow-disinfection system was devised (Kethley, Fincher and Crown, 1956), but little seems to have been heard of it.

IV. PRESERVATION AND PRESERVATIVES

A. The problem

1. *The methods of preservation*

Preservation against microbial spoilage is of increasing importance in both the health and economy of the community. It involves not only foods and drinks but also a wide range of pharmaceutical and cosmetic products as well as industrial materials such as paints, wood, textiles, paper and even concrete.

Preservation can be brought about by physical or chemical means, and its purpose is to ensure that during the normal storage life of a product it remains undamaged and unspoilt by microbial action. The sources of microbial contaminants are from the individual constituents of a product, from the equipment and containers used for its manufacture, from the operators, from the air and from the consumer whilst the product is in use. A preservative has two functions therefore: (1) in the short term, to deal with the contaminants entering the product during manufacture; (2) in the longer term, to protect the product during use. Reasonable stability in the preservative system is thus a prerequisite. Under (1) it should be axiomatic that a preservative must never be used to cover bad housekeeping; high standards of cleanliness and of good manufacturing procedures are fundamental in minimizing the level of contamination.

As a general rule chemical preservatives are more suitable for treating pharmaceutical and cosmetic preparations and for industrial materials: physical methods are more appropriate and acceptable in handling foods. The latter methods involve establishing a physical or physicochemical environment which is outside the limits of tolerance of the microbial cell, and is achieved by drying, heat (pasteurization), freezing, osmosis or irradiation, each of which either reduces the microbial content of a product to a relatively safe level or prevents its proliferation, thereby extending the safe storage period of the product.

2. *Conditions for microbial growth and spoilage*

Bacterial and fungal growth can only take place if adequate nutrients, including water, are present and other suitable cultural conditions are established. Only micro quantities of nutrients are required to allow of substantial growth, so that microbial spoilage is probably more common than might at first be expected and occurs in unexpected places and conditions. For example, waterborne bacteria can grow readily in magnesium hydroxide suspensions even though the pH value is ca. 10 and lens mountings in optical instruments can carry enough materials to support substantial mould growth; likewise large bacterial populations can develop

in a few days in distilled and purified waters and moulds can adapt themselves to grow in all sorts of conditions, including standard solutions of
mineral acids and even in a Benedict's solution containing 1·8% of copper
sulphate!

Not all organisms will grow in every situation: some are more exacting
and specific in their nutritional and other cultural requirements and so
inevitably some selection, sometimes as far as a single genus or species,
takes place, rendering some types of preparation subject to certain types of
contamination and others to different types.

The organisms involved, depending on the nature of the materials and
other conditions, may be not only pathogens or potential pathogens but
also those which cause spoilage by simple disfiguration or by altering the
characteristics of a product by fermentation, production of off-odours and
tastes, breaking an emulsion, etc. Pathogens include *S. aureus*, the pyogenic
streptococci, the salmonellae and others of the Enterobacteriaceae including
Escherichia coli type 1, *Pseudomonas aeruginosa* and the clostridia, notably
Clostridium botulinum, *Cl. tetani*, and *Cl. welchii* (*Cl. perfringens*): the non-
pathogens include waterborne bacteria of the *Pseudomonas–Achromobacter*
type, spore-forming aerobic bacteria and other Gram-positive organisms,
and moulds including *Cladosporium*, *Aspergillus*, and *Penicillium* spp.
Viruses are of no concern, except in some vaccines and antisera.

3. *The choice of preservative*

Compared with the products used in the general fields of disinfection
and antiseptics the choice of compounds as preservatives in foods, drinks,
pharmaceutical and cosmetic preparations is restricted—with industrial
products the choice is wider—the first requirements, apart from their
antimicrobial activity, being that at the concentrations employed they shall
be tasteless, odourless, nontoxic, nonirritant, compatible with the constituents of the preparation, stable and usually colourless. These are properties which can only be assessed chemically, physically or physiologically.
Microbiologically, the agent must be able to control the growth of bacteria
and moulds in a product—but what is meant by "control" in this context?

The safest way of preventing spoilage or damage is either to prepare the
product sterile (an expensive procedure and confined in practice to products
for parenteral and certain other specific uses) or to ensure that the formulation is such that the contaminating organisms in it are actually killed. This
is an ideal rarely attainable—bacterial spores, for example can only be
dealt with effectively by a limited number of chemicals—and so a compromise must be sought: nevertheless, some degree of lethal activity is
essential. In this context Sykes (1958) drew attention to the narrowness of

the margin between the effective lethal concentration and the minimum inhibitory concentration of many antibacterial agents.

As with other antibacterial formulations it is essential to assess the performance characteristics of a preservative in the preparation in which it is to be used. In this respect, susceptibility of the preparation to a particular group of organisms, pH value and the presence of suspended substances each may be important, and in creams, ointments and other products containing oils the oil/water ratio and the partitioning of the preservative between the oil and water phases are additional points to be considered: in anhydrous ointments and other dry preparations a preservative is ineffective and probably unnecessary.

B. Methods of testing

1. *Parenteral preparations*

In pharmacy and medicine, phenol at a concentration of 0.5% has been the traditional preservative and the tendency has been to use this as the basis for comparing and judging the efficacies of other compounds. Such a comparison, however, has its disadvantages, not the least of which are the different relative activities of each compound against different types of organism. There is, in fact, little point in making comparisons of this type, each agent should stand on its own merits.

In a report by a Ministry of Health Subcommittee (1957) the suggestion was made that an antimicrobial preservative used in a parenteral preparation (injections of procaine penicillin were being studied) should be effectively lethal to a wide range of bacteria, and it was proposed that selected strains of *S. aureus* and *Ps. aeruginosa* should be the organisms tested as representing the most resistant of the Gram-positive and Gram-negative bacteria. By "effectively lethal" is implied the virtual sterilization of a moderately heavy inoculum of the test organism within 24 h. Sykes (1958) supported this, but also considered that moulds should be included because often they will grow where bacteria do not. More recently the Pharmacopoeia of the United States (XVIIIth edition) (1970) introduced an "Antimicrobial Agents—Effectiveness" test in which an agent is considered to be satisfactory for use in parenteral and ophthalmic preparations "if there is no significant increase in the number of *Candida albicans* or *Aspergillus niger* organisms [two of the selected test organisms], and if the number of viable vegetative micro-organisms [*E. coli*, *Ps. aeruginosa* and *S. aureus*] is reduced to not more than 0.1% of the initial number and remains below that level for a 7-day period within the 28-day test period".

The method specified is first to prepare cultures of the five above named organisms (specific ATCC strains are quoted) on agar, incubating the bacterial cultures at 37°C for 18–24 h, the *C. albicans* at 25°C for 48 h and the *Asp. niger* at 25°C for 1 week. The resultant growths are harvested individually in saline (0·05% of Tween 80 is added to wet the spores of the mould cultures) and the concentration of cells adjusted to 50×10^6/ml. Then 5 tubes, each containing 20 ml of the test solution, are inoculated with 0·1 ml of one of the inoculum suspensions, mixed and incubated at 30–32°C. On at least two occasions, not more than 7 days apart, within a 28-days period the test solutions are examined for changes in appearance and plate counts are made to determine the numbers of surviving cells. The medium recommended for preparing the initial cultures and making the plate counts is a soya-casein digest agar.

In our laboratories we look for a much more rapid kill than is indicated in the U.S.P. test. This is because in multidose containers in particular, from which several injection doses are removed at intervals, demonstrable disinfection should have occurred in the interval between doses, so that a contamination introduced inadvertently on removing one dose is effectively dealt with by the time the next dose is removed. The following is our procedure. Five groups of cultures are used: (i) up to 10 strains of *S. aureus*, some of which are preferably freshly isolated; (ii) a similar selection of strains of *Ps. aeruginosa*; (iii) a similar selection of strains of *E. coli* type 1; (iv) a mixture of bacterial cultures isolated from various contamination situations; (v) a similar mixture of mould cultures. The cultures are grown separately on agar slopes, suspended in a diluent such as $\frac{1}{4}$ strength Ringer's solution or 0·1% peptone water (for the moulds ca. 0·1% of Tween 80 is added to wet the spores), mixed and added to a vial or tube of the test solution to give a concentration of ca. 10^5 cells/ml. The tests are kept at room temperature, or 25°C, and survivor counts determined after $\frac{1}{2}$, 1, 2, 6 and 24 h. An effective preservative will virtually sterilize the bacterial cultures and substantially reduce the mould count in 6 h. With some ophthalmic preparations, in which a more rapid action is desirable, we prefer to see the lethal period reduced to 30 min.

2. *Other pharmaceutical and cosmetic preparations*

In preparations other than those for parenteral use it is not normally necessary to specify a kill within a few hours, and subculture periods more in line with those of the U.S.P. tests are suitable. In our tests we use three mixed culture preparations: (i) a mixture of up to 40 cultures of standard strains and organisms isolated from various contamination situations, especially those from products similar to the one under test; (ii) similar mixed yeast cultures; (iii) similar mixed mould cultures. The procedure follows that described above except that the samples are kept in their intended final container and counts are made immediately before inoculation (to determine the natural level of contamination), immediately after inoculation and then after 1 day, 1 week, 2 weeks and so on up to 10 or 12 weeks. Adequate preservation is considered to have been achieved if (a) the counts have been reduced from the initial 10^5/ml to less than 10 bacteria/ml, (b) the mould count has been substantially reduced, and (c) there is no subsequent increase in either count. As a further check, and also to assess the stability of the preservative, the sample is reinoculated when two successive zero counts at weekly intervals have been obtained, and the procedure repeated.

The foregoing is applicable to all liquid preparations. With creams, the three lots of cultures are mixed evenly into the test material by stirring; in addition a light inoculum of the moulds is spread on the surface of the sample, the container is closed and growth noted macroscopically after several weeks. Alternatively the samples are exposed for several weeks in a chamber (a fish tank is suitable) kept at high humidity and containing pieces of inoculated wood and cardboard as a source of mould spores. In such conditions the surface of the test material is exposed to contamination by the spores which then have the maximum opportunity to develop, if the formulation allows them to do so.

3. Foods and soft drinks

The preservatives which may be added to foods and drinks are severely restricted by law in Great Britain. In meats, sodium nitrate is permitted in limited amounts, but only in hams, bacon and cooked meats: in fruit juices and squashes, only benzoic acid and its sodium salt, sulphur dioxide and certain sulphites may be used, but again the amounts are restricted to 800 ppm of benzoic acid and 350 ppm of sulphur dioxide (or sulphite) either singly or in combination. Sulphur dioxide may also be used in certain dried fruits, dehydrated foods, preserves and wines, the maximum amount varying with the type of product. In other countries other additives are permitted such as sorbic acid, propionic acid, sodium diacetate and certain antibiotics.

The examination of foods for spoilage and preservation is a complex and specialized procedure and those concerned should consult the many textbooks and journals, including the *Journal of Applied Bacteriology*, which deal with the subject.

Fruit squashes are now sold mainly as concentrates containing a high proportion of sugars and so have high osmotic pressures. The same applies to malt preparations. Because of this they are immune from bacterial contamination but can be susceptible to osmophilic yeasts, and sometimes moulds. In making tests on this sort of product the time factor is important because growth may take several weeks to become evident.

4. *Industrial materials*

As already stated the types of product included under this heading are diverse and include textiles, wood, paper and board, plastics (by virtue of the added plasticizers), masonry, leather and paints. Most are subject to fungal, rather than bacterial, attack and the methods used for assessing the efficacies of preservatives vary according to the type of material to be tested and the whim of the worker. All are empirical: all are carried out *in situ*: one or two merit a mention.

To assess the value of a preservative on paper, wood and similar uneven surfaced materials Howard (1954) modified a method devised earlier by Thom in which a piece of the test material, 5 cm square, is placed on the surface of a Petri plate containing Czapek's agar, pH 4·2, and the whole area sprayed with a suspension of mould spores. After 14 days at 25°C the mould growth on the material is assessed and from it the relative efficacy of the preservative determined. A similar method could be used with textiles.

Modern paints present a two-fold problem: first is the protection of the water-based products against bacterial invasion; secondly is the protection of the finished surface against (mainly) mould spoilage. Preservation in the first situation can be assessed by the test in liquids described above (Section 2), and using appropriately selected organisms. Preservation of a finished surface is a different matter. As indicated in the opening paragraphs of this Chapter, whilst the surface remains dry there is no problem: moisture is essential for growth. Equally essential is a measurable solubility in water of the preservative in the paint film if it is to exert its action. Because of this, one method for determining preservative power is by an adaptation of the agar diffusion technique (see Section I.A, 1(a)). In this, strips of filter paper, ca. 1 × 5 cm, are dipped in the paint, drained, allowed to dry and laid on the surface of an agar plate. Cultures of the chosen test organisms are streaked across the agar and paper at right angles to the length of the strip and, after incubation, the extent of inhibition observed (a) over the surface of the paper itself and (b) around the edges of the paper.

A field type test, devised by Galloway (1954), uses wood blocks or compressed wood pulp, and an adaptation has been used in our laboratories. In this, panels of hardboard, 4 × 12 in., are sprayed with a suspension of bacteria and mould spores in broth and allowed to dry. Each panel is then divided into three sections and two coats of paint applied to one of the sections. After the paint has dried and hardened the panel is reinfected and incubated in a sealed chamber kept at high humidity. Incubation for ca. 1 month at 22–25°C is sufficient to differentiate between a paint with good preservation properties and an untreated one.

5. Conclusion

In the foregoing Sections, straightforward examples of the type of test that can be applied to the assessment of preservatives in or on different products and materials are given. All are essentially *in situ* tests in which the conditions established are as near as possible to those found in practice. Obviously many variations, and even more novel techniques, could be devised, appropriate to different situations, but they are all variations on a standard theme. One word of caution—even though a preservative may be effective against the range of chosen test organisms it is not impossible that more resistant types may be found which are not suppressed to the same degree and may even be able to flourish. There is no general recipe to overcome this.

V. *IN VIVO* TESTING

A. General principles

The development of methods for the assessment of antibacterial activity *in vivo* has been essential to the discovery of agents for treatment of infectious diseases in man and his animals. *In vitro* techniques are applicable to disinfectants rather than to drugs and it was only after Domagk and his colleagues at I.G. Farbenindustrie used mice infected systemically with a haemolytic streptococcus to assess the activity of their new compounds that the first agents for the therapy of bacterial infections, the sulphonamides, were found. *In vitro* techniques take no account of toxicity or of the dynamic interplay between compound, host and pathogen *in vivo*. However activity against experimental infections in animals is not necessarily a prediction of activity in man; it is to be regarded rather as a step in the right direction, an indication that a chosen compound is active in the presence of "flesh and blood". The existing evidence indicates that all the known chemo-therapeutic antibacterials exert their effect entirely by inhibiting the infect-ing organism rather than by stimulating host defence mechanisms. As it can be shown that they reach the site of infection at concentrations which are inhibitory *in vitro* this is not surprising. Nevertheless assessment against an experimental infection measures not only the ability of a compound to inhibit an organism in the blood or tissues but its ability to reach that site in adequate concentration. This involves the pharmacological properties of the compound and these will vary between animal species.

Some work has been conducted in infected eggs or tissue cultures but these lack realism and most evaluations have been made with small rodents infected systemically to give a severe, fulminating infection. This often culminates in the death of the host within a few days and the disease bears little resemblance to that caused by the same organism in man. The activity of the compound under assessment is measured by its ability to prevent or delay death in a group of treated animals. Other models, sometimes more realistic, utilize the localization of an infection in, or on, an animal. In this type of test the effect of a compound is judged by its eradication or limitation of the infection. When a suitable infective organism is not available or, sometimes, to obtain detailed information about the mode of action of an antibacterial agent *in vivo* an indirect assessment is made. The compound is administered to an animal and the presence of antibacterial activity in tissue or body fluid sought by *in vitro* techniques using a relevant organism. A picture of the pharmacological properties of the compound can be obtained in this way.

Because of the number of variables involved it is difficult to regard the *in vivo* test as a simple extension of the *in vitro* one. Comparison of the activities of compounds between laboratories is rendered difficult because of differences in the sensitivities and infectivities of bacterial strains, differences in host animals and differences in dose regimens. It is essential to standardize conditions in any one laboratory as closely as possible but even so reproducible results in different experiments are often difficult to obtain. It is important that groups of animals be used, rather than single ones, and that the numbers in each group are large enough to ensure the desired degree of precision when computing results. Even when all these precautions are observed any assessment indicates only the activity of a compound against the infecting organism under the conditions employed.

It must be emphasized that the handling of experimental animals infected with bacteria is a hazardous undertaking, the degree of danger depending on the organism used. It is therefore important that proper precautions are enforced and that a sense of discipline and awareness is encouraged in the infected animal rooms. Safety and discipline in this context are discussed by Darlow, this Series, Vol. 1.

B. *In vivo/in vitro* assessments

It is sometimes convenient, and even illuminating, to assess the antibacterial activity of a compound by administering it to a suitable animal, obtaining samples of body fluids or tissues at intervals and measuring the antibacterial activity present. In this way a rapid impression of the pharmacodynamics of a compound or its antibacterial metabolites can be obtained. For instance Brownlee *et al.* (1948) took heart blood from dosed guinea-pigs and assessed its tuberculostatic activity *in vitro* in order to assess what range of doses may be required to show effects in the much more time consuming animal protection tests. While the technique is no substitute for tests in infected animals it is economical in animals and time. It can be used both to examine rapidly the *in vivo* potential of compounds exhibiting antibacterial activity *in vitro* and to seek reasons for the inactivity of compounds expected, from their *in vitro* properties, to be effective.

The mouse is a convenient host animal for preliminary experiments. It is possible to kill mice at intervals after dosing them and prepare whole body homogenates with a Kenwood mixer in a few minutes. This homogenate, after suitable cooling and dilution, can be assayed for antibacterial activity by an agar diffusion method. In this way an idea of the rate of inactivation of the compound in the body can be obtained. It is important to remember, however, that during the period of incubation of the assay plate inactivation can continue in the homogenate. This can be avoided to some extent if the plate is first refrigerated for a few hours to retard destruction of the active component and to allow it to diffuse away into the agar.

A rapid idea of the distribution of an antibacterial compound in the body can be obtained by killing the mice at intervals after dosing, rapidly dissecting out the organs and taking samples from these, as well as the heart blood and urine, by syringe. These again can be assessed by a surface diffusion method, the sizes of the zones of inhibition indicating the rate of absorption and excretion of a compound, and its ability to concentrate at a particular site. Again it is advisable to allow time for the compound to diffuse from the tissue sample before incubating the plate.

More accurate assessments are made by taking samples of body fluids and of tissues from groups of animals sacrificed at intervals after dosing. Microbiological assay by a diffusion technique is then employed to assay the antibacterial activity present in the fluids and in aqueous extracts of the tissues. Standard curves are constructed by diluting the test compound in serum, urine and extracts of tissues obtained from normal animals. The efficiency of the extraction procedure is first checked by adding known amounts of the compound under assessment to samples of normal tissues, carrying out extractions and assays, and then calculating the recovery obtained.

The penetration of antibacterial agents into foci of infection can also be measured in this way. For instance, samples of pus can be obtained from experimental staphylococcal abscesses induced by the injection of a virulent strain of *S. aureus* into the skin of the rabbit (Bough *et al.*, 1971).

C. Systemic infections

1. *The host animal*

Mice are the most frequently employed animals because of their cheapness and relatively small size. It is important to choose strains which exhibit genetic homogeneity and are not highly resistant to the infecting organism.

Differences in susceptibility exist between the commercially available inbred strains but these are of no great practical importance. It is important, however, that groups of the same sex are employed where comparative assessments of compounds are undertaken. It has been shown that the female mouse is more resistant to several different infective organisms, so that when infections in male and female animals are treated with sub-optimal doses of chemotherapeutic agents considerable differences in effect can exist (Wheater and Hurst, 1961). The age of the mouse also influences its response to infection, the younger the animal the greater is its susceptibility. In practice, groups of animals weighing 18–22 g are generally employed. Heavier animals involve the use of larger amounts of often valuable compounds and so are avoided on this account.

Larger species of animals are generally reserved for special evaluations. For instance rats are susceptible to pneumococci and guinea-pigs to clostridia, brucellae, corynebacteria and mycobacteria. Rabbits may also be used for evaluations with mycobacteria and pathogenic Gram-positive cocci.

2. Infecting species

A wide range of bacteria which infect mice systemically is available for the assessment of antimicrobial activity *in vivo*. However, not all species pathogenic to man can be established easily in animals and different strains of suitable pathogenic species often vary in their virulence.

Apparent resistance to infection by the chosen host is often only relative and can frequently be overcome by increasing the number of bacteria injected. But if very large doses are given death may be caused by toxaemia resulting from the liberation of endotoxins rather than by the effect of bacterial multiplication within the host. When the virulence of a test organism is low, experimental infection can frequently best be obtained by injecting organisms by the intraperitoneal route after suspending them in gastric mucin. Nungester *et al.* (1932) studied the effect of various substances on phagocytosis of the pneumococcus in the peritoneal cavity of the mouse and were the first to observe that the lethal number of bacteria was smaller when suspended in mucin than in saline. It is now established that the lethal dose can be greatly diminished; for instance the infectivity of *Salmonella typhi* for mice can be enhanced at least 10^6-fold (Rake, 1935). The early work, summarized by Olitzki (1948), did not explain the mechanism of action of mucin and it is still not clearly established whether mucin interferes with host defence mechanisms or whether it enhances the virulence of the infecting bacteria, or does both. It is interesting to note that mucin can also promote the infectivity of bacteria for grasshoppers as well as mice (Stephens, 1959).

The essential component of mucin has not yet been identified. The earliest work was carried out with dried gastric mucin sterilized by heat or with alcohol. Now special commercial preparations are available and have found widespread acceptance. Different batches can vary in their activity and it is wise to check them *in vivo* with an organism of known virulence. For use the mucin is coarsely ground in a mortar and homogenized with distilled water in a mechanical homogenizer; the suspension is sterilized by autoclaving at 121°C for 15 min and then adjusted to pH 7·3. Each preparation should be checked for sterility: if organisms are present the batch should be discarded, as repeated heat sterilization reduces the infection enhancing activity. Suspensions retain their activity for at least one month in the refrigerator.

A list of organisms which are commonly used to assess antibacterial activity in mice is given in Table I.

TABLE I

Some species of bacteria used to infect mice by the intraperitoneal route

Usually with mucin	Usually without mucin
Staphylococcus aureus	Streptococcus pyogenes
Streptococcus faecalis	Streptococcus pneumoniae
Listeria monocytogenes	Salmonella schottmuelleri
Haemophilus influenzae	Salmonella enteritidis
Pseudomonas aeruginosa	Salmonella typhimurium
Escherichia coli	Pasteurella pestis
Klebsiella aerogenes	Pasteurella multocida
Shigella dysenteriae	
Shigella sonnei	
Shigella flexneri	
Proteus vulgaris	
Proteus morganii	
Proteus mirabilis	
Salmonella typhi	
Pasteurella haemolytica	

3. *Preparation of infecting doses*

The infectivity of the culture to be employed has first to be established by injecting graded numbers of cells into groups of mice. The Median Lethal Dose (MLD), that is, the number of organisms which kills 50% of the animals in the group, is then determined by plotting numbers of dead animals against log number of organisms in the dose, or by using a suitable statistical method such as that described by Litchfield and

Wilcoxon (1949). If the virulence of the desired strains is too low it can frequently be enhanced by recovering the organism from the heart blood, spleen, or kidney of an infected animal and, after ensuring the identity of the isolate, repeating the passage a number of times, using mucin if necessary.

When making passages with organisms in mucin it is essential to identify the re-isolated organism because the injection of mucin to mice can stimulate a latent bacterial infection. Engley (1954) recovered *Ps. aeruginosa* from the heart blood of mice which had died after being injected with sterile mucin, and coliform bacteria have been recovered in our laboratories from the heart blood of sick mice after they had been infected experimentally with *Salm. typhi* in mucin.

Once satisfactory virulence has been obtained the culture can be freeze dried for future use.

Considerable variation in the MLD can occur between experiments even when bacteria from the same freeze dried batch are used. For instance although great care had been exercised to standardize the infecting dose the MLD of *Pasteurella pestis* for mice varied 50-fold in different experiments (Buckland and Treadwell, 1957).

In practice, to ensure that the majority of mice in a control, untreated group will succumb to an infection it is customary in chemotherapeutic investigations to use 100–1000 MLD. To prepare this an 18–24 h broth culture or growth washed from an agar slope is diluted to the desired concentration and assessed turbidimetrically or by opacity tube. A further dilution is made with mucin where necessary so that the infecting dose is suspended in 0·5 ml of a 4·5% mucin suspension. This is injected as soon as possible. Diluents other than broth or 0·1% peptone water should be avoided as it is well established that water and saline as suspending agents can be bactericidal. It is also important, of course, to ensure that cell division does not occur before injection.

4. *The infecting route*

The intraperitoneal route is generally employed to inject suspensions of organisms. This has to be done with care because even experienced workers have found that up to 14% of their injections deposited material outside the peritoneal cavity (Arioli and Rossi, 1970).

Where the intravenous route is employed, as in experiments with *Mycobacterium tuberculosis* in mice (see Youmans and Youmans, 1964), care must be exerted to ensure that the suspension of organisms is uniform and free from large clumps and particles. The suspending vehicle must be isotonic with blood cells and free from irritant material. The experienced worker, with an assistant and a suitable holder, can rapidly inject the tail veins of conscious mice, warming the tails to 45°C for a few minutes by immersion in a water bath to cause vasodilation (see Kingham, this Series, Vol. 5A). When a dangerous organism is employed, however, it is quicker and safer to anaesthetize the mice. We have employed an intraperitoneal dose of 0·15 ml of a 1% (w/v) solution of sodium pentobarbitone in 10% ethanol to anaesthetize 20 g mice satisfactorily.

It is possible, and sometimes advantageous, to employ the oral route with organisms such as *Salm. typhimurium*, which normally invade from the intestine.

5. *Dosing schedules*

The timing of the treatment in relation to that of the infecting dose can be critical. A compound can be administered before the infection, that is, in the prophylactic or protecting sense, or afterwards, in the therapeutic sense. In assessing the activity of a compound its pharmacology must be considered and allowance made for its excretion rate or for its slow absorption from the site of infection. Thus, although it is intrinsically much less active, a single dose of aureomycin is more effective than a similar dose of penicillin against a systemic pneumococcal infection of mice because the latter is absorbed more quickly and excreted rapidly (Klein *et al.*, 1950).

If given by the same route as the infective organism, e.g. intraperitoneally, sufficient time must elapse between giving the infection and giving the agent so that the organisms will have escaped from the peritoneal cavity, otherwise activity against the organism in the peritoneal cavity rather than against the general infection will be measured. Stimulation of host defences may also occur and this can confuse an assessment of chemotherapeutic activity. Thus, intraperitoneal injection of saline to mice before infecting them by the same route with *S. aureus* in mucin has protected some of the animals due, presumably, to stimulation of phagocytosis in the peritoneal cavity (Hale and Spooner, unpublished). Dineen (1961) has studied the timing of doses in staphylococcal infections in mice and found that a single dose of the drug 1 h after challenge was as effective as multiple doses. Variation in the duration of the interval greatly reduced the efficacy of the dose. Usually a single dose given within a few hours of infection is as effective as multiple doses spread over the ensuing days, and is economical in labour and materials. Single doses are also less toxic than are multiple ones, and they eliminate the stress to the animals, and the hazard to the operator, of frequent handling.

A variety of routes is available for the administration of the agent to be assessed. That chosen depends on a knowledge of the pharmacology and solubility of the compound. If of low solubility the compound is best given as a suspension in sterile 5% gum acacia or 1% carboxymethylcellulose solution by the intraperitoneal or oral routes. In our experience it is not necessary to sterilize the agent itself, but it should be weighed cleanly and dissolved or suspended in a sterile vehicle.

For long term therapeutic experiments a compound may be administered in the diet. A healthy 20 g mouse consumes ca. 5 g of food daily and so the dose of a drug mixed with the food can be calculated on this basis: an upper limit is set by the acceptability of the compound in the diet, because this may affect the food intake. This should be predetermined with a small group of animals. Nevertheless it is necessary to check by weight the daily intake of food to calculate accurately the amount of drug consumed. With compounds which are not efficacious against the infection it is likely that as the infection progresses the food intake, and hence the dose, will diminish. Another complication is the stability in the diet for the compound under test. It is usually economical to make up batches of diet for at least a week in advance but even in the apparently dry state some labile compounds will decompose on storage and this should be checked in preliminary experiments.

A healthy 20 g mouse consumes ca. 5 ml of drinking water per day and the dose may be administered in this. Here, again, the factors of stability and palatability must be considered. An increased water intake can be attained by sweetening the water with sucrose and this also assists in palatability.

6. *Assessment of activity*

By whichever route a drug is administered the pharmacological principle of establishing the dose which saves 50% of the number of animals in a group is that most commonly employed. This value is reported as the ED50, the PD50 or the CD50, that is, the dose which is effective, protective or curative to half the population. It is obtained by recording the response of graded doses over the likely range. It may be necessary to use widely spaced doses, perhaps 5- to 10-fold, in preliminary experiments before making more accurate assessment. Mucin-aided systemic infections kill mice in 1–3 days and survivors are usually counted at the end of one week. It is important to ensure that dead animals are removed daily before they are consumed by survivors.

The ED50 is obtained from plots of numbers of survivors at each dose level against the log of the dose using log-probit graph paper, and several simple statistical methods are available for the estimation of the ED50 and its standard error (Reed and Muench, 1938; Miller and Tainter, 1944; Litchfield and Wilcoxon, 1949) (see Boyce and Roberts, this Series, Vol. 7A).

Another method of measuring the activity of a compound is to determine its effect on the survival time of the infected mice. If all animals in the control group die during the experimental period then the mean survival time, usually in days but sometimes in hours, is calculated. If there are some survivors in the control group then the median survival time can be utilized. Statistical methods, such as that of Litchfield (1949), are available for the calculation of these values and their fiducial limits.

A useful working guide of ED50 values of 11 antibiotics for ten different bacterial infections in mice has been published (Wick et al., 1961).

D. Localized infections

An alternative to the severe test of controlling a systemic infection in animals is to employ a localized infection. The antibacterial activity of a compound is then manifest by its ability to prevent spread of the infecting organism, which otherwise may result in the death of the animal, or to diminish a lesion caused by the organism.

1. *In wounds and in the skin*

Early work on the use of experimentally infected wounds to assess potential antiseptics and disinfectants has been reviewed by Gordon (1947) and Browning (1963). It has proved difficult to control the degree of infection, amount of trauma and other variables. In particular, results from those methods involving multiple infection or deep and extensive damage to tissues have not been encouraging and infected animals were generally not saved from death by topical treatment. The controlled contact of the test compound and target organism under *in vivo* conditions has been attained by other means, but gain in standardization has resulted in considerable loss in realism. Contact in the peritoneal cavity by injection of the test compound followed by injection of the required organism gives variable results because the bacteria are unevenly dispersed and may escape from the site before coming into contact with the potential antiseptic. A subcutaneous site has therefore been employed, a small volume of culture of a suitable organism being injected subcutaneously into the centre of the abdominal wall of albino mice. This is followed by infiltration of the same area by ca. 1 ml of a solution of the test compound: several graded concentrations are needed. Mice are then killed at intervals and the severity of the subcutaneous lesion assessed; cultures are also taken from the site of the infection. This method, too, can give unreliable results because the bacteria and compound may escape contact.

Martin (1959) attempted to standardize test conditions by injecting a lethal dose of *Strep. pyogenes* subcutaneously into anaesthetized mice; 20 min later a solution of the test compound was injected through the same needle which had been left *in situ*. The survival times of treated and control groups of mice were then compared. Natural wound conditions were simulated more closely by administering the streptococci and the compound in horse blood. Jeffries and Price (1964) found that in their laboratory deaths were caused by the anaesthetic and by some antiseptics, and this confused the assessments. They modified the technique to include a group of mice to control the toxicity of the anaesthetic and extended the test to include *Ps. aeruginosa* and *Kl. aerogenes*.

The injection of bacteria into skin, instead of under it, can also be employed to assess antibacterial compounds. Miles and Burke (1957) showed that several common species of bacteria produce measurable lesions when injected intradermally into guinea-pig skin. The diameter of the mature lesions are roughly linearly related to the log of the number of infecting bacteria. *Corynebacterium ovis* gives large, regular and clearly marked lesions, and Collier and Grimshaw (1958) evaluated compounds by injecting this organism and the compounds at the same site. By comparing lesion sizes in groups of treated and control guinea-pigs a measure of antibacterial activity could be obtained, although it is necessary to exclude the possibility that the test compounds are effective because they have anti-inflammatory activity or inactivate bacterial toxins.

Burns are an important portal of entry for invasive bacteria, particularly *Ps. aeruginosa*. When administered in mucin to mice many strains of this organism give rise to a very severe infection which is difficult to control. However if the tails of anaesthetized mice are "burnt" by immersing them in water at 70°C for 5 sec and then dipped into a diluted culture of a virulent strain of *Ps. aeruginosa* a few hours later, an infection is initiated which will kill mice in ca. 1 week. The effect of systemic or topically administered therapy can then be followed (Rosenthal, 1967).

2. In muscle

Studies of gas gangrene during the 1939–45 war involved the technique of initiation of lethal infections in mice by the intramuscular injection of clostridia. From this was developed the use of *S. aureus* (Selbie and Simon, 1952) and *M. tuberculosis* (Selbie and O'Grady, 1954) to obtain measurable swellings in the thigh of the mouse. The infection remains localized and can be used to assess the activity of chemotherapeutic agents. Infection is initiated in groups of animals by injecting 0·2 ml of culture into the posterior aspect of the left thigh. At daily intervals the swollen thigh is measured with calipers and the width of the other thigh subtracted to give the degree of swelling caused by the infection. The effect of antibacterial compounds is assessed by dosing groups of animals and comparing their mean lesion diameters with those of undosed animals. The use of a strain of hairless mouse facilitates the measurements.

3. In corneal ulcers

Robson (1944) produced corneal ulcers in the eye of the rabbit by inject-ing small volumes of cultures of staphylococci, streptococci, pseudomonads or diphtheroids. This technique has been used in mice to assess the activity of compounds, administered systematically or topically, against *M. tuber-culosis* and *M. lepraemurium*. It allows direct observation of the developing lesions throughout the experiment, the effect of treatment being assessed by means of an arbitrary numerical scale applied to treated and control groups of animals.

(a)

(b)

Fig. 14. Lung lesions in mice infected intravenously with *Mycobacterium tuber-culosis*. (a) Lungs from untreated animals, (b) lungs from mice given isonicotinic acid hydrazide (6 mg/kg/day) in the diet.

The causative agent of leprosy, *M. leprae*, cannot be used in this way because hitherto it has not been cultured *in vitro* or transmitted to animals. However, significant, if slow and limited, multiplication of organisms from human lesions has now been obtained in the footpad of the mouse (Shepard, 1960) and this at last provides the basis of a laboratory technique for the assessment of antileprosy activity.

4. By intravenous injection

Most mouse pathogenic bacteria are less lethal to mice when injected intravenously. For instance, Dutton (1955) found this to be true with 8 of 10 species from 8 genera. Often the intravenously injected bacteria are rapidly cleared from the blood stream and set up infection in specific organs. Examination of the lesions produced and the enumeration of the organisms present provides parameters for the assessment of the anti-bacterial activity of administered compounds. This allows the effect of a compound to be studied in some detail because the infection develops more slowly than those produced by intraperitoneal injection.

Mycobacterium tuberculosis injected intravenously into mice causes marked changes in the lungs and the effect of drugs on this change can be assessed (Youmans, 1949) (Fig. 14). Similarly, *S. aureus* causes characteristic lesions in the kidneys (Gorrill, 1951) and by removing these infected organs, homogenizing them and counting the colony-forming units by conventional techniques the effect of a chemotherapeutic agent can be measured. It is important, of course, to ensure that no residual agent in the tissue of the dosed animal interferes with the enumeration. It is interesting to note that even highly effective agents such as penicillin do not completely eradicate *S. aureus* from the kidneys although persisting bacteria are found to be fully sensitive to the antibiotic *in vitro* (McClune *et al.*, 1960).

5. In the urinary tract

Although some species of Gram-positive cocci localize in the kidneys of animals after intravenous injection it is more difficult to establish Gram-negative bacteria at this site even though these organisms, particularly *E. coli*, are the major cause of infections of the urinary tract in man. During the last 15 years there have been many studies of experimental pyelo-nephritis, but the methods used to obtain suitable infections have involved the surgical implantation of foreign bodies in the bladders, kidney massage or similar rather specialized techniques. Their use to evaluate antibacterial compounds was summarized by Jackson (1965).

A simpler technique is to inject *E. coli* directly into the kidney. In our laboratories it has been found possible to inject 0·02 ml of a suspension containing 10^8 organisms/ml directly through the body wall of anaesthet-ized 200 g albino rats (L. J. Hale, pers. comm.). Viable counts after homo-genization of each kidney in individual disposable hypodermic syringes without needles, suggested to us by Dr. D. M. Harris, show that *E. coli* can be recovered from most of the injected kidneys for about a month, although the infection slowly regresses. Lesions are visible macroscopically and sections of the kidneys exhibit congestion and acute inflammation with necrosis. The organisms are also recoverable from the bladder and the uninjected kidney. A method has recently been described for the assessment of the activity of antibacterial compounds against an infection initiated in this way (Burrows and Cawein, 1969).

Limited multiplication of *E. coli* can also be obtained in urine instilled into the bladders of catheterized rabbits or dogs. The organisms are rapidly eliminated, but counts made on urine samples obtained at intervals up to 8 h after "infecting" can indicate the effect of an antibacterial compound administered orally or parenterally.

6. *In the gastro-intestinal tract*

Information on the activity of an antibacterial compound in the intestine can be obtained simply by observing its effect on the flora of the gastro-intestinal tract. The compound is administered orally by intubation, in the diet or in the drinking water, and the organisms in freshly voided faeces are submitted to differential counts using a variety of selective media. Suitable techniques have been described by Smith and Crabb (1961) who performed such counts on the faeces of 11 species of animals, including some laboratory rodents, at different ages. In studies with mice, Dubos *et al.* (1963) noted the profound changes caused by the inclusion of a variety of antibiotics in drinking water. Again it is important to ensure that inhibitory amounts of antibacterial compounds are not carried over with the faeces to interfere in the bacteriological assessment.

A variation of this technique is to administer a target organism orally to animals and to assess the activity of a compound in eliminating it from the intestine by using a suitable selective medium for faecal counts. For example, Lindh (1959) described a method for evaluating antifungal drugs against *Candida albicans* administered to mice in drinking water, the drugs being included in the diet. As the organisms appear to pass straight through the intestine it cannot be claimed that the activity of a compound against an infection is being measured, but a method of this type provides some information about the activity of a compound when a suitable infection cannot be obtained. However, this sort of evidence is no guarantee that the compound will eradicate organisms located in pathological processes in the mucosa resulting from infections such as bacillary dysentery.

Dysentery caused by species of *Shigella* appears to be restricted to primates; it has not been possible to obtain a realistic model, for example, in laboratory rodents by administering shigellae *per os* or *per rectum*. It appears that the normal intestinal flora of laboratory animals prevents the establishment of the shigellae. In recent years some success has been achieved in America by suppressing the normal flora before infection, thus allowing the pathogens sometimes to become established and give rise to characteristic lesions (Freter, 1956; Formal *et al.*, 1958). Germ-free guinea-pigs have also been employed, but here the shigellae spread from the intestine to other organs (Formal *et al.*, 1963). While more relevant than mucin-aided systemic infections for the *in vivo* evaluation of drugs against *Shigella* the necessary suppression of the normal flora renders the conditions still artificial—probably more meaningful results would be obtained in monkeys, which are natural hosts.

E. Skin disinfection

Techniques devised for determining the effect of antibacterial compounds on the flora of the skin of human volunteers have been reviewed by Sykes (1965) and Gibbs and Stuttard (1967). They consist either of the enumeration of the normal resident flora of the skin before and after treatment or of the assessment of the number of "transient" bacteria surviving after treatment of a small area of deliberately infected skin. It is again important to ensure that small amounts of the test agent on the skin do not interfere with the viable counts by employing adequate dilution or inactivators.

1. *Hand washing tests*

The original hand washing method was published by Price (1938). In this, the hands and arms are scrubbed with an ordinary soap and then rinsed with a number of samples of water. The bacterial content of each rinse is obtained by viable counts and the whole procedure is repeated after treating the hands with the test material. Its effect is assessed by comparing numbers of bacteria removed at each rinsing. The procedure is tedious, even when less rinses are used (Cade, 1950), as a number of volunteers must be employed to obtain significant results. One variation, the "split-use procedure" (Quinn *et al.*, 1954) utilizes one hand of each subject as a control. It is covered with a rubber glove while the other hand is treated with the preparation under evaluation. The percentage reduction in viable counts on one hand compared with the control is then calculated.

In another method (Bowen, 1958) rubber gloves are donned after washing the hands. Two hours later the gloves are removed, the insides carefully rinsed with water and the bacteria in the water counted. The procedure is repeated after treatment of the hands and the effect of the treatment on the skin thus assessed.

2. *Direct swabbing*

Tests involving direct swabbing are simpler and quicker than hand washing methods. Sykes (1965) describes a method which is useful for assessing the retention of the activity of a disinfectant on the skin. A measured amount of each of a number of dilutions of the test compound is spread on ca. $\frac{1}{2}$ in.^{2} of the basal internodes of the backs of the fingers and allowed to dry. After a fixed time a loopful of a diluted culture of *S. aureus* is spread within each area. Exactly 10 min later the area is thoroughly swabbed and the organisms on the swab enumerated. The results can then be referred to those for a known standard germicide. Story (1952) describes a similar method in which the culture is applied to the skin before the disinfectant.

3. Replica methods

When contact impressions from an area of skin are taken on to nutrient agar the resulting growth gives a convenient, if approximate, indication of the number and type of bacteria on the skin. This technique provides a simple method for assessing the effect on the skin flora of antibacterial compounds and some of the methods have recently been summarized by Croshaw, Hale and Spooner (1969).

Before and after treatment of the chosen area of skin of volunteers impressions are transferred to nutrient agar with velvet (Vardon, 1961) or adhesive tape (Woodworth and Wengard, 1963; Updegraff, 1964) or made directly on to agar (Laurie and Jones, 1952). The degree of reduction of the numbers of colonies on the replicas is taken as an indication of the activity of the test compound. It is important to ensure that the actual taking of the first replica does not significantly diminish the flora and that the residual antibacterial compound is not carried over in amounts sufficient to inhibit the recovery growth of surviving organisms.

The substantivity of an antibacterial compound, that is, its ability to remain on the skin in spite of subsequent washes, has been assessed by taking impressions from the finger tips directly on to agar inoculated with a sensitive strain of *S. aureus* (Vinson *et al.*, 1961). Volunteers immerse their finger-tips in a 10% solution of test preparation, usually a soap formulation of an active agent, at 40°C for 30 sec. After rinsing and drying the finger-tip impressions are then taken. The zones of inhibition around the contact areas after incubation indicate the degree of substantivity of the test compound. It should be remembered, however, that, as size of inhibition zone is partly a function of the diffusion rate of the compounds in agar, the size of zones is not necessarily a measure of the total or relative activity of the compound.

F. Oral disinfection

The most meaningful tests for oral disinfection are obtained from *in vitro* observations. They are therefore discussed under *In Vitro* Testing, Section B, 5.

REFERENCES

Arioli, V., and Rossi, E. (1970). *Appl. Microbiol.*, **19**, 704.

Association of Official Agricultural Chemists (1955). Official Methods of Analysis of the A.O.A.C., 8th edn. Washington, D.C.

Association of Official Agricultural Chemists (1960). Official Methods of Analysis of the A.O.A.C., 9th edn. Washington, D.C.

Bough, R. G., Everest, R. P., Hale, L. J., Lessel, B., Mason, C. G., and Spooner, D. F. (1971). *Chemotherapy*, **16**, 183

Bowers, A. G. (1950). *Soap*, **26**, 36.

British Standards Institution (1934). "Technique for Determining the Rideal–Walker Coefficient of Disinfectants". B.S. 541 : 1934.

British Standards Institution (1938). "Specification for the Modified Chick–Martin Test for Disinfectants". B.S. 808 : 1938.

British Standards Institution (1960). "Method for the Laboratory Evaluation of Quaternary Ammonium Compounds by Suspension Test". B.S. 3286 : 1960.

Browning, C. H. (1964). *In* "Experimental Chemotherapy", Vol. 2, Eds R. J. Schnitzer and F. Hawking. Academic Press, London.

Brownlee, G., Green, A. F., and Woodbine, M. (1948). *Br. J. Pharmacol.*, **3**, 15.

Buckland, F. E., and Treadwell, R. H. (1957). *J. Hyg., Camb.*, **59**, 49.

Burrows, S. E., and Cawein, J. B. (1969). *Appl. Microbiol.*, **18**, 448.

Cade, A. R. (1950). *Soap*, **26**, 35.

Chabbert, V. A., and Patte, J. C. (1960). *Appl. Microbiol.*, **8**, 193.

Chick, Harriette, and Martin, C. J. (1908). *J. Hyg., Camb.*, **8**, 654.

Cobb, R., Crawley, D. F. C., Croshaw, Betty, Hale, L. J., Healey, D. R., Pay, F. J., Spicer, A. B., and Spooner, D. F. (1970). *In* "Automation, Mechanization and Data Handling in Microbiology", Eds A. Baillie and R. J. Gilbert. Academic Press, London.

Collier, H. O. J., and Grimshaw, J. J. (1958). *Br. J. Pharmacol.*, **13**, 231.

Coulthard, C. E., Michaelis, R., Short, W. F., Sykes, G., Skrimshire, G. E. H., Standfast, A. F. B., Birkinshaw, J. H., and Raistrick, H. (1945). *Biochem. J.*, **39**, 24.

Croshaw, Betty, Hale, L. J., and Spooner, D. F. (1969). *In* "Isolation Methods for Microbiologists", Eds D. A. Shapton and G. W. Gould. Academic Press, London.

Dickinson, J. C., Wagg, R. E., and Litchfield, Susan (1970). *J. appl. Bact.*, **33**, 566.

Dineen, P. (1961). *J. infect. Dis.*, **108**, 174.

Dodd, A. H. (1969). "The Theory of Disinfectant Testing", 2nd edn. Swifts, London.

Dubos, R., Schaedler, R. W., and Stephens, M. (1963). *J. exp. Med.*, **117**, 231.

Dutton, A. A. (1955). *Br. J. exp. Path.*, **36**, 128.

Engley, F. B. (1954). *Tex. Rep. Biol. Med.*, **12**, 64.

Formal, S. B., Dammin, J. G., LaBrec, E. H., and Schneider, H. (1958). *J. Bact.*, **75**, 604.

Formal, S. B., Dammin, G., Sprinz, H., Kundel, D., Schneider, H., Horowitz, R. E., and Forbes, M. (1961). *J. Bact.*, **82**, 284.

Freter, R. (1956). *J. exp. Med.*, **104**, 411.

Galloway, L. D. (1954). *J. appl. Bact.*, **17**, 207.

German Society for Hygiene and Microbiology (1959). "Instructions for Testing Chemical Disinfectants". Fischer, Stuttgart.

Gibbs, B. M., and Stuttard, L. W. (1967). *J. appl. Bact.*, **30**, 66.
Gordon, J., McLeod, J. W., Mayr-Harting, A., Orr, J. W., and Zinneman, K. (1947). *J. Hyg., Camb.*, **45**, 297.
Gorrill, R. H. (1951). *Br. J. exp. Path.*, **32**, 151.
Hale, L. J., and Inkley, G. W. (1965). *Lab. Pract.*, **14**, 452.
Howard, E. (1954). *J. appl. Bact.*, **17**, 219.
Hoy, W. A., and Clegg, L. F. L. (1953). *Proc. Soc. appl. Bact.*, **16**, i.
Ingram, G. I. C. (1962). *Immunology*, **5**, 504.
Jackson, G. G. (1965). *In* "Process in Pyelonephritis", Ed. E. A. Kass. Davis, Philadelphia.
Jeffries, L., and Price, S. A. (1964). *J. clin. Path.*, **17**, 504.
Kavanagh, F. (1963). "Analytical Microbiology". Academic Press, New York and London.
Kelsey, J. C., Beeby, M. M., and Whitehouse, C. W. (1965). *Mon. Bull. Minist. Hlth.*, **24**, 152.
Kelsey, J. C., and Maurer, I. M. (1966). *Mon. Bull. Minist. Hlth*, **26**, 110.
Kelsey, J. C., and Sykes, G. (1969). *Pharm. J.*, **202**, 607.
Kethley, T. W., Fincher, E. L., and Cown, W. B. (1956). *Appl. Microbiol.*, **4**, 1.
King, M. B., Knox, R., and Woodroffe, R. C. (1953). *Lancet*, i, 573.
Klein, M., Schoor, S. E., Tashman, S., and Hunt, A. O. (1950). *J. Bact.*, **60**, 159.
Knox, R., and Collard, P. (1952). *J. gen. Microbiol.*, **6**, 369.
Krönig, B., and Paul, T. (1897). *Z. Hyg. InfektKrankh.*, **25**, 1.
Lawrie, P., and Jones, B. (1952). *Pharm. J.*, **168**, 288.
Lederberg, J., and Lederberg, E. M. (1952). *J. Bact.*, **63**, 399.
Lindh, H. F. (1959). *Antibiotics Chemother.*, **9**, 226.
Lisboa, N. P. (1959). *Proc. XVth Int. Diary Congress*, **3**, 1816.
Litchfield, J. T., and Wilcoxon, F. (1949). *J. Pharmac. exp. Ther.*, **96**, 99.
Litchfield, J. T. (1949). *J. Pharmac. exp. Ther.*, **97**, 399.
Mahouny, D. E., and Chadwick, P. (1965). *Can. J. Microbiol.*, **11**, 829.
Martin, A. R. (1959). *J. clin. Path.*, **12**, 48.
McCune, R., Dineen, P., and Batten, J. C. (1960). *J. Immun.*, **85**, 447.
Miles, A. A., Miles, E. M., and Burke, J. (1957). *Br. J. exp. Path.*, **38**, 79.
Miller, L. C., and Tainter, M. L. (1944). *Proc. Soc. exp. Biol. Med.*, **57**, 261.
Ministry of Health Subcommittee (1957). "Report on Bacteriostatics for Parenteral Injections of Procaine Penicillin". Ministry of Health, London.
Nungester, W. J., Wolf, A. A., and Jourdonais, L. F. (1932). *Proc. Soc. exp. Biol. Med.*, **30**, 120.
Olitzki, L. (1948). *Bact. Rev.*, **12**, 149.
Parry, J. (1961). *J. appl. Bact.*, **24**, 218.
Pharmacopoeia of the United States, XVIIIth Edition (1970). Washington, D.C.
Price, P. B. (1938). *J. infect. Dis.*, **63**, 301.
Pulvertaft, R. J. (1952). *J. Path. Bact.*, **64**, 75.
Quastel, J. H. (1966). *J. gen. Microbiol.*, **45**, XIV.
Quinn, H., Voss, J. G., and Whitehouse, H. S. (1954). *Appl. Microbiol.*, **2**, 202.
Rahn, O. (1945). *Biodynamica*, **5**, 1.
Rake, G. (1935). *Proc. Soc. exp. Biol. Med.*, **32**, 1523.
Reed, L. J., and Muench, H. (1938). *Am. J. Hyg.*, **27**, 493.
Rideal, S., and Walker, J. T. A. (1903). *J. R. sanit. Inst.*, **24**, 424.
Robson, J. M. (1944). *Br. J. Ophthal.*, **28**, 15.
Rosenthal, S. M. (1967). *Ann. Surg.*, **165**, 97.

Ruehle, G. L. A., and Brewer, C. M. (1931). "United States Food and Drug Administration. Methods of Testing Antiseptics and Disinfectants". U.S. Dept. of Agric., Circular 198.

Selbie, F. R., and Simon, R. D. (1952). *Br. J. exp. Path.*, **33**, 315.

Selbie, F. R., and O'Grady, F. (1954). *Br. J. exp. Path.*, **35**, 556.

Shepard, C. C. (1960). *Am. J. Hyg.*, **71**, 147.

Showacre, Jane, Hopps, H. E., DuBuy, H. G., and Smadel, J. E. (1961). *J. Immun.*, **87**, 153.

Smith, H. W., and Crabb, W. E. (1961). *J. Path. Bact.*, **82**, 53.

Statutory Instrument 1372 (1970). H.M.S.O., London.

Stephens, J. M. (1959). *Can. J. Microbiol.*, **5**, 73.

Streitfield, M. M., and Suslaw, M. S. (1954). *J. lab. clin. Med.*, **43**, 946.

Story, P. (1952). *Br. med. J.*, **2**, 1128.

Stuart, L. S., Ortenzio, L. F., and Friedl, J. L. (1953). *J. Ass. off. agric. Chem.*, **36**, 466.

Sykes, G. (1958). "Disinfection and Sterilization", Spon, London.

Sykes, G. (1962). *J. appl. Bact.*, **25**, 1.

Sykes, G. (1965). "Disinfection and Sterilization", 2nd edn. Spon, London.

Szybalski, W., and Bryson, V. (1952). *J. Bact.*, **64**, 489.

Temperly, H. N. V., and Blyth, G. E. K. (1968). *Nature, Lond.*, **219**, 1218.

Updegraff, D. M. (1964). *J. invest Derm.*, **43**, 129.

Verdon, P. E. (1961). *J. clin. Path.*, **14**, 91.

Vinson, L. J., Ambye, E. L., Bennet, A. G., Schneider, W. C., and Travers, J. J. (1961). *J. pharm. Sci.*, **50**, 827.

Wheater, D. W. H. and Hurst, E. W. (1961). *J. Path. Bact.*, **82**, 117.

Wick, W. E., Streightoff, F., and Holmes, D. H. (1961). *J. Bact.*, **81**, 233.

Woods, D. D. (1962). *J. gen. Microbiol.*, **29**, 687.

Woodworth, H. H., and Newgard, P. M. (1963). "Measurement of Skin Contamination". Stanford Research Institute, California.

Youmans, G. P. (1949). *Ann. N.Y. Acad. Sci.*, **52**, 662.

Youmans, A. S., Youmans, G. P., and Doub, L. (1954). *Antibiotics Chemother.*, **4**, 521.

Youmans, G. P., and Youmans, A. S. (1964). *In* "Experimental Chemotherapy", Vol, 2, Eds R. J. Schnitzer and F. Hawking. Academic Press, New York.

CHAPTER V

Photomicrography and Macrophotography

LOUIS B. QUESNEL

Department of Bacteriology and Virology, University of Manchester, Manchester, England

I. INTRODUCTION

The microbiologist often requires a permanent record of specimen material that is either small or microscopic, and such records can be made using the techniques of macrophotography and photomicrography. Obviously, before one can take good photomicrographs one must be able to set up the microscope so that it gives the best possible performance within its design limitations and the strictures of the experimental situation, and it will be assumed that the reader is familiar with the design and use of the microscope, at least in all the aspects covered by Chapter 1 in Volume 5A of this Series. The finest camera in the world will not improve upon a bad microscope image and micrographs tend, in addition, to accentuate defects in the image for which the human eye has compensated and which have, subconsciously, been rectified or disregarded during visual observation.

II. PRINCIPLES

A. Photomicrography and macrophotography

Any photographic image on film which is larger in dimension than the object photographed, i.e. has been magnified, is a "micrograph" or a "macrograph". There is some dispute as to where the boundary should be drawn but for our purposes we can consider magnifications up to about 20 times to be macrophotographs and magnified images greater than this to be photomicrographs. The big difference between the two is that macrophotographs may be made simply with a camera while photomicrographs require the use of a microscope.

It should be noted that (in Britain and the U.S.A.) the term *Microphotography* means the opposite of photomicrography, that is, it is the production, by optical reduction, of very small photographs from much larger natural objects, e.g. images of book pages reduced to say $\frac{1}{2}$ in^2 in "microfile" libraries.

B. Macrophotography

1. *Supplementary lenses*

In normal photography the object size is considerably greater than the image size while for macrophotography the magnification (M) must be 1 : 1 or greater. This situation may be achieved in a number of ways. It was shown in Volume 5A, Chapter 1 of this Series, how a magnifier worked by enabling the object to be brought within the "least distance of distinct vision" and so enabled a greater "visual angle" to be obtained with consequent increase in the defined image size, that is, magnification of the image on the retina. Much the same applies to the camera when used with an appropriate supplementary lens to permit close-up work. In this arrangement the camera lens is focused at infinity and the supplementary lens represents the magnifier. The degree of magnification is given by:

$$M = \frac{f_{\text{cam}}}{f_{\text{supp}}}$$

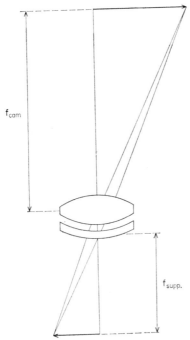

FIG. 1. Use of a supplementary lens in conjunction with a camera lens of long focal length, focused at infinity.

so that magnification increases as the focal length of the supplementary lens decreases, and for an image scale greater than 1 : 1 the focal length of the camera lens must exceed the focal length of the supplementary (Fig. 1).

All major camera manufacturers supply such supplementary lenses and their power is given in diopters, the diopter being equal to the reciprocal of the metre, and:

$$\text{Focal length in cm} = \frac{100 \text{ cm}}{\text{power in diopters}}$$

However, in order that the supplementary lens be relatively free from error it must be made with a fairly long focal length (e.g. 200 cm) which means that for image scales even to approach 1 : 1 a long-focus camera lens must be used. For this reason the use of supplementary lenses for macrophotography are not recommended.

2. Camera with extension tube or bellows

In photography, as stated above, the object distance is usually much greater than the image distance, that is the distance from lens to film plane is much shorter than the distance from lens to object, so it is not surprising that when the lens is reversed and used in conjunction with an extension tube or bellows it is possible to make the image distance (v) greater than the object distance (u) and so $M > 1 : 1$ (since $M = v/u$). The degree of magnification which is possible with such systems depends upon the focal length of the lens used and upon the extension of the bellows. Also, since the image is larger than the object, light leaving the object area has to fill an image area which is greater and normal light-meter readings cannot be used. The increase in exposure which is needed depends on the bellows extension, i.e. on the magnification, and tables are provided of the exposure factors which are required. These may be considerable since they are dependent on the "inverse square ratio" law of illumination. When the camera is designed with a built-in through-the-lens light metering system of sufficient sensitivity this may be used directly for exposure determination once its latitude has been checked by a preliminary calibration "experiment".

3. Cameras for macrophotography

The best results in macrophotography will be obtained when special lenses designed for the purpose are used. These lenses are specifically corrected for short working distances and will enable a wide range of magnification depending on the focal length of the lens and the image distance which is possible. The latter will depend on the means provided for camera extension which is normally an extensible bellows.

The various parameters such as object size (O), image size (I), object distance (u), image distance (v), magnification (M), and focal length (f) are related by equations from which they may be calculated. When the magnification is $1 : 1$ then:

$$M = \frac{v}{u} = 1, \text{ and } v = u = 2f$$

For image size

$$I = \frac{O \times v}{u}$$

For object size

$$O = \frac{u \times I}{v}$$

For image distance

$$v = \frac{I \times u}{O}$$

For object distance

$$u = \frac{O \times v}{I}$$

When a specific magnification factor is required and the focal length of the lens is known then the image distance (bellows extension) can be obtained from:

$$v = f + (M \times f)$$

and the object distance from

$$u = f + \frac{f}{M}$$

We may therefore make a number of general statements about the use of lenses for macrophotography. For a lens of any given focal length, the greater the bellows extension the greater the magnification. For any given bellows extension, the shorter the focal length of the lens the greater the magnification. For a lens of any given focal length, the object distance, and more practically the working distance (from specimen to front edge of lens mount), becomes shorter as the magnification increases. For any given magnification, the shorter the focal length of the lens the shorter the working distance. The total distance from object to image is at a minimum when the magnification factor = 1, and this distance = $4 \times f$; magnification or reduction of image scale increases the total distance. For a given magnification the longest total distance is required by the longest focal length lens.

If camera extension tubes are used that distance will be fixed and magnification may be altered in discrete steps by using lenses of different focal length, whereas the use of a bellows extension enables continuously variable magnification to be achieved within the limits of extension of the bellows. Continuous curves may, therefore, be plotted to relate the parameters. A series of curves for the Summar and Milar macrophotographic lenses manufactured by Leitz to cover the focal length range from 120 mm to 24 mm are given in Fig. 2. In general if the desired magnification can be obtained with several lenses, the lens of the longest focal length will give the best result.

In practice it is often most convenient to obtain magnification values by photographing an accurately ruled calibration graticule under the identical conditions used to photograph the specimen. The grid distance is known and the grid distance in the photographic image may be measured; the ratio of the latter to the former gives the magnification. (A sheet of graph paper makes a convenient 'graticule'.)

4. Using the microscope for macrophotography

It is possible to obtain low magnification photographs with the aid of a microscope, but these may be very poor unless proper precautions are taken. The proper combination of objective and eyepiece must be used

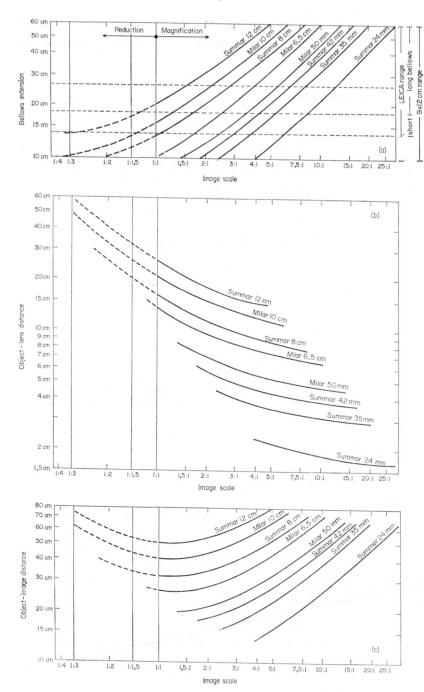

since the latter is often designed to correct residual errors in the image from the objective alone and attempts to obtain low magnification photographs by omitting the eyepiece will be of poor quality. For best results low-power flat-field objectives must be used in conjunction with low-power eyepieces, and the specimen must be adequately illuminated. The latter requirement cannot be met with normal condensers used for the higher power objectives since the specimen field which is photographed may be several times greater than the field adequately illuminated by the Kohler system of illumination. Special simple spectacle-lens condensers must be used, which give a wide aperture beam of rays illuminating a large area of the field. The condenser is adjusted so that the exit pupil of the condenser lies in the entrance pupil of the objective. (The entrance and exit pupils determine the brightness of the image. If we imagine a two-component camera lens with the f-stop diaphragm in the middle such that the f-stop aperture is the smallest aperture in the lens system, then the entrance pupil is the image of that aperture as viewed from the object and the exit pupil is its image as viewed from the image point. The entrance pupil is thus the "new aperture" (diameter of image of the f-stop) and is the aperture which subtends the smallest cone angle of rays from the axial object point. In the same way, the exit pupil is the image of the f-stop formed by the optical element which follows it, and this subtends the smallest cone angle at the axial image point.) If the wide aperture spectacle-lens condenser is not properly adjusted the field will show a difference in the intensity of illumination across its diameter. This may not be obvious to the eye but may easily be seen on photographs as a progressive darkening towards the periphery of the field.

Condensers which have removable top elements can be used on the principle of diascopic illumination when the long-focus low element is so adjusted that the image of the source is focused in the plane of the objective to give a wide-diameter illuminating cone, Fig. 3.

5. *Illumination for macrophotography*

As mentioned under 4 above the wide field area used in macrophotography poses problems of even illumination for transilluminated specimens which also apply to the bellows camera systems. For adequate

---

Fig. 2. Graphs for macrophotography with Leitz Summar and Milar lenses. To set the optical system for a given magnification, the object-lens mount distance is read from graph (b) and the bellows extension from graph (a). Slight final focusing of the lens will be necessary. Exact image scales must be obtained in one of the ways described in Chapter 1, Volume 5A of this Series. (Courtesy E. Leitz, Wetzlar.)

transillumination the bellows camera should be used with the diascopic illuminator and stage designed for it by the particular manufacturer. These dia-illuminators are supplied with an appropriate lamp and a series of condenser lenses of focal length and diameter designed to "match" the

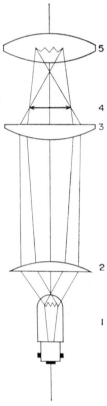

Fig. 3. The diascopic principle of illumination. (1) Light source, (2) lamp collector lens, (3) substage condenser, (4) object plane, (5) objective and filament image.

various macrolenses of the camera (see below). A general diagram of the optical path within such a system is given in Fig. 4.

Of course, incident light illumination will be required for opaque and three-dimensional specimens, or, a mixture of dia- and epi-illumination may be used. The ability to use a particular epi-illumination system will depend to a large extent on the focal length of the lens used as the proximity of the lens to the specimen may obstruct the illumination. For such specimens ring illuminators are particularly helpful since they may be fitted around the lens. Some lenses are manufactured with built-in "ring-flash", such as the Micro-Nikkor Medical. A comprehensive range of epi-illumination techniques may be found in "Photography for the Scientist" Ed. C. E. Engel (Academic Press).

While specimen illumination creates difficulty the determination of exposures for macrophotography creates its own special problems and these are dealt with in a later Section.

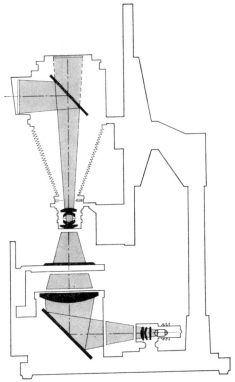

FIG. 4. Ray path in the Nikon Multiphot camera set up for macrophotography with a diascopic illuminator. (Courtesy Nippon Kogaku, Tokyo.)

C. Photomicrography

It will be remembered from the discussion in Volume 5A of this Series (Chapter 1, Quesnel) that the intermediate image of the microscope is formed at the focal distance from the eyelens. This means that the rays of light leaving this image will emerge from the eyelens in parallel bundles and therefore, in effect, the microscope image is formed at infinity. However, by placing the eye at the eyepoint the rays entering the eye will be

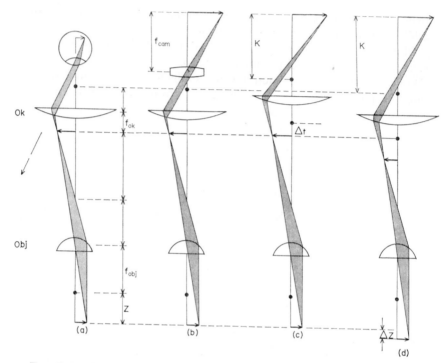

FIG. 5. Methods for producing the real image required for photomicrography. (a) Image formation by the eye. (b) Use of camera with lens focused at infinity. (c) Objective focus fixed but eyelens raised so that intermediate image falls outside f_{ok}. (d) Eyelens fixed but objective focus changed (increased) so that intermediate image falls outside f_{ok}. Ok, eyelens; Obj, objective; f_{ok}, focal length of eyelens; f_{obj}, focal length of objective; f_{cam} focal length of camera lens; K, camera length; Δt, distance from intermediate image to lower focal plane of eyelens; Z, distance from object to lower focal plane of objective; ΔZ, increase in objective working distance.

caused to form converging image-forming pencils due to the curvature of the eyelens. In this way the image is formed on the retina (Fig. 5(a)) and it appears to be derived from a virtual image at the least distance of clear vision, taken to be 250 mm. For photographs to be taken the image must be formed at a fixed distance and at this point the film must be placed, viz. in the focal plane of the camera.

There are at least three ways in which this situation may be achieved.

1. *A camera with its lens focused at infinity*

The camera is placed with its lens at the eyepoint of a microscope correctly adjusted for visual observation, when the microscope image will be re-formed in the film plane (Fig. 5(b)). Where possible the camera should be positioned so that the exit pupil of the microscope coincides with the entry pupil of the camera lens system. The magnification (M) at the film then depends upon the microscope magnification (M_{mic}) and the focal length of the camera lens (f_{cam}) in mm:

$$M = M_{mic} \times \frac{f_{cam}}{250}$$

It may not always be possible to couple the two instruments adequately as this may depend on the type of mount in which the camera lens is fixed. In addition, the "angle of view" of the eyepiece must at least match that of the camera lens. A standard camera lens will cover an angle of about 50–60° and the eyepiece lens must cover at least this angle, so a wide-angle eyepiece should be used wherever possible. If the eyepiece does not cover the angular aperture of the camera lens then vignetting will take place and the exposed photograph will appear as a circular area in the middle of the film format. In such cases field limitation will have been brought about by the eyepiece aperture and not by the camera. If another eyepiece is not available it may be possible to overcome the problem by using a long focus camera lens with lower angle (90 mm, 30°). With this arrangement a light-masking sleeve should be placed around the junction of camera and microscope. The great advantage of the system is that the camera is not physically attached to the microscope and so will not affect focus, for example. The camera itself must be rigidly fixed to an appropriate stand.

2. Use of camera without camera lens

In the second method the camera lens is removed, the microscope objective focus is maintained at the normal position as before, but the eyepiece is drawn away from the objective so that the intermediate image now falls outside the focal length of the eyelens. In this position the rays leaving the eyelens are no longer parallel and will focus an image at a distance determined by the amount which the eyelens has been retracted. The arrangement is shown in Fig. 5(c).

Adjustable eyelenses of this sort are supplied by manufacturers. When used in this way the overall focal length of the microscope is altered and the magnification of the final image changed. The overall magnification is now given approximately by:

$$M = M_{mic} \times \frac{K}{250}$$

where K is the camera extension in mm when the image is in focus on the filmplane. This equation does not, however, take into account the slight change in focal length due to the retraction of the eyepiece and again, in practice, it is probably easiest and best to photograph a known scale under the same conditions in order to derive the true image scale.

3. *Focusing by changing the objective working distance*

In the third method the distance between eyelens and objective is retained at the normal value for visual observation, but the distance between objective and specimen is changed to a longer working distance. With the object now further beyond the focal length of the objective the conjugate plane in which the intermediate image is formed will be nearer to the objective and further from the eyelens (Fig. 5(d)). As before, with the intermediate image now lying outside the focal length of the eyelens a real inverted image of it will be formed at a plane beyond the eyelens—the film plane. In this arrangement the intermediate image is formed at a distance which is shorter than that computed for the minimum-error functioning of the objective and high N.A. objectives, in particular, will exhibit under-correction. This is compensated for by the use of special photo-eyepieces which restore the computed "correctness" of the objective.

D. Focusing

The means of focusing will depend on the photomicrographic system being used and the provisions made in the equipment itself. In the case of the first method above (Fig. 5(b)) the camera lens is set at infinity and the microscope is adjusted so that the image is produced at infinity. This will be so when the microscope is focused with a relaxed eye. If difficulty is experienced the image can be viewed by focusing the image with the aid of a pocket telescope which has previously been adjusted "at infinity" and placed on top of the eyepiece. Remove the telescope, bring the camera into place and the image should now be focused on the film plane.

If a ground glass screen can be placed in the film plane for viewing the final image then this should be used. Alternatively, with a through-the-lens reflex camera the normal viewing window of the camera should be used and the image brought into sharp focus by adjusting the microscope while the camera lens is set at infinity.

In most photomicrographic apparatus focusing is performed with the aid of a ground glass screen or focusing telescope; normally, for attachment cameras the ground glass system is used for the larger formats and focusing telescopes for 35 mm.

Ground glass screens are not always satisfactory when performing critical focusing of fine detail, especially when using high power objectives (when depth of focus is very short), and especially if the "frosting" has a large grain. In these cases it is far better to make a clear-glass area in the screen and to use a focusing magnifier to view the aerial image. To do so Ruthmann (1970) suggests that fine India ink marks should be made on the lower surface of the screen, a drop of mounting fluid applied, and a cover-slip fixed on. Since the refractive index of mounting fluid can be chosen to be the same as glass the frosting in that area will "disappear" leaving the black marks to identify the film plane. To use this system the focusing magnifier is "fixed" in place and adjusted to focus on the India ink marks; then the specimen is brought to focus while looking through the magnifier. Some manufacturers supply equipment whose glass screens already have clear-glass areas, e.g. the Nikon and Leitz bellows cameras have clear diagonal bands.

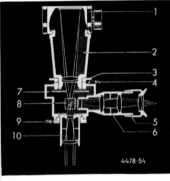

Fig. 6. Leitz Mikas micro-attachment. (1) Leica 35 mm camera, (2) $\frac{1}{3} \times$ adapter, (3) clamping ring, (4) shutter setting knob, (5) rotating eyepiece mount, (6) stop with graticule of the focusing telescope, (7) synchronized central shutter, (8) deflecting prism, (9) knurled clamping screw, (10) eyepiece. (Courtesy E. Leitz, Wetzlar.)

Effectively the same arrangement is found in the Type C viewing screen of the Nikon F camera (which the author recommends for use in photo-micrography). It is a frosted screen with a central clear-glass spot and an etched cross.

A clear-glass screen with etching is the standard method for focusing an image when an attachment camera with built-in focusing telescope is used. A typical system is shown above (Fig. 6), the Leitz Mikas attachment. In the section diagram, Fig. 6, the camera back (1) is attached to a $\frac{1}{3} \times$ adaptor sleeve (2) which is coupled to the top of the focusing attach-ment by means of the clamping ring (3); the shutter setting knob (4) is on top of the shutter housing (7). The telescope viewing eyepiece is fixed in a

rotating mount (5) so that it may be independently focused on the etching
on the clear-glass screen (graticule) which is fixed in the position of the
stop (6). The deflecting prism (8), when inserted into the light path, directs
25% of the light to the telescope and 75% is transmitted to the camera;
the image may therefore be viewed while taking the picture. When swung
out of the optical path all the light from the microscope eyepiece (10) passes
on to the camera. The whole attachment is rigidly fixed to the microscope
by means of the clamping-screw (9).

By design the image will be in focus on the film when it is in focus in
the plane of the graticule (at 6). But whether or not the image is in the
plane of the graticule will depend upon the distance between the specimen
and the objective since this affects the positions of the conjugate planes of
all the images involved (see above concerning Fig. 5(d)). The first step,
therefore, is to focus the viewing telescope eyepiece precisely so that a
sharp image of the graticule is formed for the particular observer. Do this
against a plain illuminated background. Any image which is made sharp
in the plane of the graticule will obviously now also appear sharp in the
viewing telescope, so the microscope is adjusted until the specimen is in
sharp focus on the field of the telescope. It will now be in the plane of the
graticule since the eyepiece is focused on the graticule for the particular
observer, and so the image will also be sharply focused on the film. The
exposure can now be made.

E. Depth of field and depth of focus

When a lens is focused on an object (O, Fig. 7) other objects within the
zone ZY will also be rendered acceptably sharply defined. The distance
before and behind the object yielding sharp definition is called the depth
of field of the lens. As a rule the near zone (OY) is about one half the depth
of the farther zone (OZ). Acceptable sharpness depends on the size of the
"circle of confusion" of the image "point". It can be seen from Fig. 7

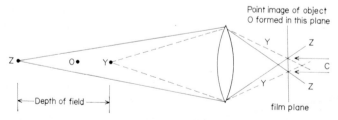

Fig. 7. Depth of field. O, focused object; Z, farthest object reproduced sharply;
Y, nearest object reproduced sharply; C, diameter of the image disc of Z and Y
in the plane of the film. C represents the limit of the circle of confusion, objects
outside ZY are judged "unsharp".

that if the object point is at Z its image will be formed closer to the lens than the image of the point were it at position Y. In the former case the image-forming cone of rays forms a diverging cone as the rays pass through the image point; this cone intersects the converging cone moving toward the formation of the image of Y. Where the two cones "intersect" there will be a disc of image forming rays, one set having just formed the image of Z the other about to form the image of Y. This disc is called the circle of confusion (C). The image of the point at O would be formed in the middle of this circle. For image sharpness in photography the standard set is that the diameter of the circle of confusion should not exceed $D/1000$, where D is the viewing distance. For a photograph at the conventional limit of clear vision, 250 mm, this would be 0·25 mm. An image "patch" with a wider diameter than this would be unacceptably out of focus.

It should be noted that the circle of confusion is that which is "viewed" in the final image—the print—and is therefore a function of the total magnification. If the print was derived by $10\times$ magnification of a negative, the circle of confusion in the negative plane at the time of photographing would have to be correspondingly $10\times$ smaller. Note, too, that sharpness depends on the viewing distance. An image which is unsharp when viewed at 250 mm may appear sharp when viewed at a greater distance in the same way that the cinema image appears blurred when viewed from the cheapest seats in the cinema but sharp when viewed from the rear stalls.

Similarly, the depth of focus is the distance through which a film emulsion may be moved in relation to a particular lens system and object, and yet yield an image which is sharp within the standard of sharpness determined by the diameter of the circle of confusion.

It will be realized, too, that the size of the circle of confusion, and hence sharpness, will depend on the effective aperture of the lens which determines the angles of the image-forming ray-cones. A decrease in aperture, effected by restricting the condenser iris diaphragm in photomicrography will increase the depth of field; but will also affect the resolution of the system (see Quesnel, this Series, Vol. 5A). Since the depth of field in photomicrography may be very small indeed a compromise should be struck in which some gain in depth of field is made at the expense of a slight loss in resolution.

The important principle to remember is: the greater the magnification the smaller the depth of field. As pointed out in Volume 5A this distance may be as little as 0·15 μm for high N.A. lenses, well below the diameter of a small bacterium. Very critical focusing is therefore required for microbiological photomicrography.

In macrophotography, where three-dimensional objects are to be photographed, the depth of field requirement may be particularly important.

For the best result due consideration must be given to the image magnification, the focal length of the lens to be used, its aperture and the viewing coefficient, i.e. the ratio of the eventual negative enlargement to the viewing distance of the final image (print) in metres. These methods have recently been described by Heunert (1968) and Martinsen (1968) to whose work the reader is referred for details.

A general impression of the depth of field available for photomicrography with a selection of objectives used for varying magnifications in the useful range is given in Table I.

TABLE I

Depth of field in μm for photomicrography with a range of Zeiss (Jena) objectives for final magnifications (M) within the useful range. Circle of confusion: 0·143 mm

Image scale M : 1	"Power" and N.A. of achromatic and apochromatic objectives					
	8 × 0·20	10 × 0·30	20 × 0·40	20(40) × 0·65	40 × 0·95	HI 90 × 1·30
100	7·14					
125	5·72					
160	4·46	2·98				
200	3·57	2·38	1·79			
250		1·91	1·43			
320		1·49	1·12	0·69		
400			0·89	0·55		
500				0·44	0·30	
630				0·35	0·24	0·27
800					0·19	0·21
1000					0·15	0·17
1250						0·13

From Table I a number of working rules may be deduced. In general, the higher the N.A. of the lens the less the depth of field. For a given final magnification the greatest depth of field is obtained by using a low power objective with "high" power eyepiece within the useful range, e.g. at M : 1 = 200 an 8 × objective with 25 × eyepiece will give twice the depth of field of a 20 × objective with 10 × eyepiece. Also it may be better to photograph at lower magnification but greater depth of field and subsequently to increase magnification by enlargement from the negative. Thus, using a 40 ×, N.A. 0·65 objective with 12·5 × eyepiece to give M : 1 = 500, an image with 0·44 μm depth of field is obtained while a 90 × objective with 15 × eyepiece would give high magnification, M : 1 = 1350, but very little depth of field < 0·13 μm. In the former case the depth of field is over three times as great, and by photographic enlargement the print image could be easily made at a correspondingly high final magnification. No process can improve the depth of field in the latter case so magnification should not be sought for its own sake. On the other hand where resolution is of great importance, the loss in resolution from using the low power lens should be considered.

F. Magnification and useful scale of reproduction

When the microscope is used to produce a "fixed" image, as by photomicrography, drawing or projection, the ratio of the size of the image to the size of the object is referred to as the image scale rather than as the magnification. A photomicrograph reproduced at an image scale of 100 : 1 means that the picture records the object 100 times larger than it, in fact, is.

In photomicrography the image scale is determined from:

$$\text{Image scale (negative)} = M_{\text{ob}} \times M_{\text{oc}} \times \frac{K}{250} \times e$$

where K is the camera extension (mm) and e the eyepiece tube magnification factor, if any, e.g. the phototube for a particular camera attachment may involve a magnification factor. M_{ob} and M_{oc} are the magnification of objective and eyepiece respectively.

As mentioned in the Chapter on Microscopy (Quesnel, this Series, Volume 5A) the magnifying powers inscribed on objectives are approximate so that calculation from a formula gives an incorrect, though usually adequate, answer. The true image scale may be easily obtained by photographing a stage micrometer under the identical conditions used for the specimen. Since the graticule interval is known, by measuring the same interval on the print and taking the ratio, the true image scale is obtained.

When the final print is to be viewed at the conventional visual distance (250 mm), contact prints from $2\frac{1}{4} \times 3\frac{1}{4}$ in. or 4×5 in. negatives may be quite suitable so long as the image scale falls within the limits for useful magnification. These limits are from $500 \times \text{NA}$ to $1000 \times \text{NA}$ for the objective used (see Volume 5A). A moderate degree of enlargement, to about half-plate ($4\frac{3}{4} \times 6\frac{1}{2}$ in.) may however, be preferred and may be necessary from small-format negatives, but greater enlargement than this is unlikely to increase detail and is unnecessary. When prints or projected images are to be viewed at further distances the scale of useful magnification must be extended. The increase may be determined from the formula:

$$\text{Extra magnification } m = \frac{L}{250}$$

Where L is the viewing distance in mm.

The maximum practical (useful) magnification now becomes:

$$M_{max} = \frac{L}{250} \times 1000 \text{ N.A.}$$

and the minimum practical magnification is:

$$M_{min} = \frac{L}{250} \times 500 \text{ N.A.}$$

It should also be noted that since the final magnification must fall within the useful scale and a reasonable degree of enlargement from a 35 mm negative may be required even for viewing at the conventional visual distance, it is possible that the magnification in the negative may fall below the minimum of the useful range. This does not necessarily impair the image quality since the resolving power of photographic emulsions is usually greater than that of the eye.

As for photomicrography, it is possible to calculate the image scale for microprojection:

$$\text{Microprojection scale} = M_{ob} \times M_{oc} \times \frac{Lp}{250}$$

where Lp is the projection distance in mm.

And the image scale in cinemicrography is obtained from:

$$\text{Image scale (cine)} = M_{ob} \times M_{oc} \times \frac{f}{250}$$

where f is the focal length of the cinecamera lens in mm.

III. CAMERAS

A very wide range of cameras suitable for macrophotography and photomicrography is available and it is not possible to make a complete list here. An attempt will be made, however, to give a brief description of a selection of representative types. Full details of the various instruments available are easily obtainable from the manufacturers.

A. For macrophotography

1. *Camera and bellows or extension tubes*

Many high quality cameras have a removable lens and provision for the insertion of extension tubes (or rings) or bellows which increase the lens-to-film distance and permit image scale reproductions up to 1 : 1 or greater if the lens can be reversed (see Section II.B). Some manufacturers have computed lenses especially for use at very short working distances. There are many microbiological applications for cameras of this type, an obvious one being the recording of colonies on agar plates.

A good example of a versatile camera system for close-up and macro-work is the Nikon system, with Photomic T finder for through-the-lens exposure determination. The specially designed close-up lens, the Micro-Nikkor Auto 55 mm f/3·5, is ideal. It is computed for optimum performance at a focused distance of just over 61 cm when the image-scale is 1 : 10. (Most general purpose photographic lenses are designed for best performance when infinity focused.) The Micro-Nikkor Auto has exceptionally high resolution and full-field crispness of details even at its closest focusing distance of $9\frac{1}{2}$ in., when the reproduction ratio is 1 : 2. By inserting the M-ring, the lens can now be made to give a 1 : 1 image scale. With the M-ring the effective maximum aperture is raised from f/3·5 to f/7·1. Because of the slight fall-off in image quality and the great loss in depth of field, lenses used under these conditions should be "stopped down" to at least f/8. The M-ring specially designed for this lens enables the auto-coupling between lens and camera to be maintained and exposure can still be determined on the built-in meter by adjusting the diaphragm, or the shutter-speed setting until the meter needle pointer moves to the centre of the scale (marked by a small "o"). The working distance (from lens-mount outer rim to object) when the M-ring is used is just over $3\frac{1}{4}$ in.

To obtain higher image scales the lens must be used with a bellows attachment inserted instead of the M-ring. When the coupling is with the lens in the normal position the image scale ranges from 3·8 : 1 with bellows fully extended to 0·8 : 1. When the lens is reversed and coupled to the bellows by means of the BR-2 Macro-Adaptor ring the range is increased to 4·2 : 1 to 1·7 : 1. It should be taken as a general rule that for ratios greater than 1 : 1 the lens should be used in the reverse position; the working distance for maximum reproduction ratio in the latter case is 50 mm while it is only 17 mm in the "normal" position. Epi-illumination of the specimen is virtually impossible at such short working distances.

The rail of the bellows attachment is calibrated to give the reproduction ratios for different bellows extensions. Since the ratio for a particular extension will depend also on the focal length of the lens only one or two scales can be provided. In the case of the Nikon bellows the rail is calibrated for use with Nikkor 50 mm f/2 lenses and for any other lenses the image scale for a particular extension must be read from a table provided with the attachment. The method of assembling for macro use is shown in Fig. 8.

With the Nikon system the maximum image scale (8·8 : 1) is obtained with the short focal length Nikkor H Auto 28 mm f/3·5 lens used reversed with bellows. Other lenses give a variety of scale ranges and full data for

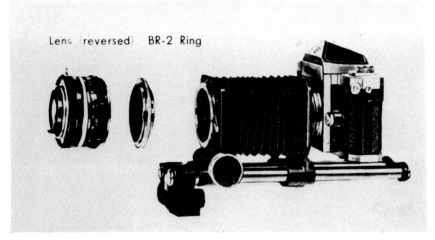

Lens (reversed) BR-2 Ring

FIG. 8. Nikon F camera fitted with macro-bellows extension; coupling ring BR.2, and lens in reverse position. (Courtesy Nippon Kogaku, Tokyo.)

systems of this kind are readily obtainable from manufacturers or their representatives.

Exposure determination in close-up and macro-work requires special attention and will be referred to in the appropriate Section below.

Although not strictly within the brief for this Chapter it is worth mentioning that a number of manufacturers can supply slide copying attachments which couple in front of a lens coupled with a bellows to the camera. If the opal glass of the attachment is "back-lighted" with quartz-iodine

FIG. 9. Nikon F camera with slide-copying attachment. (Courtesy Nippon Kogaku, Tokyo.)

lamps rated at the correct colour temperature it is possible to reproduce colour slides directly with correct colour rendering on daylight or "tungsten" emulsion depending on the mired value of the light source (see later IV.B, 3). The Nikon slide-copying system is illustrated in Fig. 9.

2. Bellows macro-camera

This type of camera with lenses primarily designed for close-up work is the most useful instrument for routine macrophotography. A good example is the Leitz Aristophot which may be equally well applied to photomicrography. The apparatus consists essentially of a rigid stand with a central prismatic guide-bar to which the camera is fixed and which allows vertical adjustment of the camera by means of a control knob. The base plate is fitted with a central disc for speedy location of a Leitz microscope or macro-dia apparatus for transmitted-light macrophotography.

Different cameras may be used on the basic stand to enable the use of different formats. The 4×5 in. bellows camera has a rotating darkslide frame and international back, the former enabling easy selection of picture area for composition within the format. The international back enables the use of Polaroid Land equipment. A 9×12 cm format bellows camera

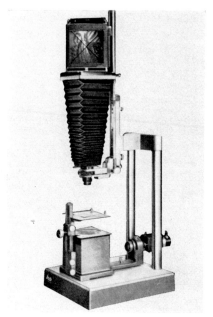

FIG. 10. Leitz Aristophot 4×5 in. bellows camera with macro-dia apparatus. The "X" which is visible on the viewing screen is the clear glass area. (Courtesy E. Leitz, Wetzlar.)

with mirror attachment and adaptors for $6\cdot5 \times 9$ cm format can also be supplied, and a separate bellows enables the use of a Leica 35 mm camera. The 4×5 in. camera is also available with fully automatic exposure control. The shutter is fixed in an anti-vibration mount within the lower bellows support.

The macro-dia apparatus provides a suitable flat field of illumination for large specimen areas and can be used with any of the cameras. This accessory consists of an illuminating lamp attached to a housing with built-in mirror and support for the various condenser lenses to be used. The large platform specimen stage above the condenser is adjustable vertically by means of a rack and pinion; a number of masking inserts (see Table II) are provided with the stage. The equipment is illustrated in Fig. 10.

When incident light is considered to be the more appropriate form of illumination the macro-ring illuminator may be used (Fig. 11). This is adjustable on an auxiliary bar fixed to the baseplate.

As mentioned above (Section II.F) the choice of equipment will depend on the image scale required and the relationships between magnification, working distance, extension and focal lengths of lenses in the special Milar and Summar range are given in the graphs in Fig. 2. Also, since in transmitted light macrophotography the condenser and lens apertures must be matched, a series of condensers will be required and Table II shows the combinations which are necessary to cover the image scale range from 1 : 1 to 25 : 1.

TABLE II

Reproduction scales obtained with the macro-dia apparatus together with the associated stage inserts and condenser lenses

Lens name and focal length		Reproduction scale*	Stage insert	Condenser
Summar	Milar			
f = 12 cm		1 : 1– 4 : 1	80 mm	120
	f = 10 cm	1 : 1– 5 : 1	70/60/50 mm	100
f = 8 cm		1 : 2– 7 : 1	60/50/40 mm	80
	f = 65 mm	1 : 1– 9 : 1	45 mm	65
	f = 50 mm	1·4 : 1–12 : 1	40 mm	42
f = 42 mm		1·7 : 1–13 : 1	30 mm	42
f = 35 mm		2 : 1–16 : 1	25 mm	35
f = 24 mm		4 : 1–25 : 1	15 mm	24

* These scales are for the 9×12 cm camera. Reproduction scales for the Leica 35 mm camera can be derived from Fig. 2.

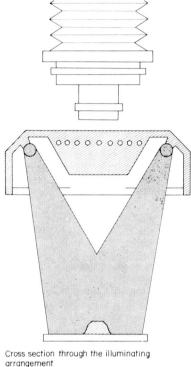

Cross section through the illuminating
arrangement

FIG. 11. Diagram of Leitz macro-ring illuminator. (Courtesy E. Leitz, Wetzlar.)

3. Macrophotography with the stereo microscope

Most stereo microscopes can be fitted with a coupling ring to enable the use of an attachment camera. This arrangement is very suitable for making photographs of bacterial colonies at fairly low magnification. Alternatively, the stereo binocular head may be replaced by a monocular tube or with a phototube for trinocular assembly. Wild Heerbrugg, for example, supply stereophototubes for Alpa, Exakta, Edixa and Praktika (general purpose cameras) in addition to the phototube with R.M.S. (Royal Microscopical Society) standard diameter. Fig. 12 illustrates some of these arrangements. A low magnification (16 × on print) photograph of bacterial colonies using a combination of epi- and dia-illumination is reproduced in Fig. 13.

4. Macrophotography with the microscope

There are also a number of special devices which enable the microscope to be used for transmitted light macrophotography. Such a device is the macro-bellows attachment made for use with Nikon 35 mm camera and

(a)

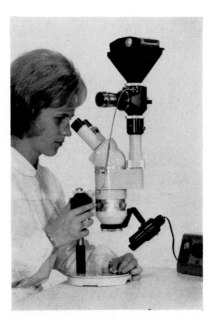

(b)

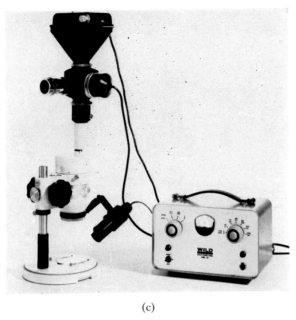

(c)

FIG. 12. (a) Wild M5 stereomicroscope with phototube and adapter for Exakta camera. (b) M5 with phototube for trinocular assembly, attachable camera and roll film magazine. (c) M5 with phototube and Photoautomat MKa4 and motor-driven 35 mm camera magazine. (Courtesy Wild, Heerbrugg, Switzerland.)

microscope (Fig. 14). The bellows attaches directly to the microscope in place of the eyepiece body so that the low power microscope objectives are used as photographic lenses in a variable length camera. The camera can now be rotated through 360° for any desired orientation of specimen within the format.

B. For photomicrography

There is no easy way to classify cameras used for photomicrography. They differ greatly in design and in format. In general the largest formats are found in bellows cameras with the shutter assembly fixed into the bellows support, but there are rigid large-format attachments which can be fixed on to shutter units made to carry small format backs as well. Also, there seems to be no particular advantage to be gained in talking about attachment cameras as a special class, as "degrees of attachment" differ and the position on the microscope stand to which they are attached varies considerably.

A camera may be attached to an eyepiece tube by a rigid physical connection or it may simply sit over the eyepiece tube without making physical

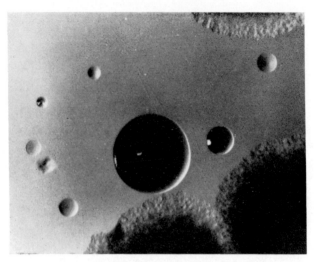

Fig. 13. Macrophotograph of colonies of a paromomycin-dependent strain of *Bacillus cereus* grown on nutrient agar after growth in 3% alcohol broth. Photographed through a Nikon SMZ-2 stereomicroscope; agar plate specimen illuminated by a mixture of epi- and dia- illumination. Reproduction scale 10 : 1.

Fig. 14. Nikon F camera and bellows in use for macrophotography through the microscope. (Courtesy Nippon Kogaku, Tokyo.)

contact with it. There is no specific reason, as a rule, why the camera should be physically attached to the microscope, so long as it is possible to bring the camera assembly and any compensating or "field lens" it may have, sufficiently close to the eyepiece and the right distance away from the eyelens to match the entry and exit pupils. Also the camera must be made co-axial with the eyelens, and this is usually determined by some sort of sleeve or collar device which fits over the eyepiece tube.

Two points are worth noting in this connection. It is better, if possible to avoid physical contact between microscope and camera since vibrations from and manipulations of the camera will affect the focus of the microscope, and, if the camera is supported on the eyepiece tube it may place a strain on the focusing mechanism in microscopes with moveable limbs. To avoid loss of focus and strain problems most modern research microscopes are of the fixed limb type and both coarse and fine focusing movements operate on the stage/condenser assembly. Microscopes with a moving limb may be provided with a "clamp" to make the limb "rigid" for photomicrography with an attached camera, as in the case of the Wild M20; the fine focus must work on the stage and be "unclamped" for such a system to be effective.

In other instruments the light to be used for photography and photometering is diverted by a beam-splitter or prism into the horizontal cross member of the limb and the camera is attached at the top of the limb at the opposite end from the eyepiece. In such arrangements there is a lens seated in the limb which replaces, but is equivalent to, the photo-eyepiece. In other designs not only is the photographic light path within the limb, but the camera itself is built into an enlarged limb as in the case of the Zeiss (Oberkochen) photomicroscopes.

Again, some manufacturers refer to the device containing the shutter which links the microscope and the camera "box" as the camera, although such a "camera" would be useless without the light-tight box whatever may be its size or format. Also, shutter types vary; they may be mechanical or electromagnetic, and varying degrees of automation may be built in.

In view of this plethora of designs it is almost impossible to categorize them into defined classes, and the only simple distinction is between cameras with extensible bellows and those with rigid construction, and so, fixed camera length. The main advantage of the bellows type is that it permits variation of the image scale and enables the long and cumbersome camera required for large format and great image scale to be packed away in a much smaller space. The drawback of the bellows camera is that it is more liable to damage and pin-hole breaks in the bellows may appear at the corners of the folds after much use.

This is not to say that the fixed length camera can only be used for reproductions at a fixed image scale. Obviously, by varying the objective and eyepiece powers, different (film negative) image scales are possible and, in addition, any negative may be enlarged in printing to increase the final image scale, but this should be done within the limits of useful magnification as discussed in Section II.F.

1. *Bellows cameras*

As examples of this class we may use the Aristophot systems (Leitz, Wetzlar) already referred to for macrophotography in Section A, 2 above. For use with the microscope the macrolens is replaced by a connecting collar which forms a light-tight, but not physically adherent, link with the microscope. A collar attachment on the eyepiece tube is marked with a ring to denote to what extent the camera must be lowered to achieve the optimum optical relationship between camera and photo-eyepiece. Focusing is best done with the aid of a (Leitz) focusing magnifier using the clear-glass area of the screen (Section II.D). Exposure determination may be achieved by means of a sensitive meter of appropriate design such as the Leitz Microsix SL which is described later (Section VII.B, 1). Alternatively, the large format Aristophot camera is available as a fully automatic camera with electronically controlled photometering and exposure. The light path in the large format camera as used for photomicrography, is shown diagrammatically in Fig. 15.

2. *Fixed length cameras*

Various fixed length cameras are to be found, ranging from the purely manual with no facility for light metering, through the semi-automatic to the fully automatic variety. Outwardly the manual and automatic cameras may appear very similar. A good idea of the type of product available for attachment to almost any microscope can be obtained from a brief survey of the three ranges which follow.

(a) *Vickers*. This manually operated camera consists essentially of a reflector body with focusing telescope which is "linked" to the microscope tube by the appropriate light trap adaptor; on top of the body is the shutter and on to this various camera backs for the range of formats from 35 mm to 9×12 cm and polaroid $3\frac{1}{4} \times 4\frac{1}{2}$ in. may be fixed. Fig. 16(a) shows the 35 mm form.

The Vickers semi-automatic camera is similar, but, instead of the mechanical shutter, the camera back and reflector body are linked by an electromagnetic shutter unit. By means of a prism 20% of the light can be deflected to the side tube designed to take a photometer cell (cadmium sulphide) which, with the photometer enables exposure times to be deter-

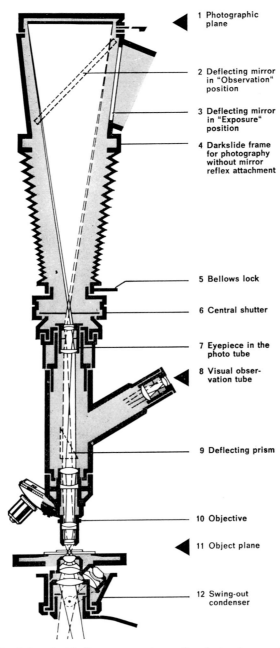

1 Photographic plane

2 Deflecting mirror in "Observation" position

3 Deflecting mirror in "Exposure" position

4 Darkslide frame for photography without mirror reflex attachment

5 Bellows lock

6 Central shutter

7 Eyepiece in the photo tube

8 Visual observation tube

9 Deflecting prism

10 Objective

11 Object plane

12 Swing-out condenser

Fig. 15. Leitz Aristophot bellows camera in use for photomicrography. (Courtesy E. Leitz, Wetzlar.)

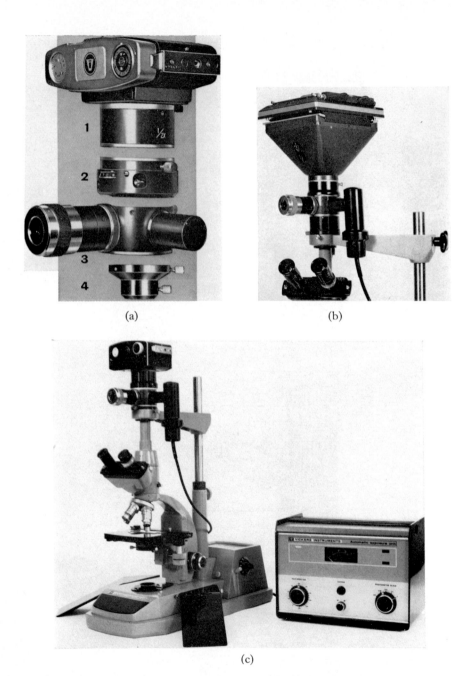

(a)

(b)

(c)

FIG. 16. (a) Vickers 35 mm camera and intermediate body (1); cable release shutter unit (2); reflector body with focusing eyepiece and tube to take photometer cell (3); adapter tube (4). (b) Vickers semi-automatic camera with large format back. (c) The autowind 35 mm camera and J35 exposure unit with the M15c microscope. The photomultiplier head is visible to the right of the trinocular vertical tube. (Courtesy Vickers Instruments Limited, York.)

mined (see Section VII). Using the J35 photometer with photomultiplier detector it becomes automatic except for film wind-on (Fig. 16(b)).

The Vickers fully automatic system consists of a 35 mm motorized autowind camera back, electromagnetic shutter unit and reflector body with J35 photomultiplier and automatic exposure unit (Fig. 16(c)).

(b) *Nikon* (*Nippon Kogaku K.K.*). The differing features of the Nikon range of attachable cameras are summarized in Table III, the main difference being in the degree of automation. A number of features are common to all three attachments. All may be used with an ocular finder (or viewing telescope) for focusing purposes; its screen is marked with the limits of the various picture formats and centre double crosslines for focusing. Alternatively the telescope may be replaced by a small screen

TABLE III

Nikon camera bodies for photomicrography

Model	PFM Manual	EFM Semi-automatic	AFM Automatic
Exposure meter	None	Coupled CdS	Coupled CdS meter and electronic shutter.
Meter reading	None	Exposure computed from integrated full-format reading taken in plane conjugate to film. Meter coupling between both shutter speed and film speed ASA settings	Integrated full-format as EFM; coupled through transistorized circuit to shutter operation
Exposure range	None	Photomicro and single-frame cinemicro: 1/250 to 32 sec B and T.1/250 to 1 sec coupled to meter. Cine: framing speeds 64 to 1 fps	Automatic setting range: 1/100 sec to 10 min. Manual setting range: 1/125 to 8 sec and T
Film speed setting range	None	Photomicro: ASA 12, 25, 50, 100, 200, 400, 800, 3200; each increment in 3 steps.	ASA 12 to 3200 as EFM but set on transistorized control box.
Meter power supply	None	1·3 V mercury battery	4 × 1·5 V penlite batteries
Shutter		Leaf type shutter equidistantly spaced click stops: T, B, 1/2, 1/4, 1/8, 1/15, 1/30, 1/60, 1/125, 1/250 sec and x-synch.	Electronic, built-in, controlled through transistorized circuit. Manual settings from 8–1/125 sec and T: x-synch.

finder with format-etched ground glass and used with a $7\times$ focusing magnifier. A different relay lens is required depending on the film format but each camera body can be adapated for 35 mm roll film use in 6×9, 6×6 or 6×4 cm formats, sheet film or plates $2\frac{1}{2}\times3\frac{1}{2}$ in., or Polaroid $3\frac{1}{4}\times4\frac{1}{4}$ in. For 35 mm format the relay lens gives the camera a 0·5 factor; for the other formats a lens giving a 1·3 factor is built in to the adaptor cone.

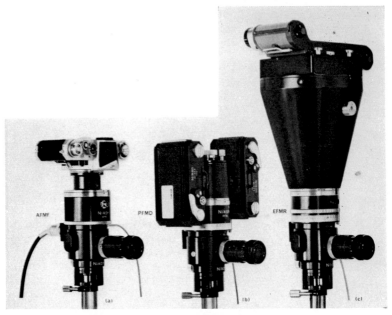

Fig. 17. Nikon attachable cameras. (a) Automatic with Nikon F camera (but without autowind). (b) Manual with double head. (c) Semi-automatic with roll-film back. For further details see text. (Courtesy Nippon Kogaku, Tokyo.)

Eyepiece adaptor collars enable any of the camera combinations to be used on microscopes with tubes of diameter 25 mm, 32·5 mm, 33 mm or 36 mm (stereo-microscopes).

Representative assemblies are shown in Fig. 17. The double-head attachment is particularly noteworthy. One camera back may be loaded with black and white, the other with colour, and selection made by rotation of a prism which directs light to one or the other as required.

(c) *Wild Heerbrugg.* In the Wild range all the camera bodies can be used with adaptors to give a range of formats from 35 mm to 4×5 in. Manual, semi-automatic and automatic versions are available. They can be attached to microscopes with eyepiece tube diameters 25 mm or 33 mm.

The attachable camera has a manually operated Synchro-Compur shutter with settable speeds from 1/500 to 1 sec and B setting for time exposure. Focusing is by the usual type of telescope but uses full light intensity deflected by a prism. On pressing the shutter the prism is automatically swung out and the full light intensity falls on the film. The camera factor is 0·5 except when the 4 × 5 in. attachment is used when the camera factor is 1·0.

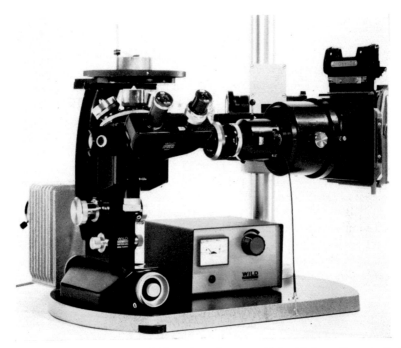

FIG. 18. Wild camera MKa2 (for trinocular assembly) with large format attachment and film holder with Polaroid-Land sheet film magazine, 4 × 5 in. The microscope is the Wild M50 metallurgical. (Courtesy Wild, Heerbrugg, Switzerland.)

Camera MKa1 (see Fig. 12) is similar but has a fixed beam-splitting prism deflecting 25% of the light to the receiver; the shutter is self-cocking, speeded from 1/125 sec to 1 sec with T and B settings, and there is a swing-out photocell to which an appropriate meter for exposure measurement may be connected. A variant, MKa2, is designed for trinocular assembly and has no viewing telescope (Fig. 18).

The Photoautomat MKa4 (Fig. 12) is a fully automatic camera with the following additional features. On the side of the body is a drum which houses a photo-resistor, the current through which varies with the amount of light falling upon it. A small proportion of the image light is diverted for this purpose. From this variation a correct exposure time is determined by the electronic control on which the film speed is set in advance. Compensation can be made for the technique in use (bright-field, dark ground, etc.) by turning a milled ring in front of the photoresistor drum. Another milled ring controls the compensation for eyepiece magnification.

A motor-driven 35 mm magazine enables the films to be automatically wound on after exposure; a warning light shows when the film is moving and when the end of the film has been reached. Most important of all (in some senses) is the safety switch which prevents the film transport mechanism from operating if the magazine slide is not fully pulled out. Many an eminent scientist has "exposed" a roll of film without having pulled out the slide. It couldn't happen with this camera.

The MKa5 is similar but designed for use with the Wild trinocular assembly and has no viewing telescope.

3. *Automatic cameras*

Reference has already been made to automatic cameras above and a wide variety of instruments is now available. Cowen (1969) has recently tabulated the main features of a range of automatic cameras and these are listed in Table XI. The definition he uses for an automatic camera is that it should "automatically determine the exposure time from the quantity of light passing into the camera. In other words it is only necessary to set the film speed on the instrument and to actuate a control which opens the shutter. The camera shutter then closes automatically after the film has been exposed for the correct period of time." This definition is a little inexact since some devices determine exposure from a small fraction of the light leaving the microscope and this fraction may never pass into the camera, but the intention is, I think, clear. Automatic wind-on is not considered essential. Some manufacturers refer to the former type as semi-automatic; and a camera with automatic wind-on as well is fully automatic.

The Reichert Photo-Automatic system is slightly different from all the others as the exposure is selected by turning the control knob until a green light denoting optimum exposure is illuminated; on either side of the optimum range red lights are illuminated.

Cowen's table although a little out of date now is nevertheless included as it gives at a glance the various features which may be sought in an automatic system. Unfortunately, the Watson apparatus must now be deleted as production has been discontinued. The Zeiss Photomicroscope I has been superseded by the Photomicroscope II and a number of additions could be made. Among these are the Reichert Kam ES—an electronic camera system which may be fitted to any microscope and adaptable for use with various "backs" to give formats up to $3\frac{1}{4} \times 4\frac{1}{4}$ in. Polaroid. The shutter is novel in that it is at the same time a fully silvered mirror reflecting all the light which would enter the camera box, on to the sensitizing photo-resistance. A beam-splitter before the shutter deflects 20% of the light to a telescope for viewing during exposure. Exposure time read-out and determination is similar to the Photo-Automatic system and easily permits deliberate exposure time deviation to allow for reciprocity failure or other special conditions.

The Gillet and Sibert Autolynx automatic photohead for attachment to GS microscopes is a very compact design permitting automatic exposure and film transport of 35 mm film. The exposure rate may be as rapid as one exposure every 6 sec. There is a choice of reflex head or beam splitter, in the former case the light is reflected either to binocular eyepieces or totally to the Autolynx; with the beam-splitter the image can be viewed while the exposure is made. These "heads" have a magnification factor of 1·2, the Autolynx unit itself has a factor of 0·5. A useful accessory not normally provided for 35 mm cameras is the ground glass focusing screen which can be placed in the film plane and used to ensure that the film image is in focus when the eyepiece-viewed image is in focus. The photo unit monitors and integrates light values in relation to area and time, and the opening interval of the electromagnetic shutter is controlled by a solid state circuit. The user selects the film speed within the range 12–160 ASA and makes any correction (e.g. by filter) for colour temperature.

One of the very useful features of the GS system is the continuously variable neutral density filter which may be placed in the filter well in the base of the GS conference microscope stand. It enables a very flexible control of the light intensity to the camera (overall ratio 1 : 500) by simply rotating a knurled ring. An iodine-quartz lamp is standard on this instrument. The arrangement is illustrated in Fig. 19.

Another very useful instrument, on which Cowen was unable to obtain information is the Olympus Photomicrographic Exposure Meter Model EMM-V. This instrument not only permits automatic exposure determination but allows the rapid determination of, and correction for, colour temperature when using colour stock. It is described in greater detail below (Section VII.B, 7).

4. Photomicroscopes

A photomicroscope may be loosely defined as an instrument designed with built-in photomicrographic facilities. In the well-known photomicroscopes manufactured by Zeiss (Oberkochen) the light used for photography is retained totally within the microscope "tube" and "stand"

and the camera is accommodated within the instrument by greatly enlarging the limb. Cross-sectional diagrams are given in Fig. 20(a) and (b).

Other instruments designed with special consideration for photomicrography may have cameras attached to the limb or stand such as the Vickers M41 Photoplan (Fig. 21) which has built-in facilities for directing light as required to the eyepiece, or the camera, or to both, or to the exit ports for the photomultipliers which may be linked to automatic exposure control

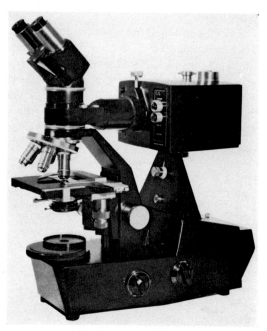

FIG. 19. Gillett and Sibert Autolynx automatic photo head mounted on the GS Conference microscope. The exposure is automatically computed and the film wound on after exposure. Note the rotating continuously variable neutral density filter mounted on the base. (Courtesy Gillet and Sibert, London.)

units or high sensitivity photometer timers. The various possibilities are demonstrated in Fig. 22. In sophisticated instruments of this kind it is also usually possible to illuminate the specimen by both dia- and epi-systems (transmitted and incident illumination) at the same time (Fig. 23). The special advantage of this for the biological scientist is the ability to combine two different types of technique; for example, the dia-illumination system may be arranged for phase contrast microscopy, while the epi-illumination may be used to provide ultraviolet light for fluorescence illumination of an appropriately stained specimen. The resulting "picture"

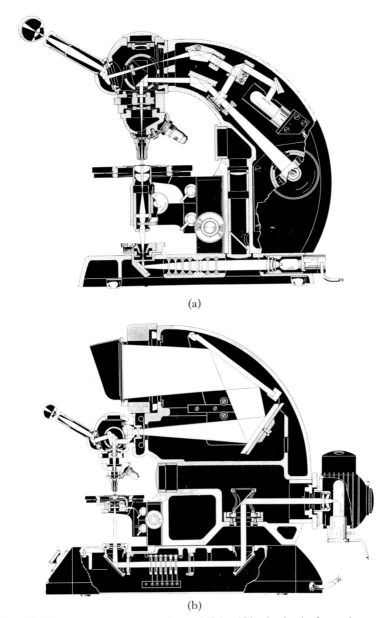

(a)

(b)

FIG. 20. Diagrammatic cross-sections of Zeiss (Oberkochen) photomicroscopes. (a) Photomicroscope II for 35 mm, and (b) Ultraphot III for up to 9 × 12 cm, automatic photomicrography. Note the photomultiplier tube housed in the enlarged limb above the film transport mechanism (a). (Courtesy Carl Zeiss, Oberkochen.)

may be easily photographed with the aid of the high sensitivity photometer with its ability to determine exposure from 10%, 1% or 0·5% of the total photographic field area.

5. *Simple tube connectors*

Despite the complexity and refinement of the photomicroscope and the various other attachments produced for easing the task of the photomicrographer, it is possible to produce excellent results with a cheap and simple tube connector or adaptor so long as one has a suitable camera. Such a

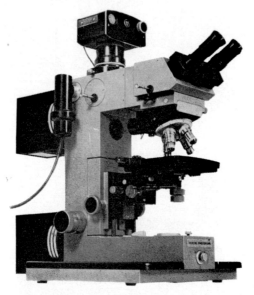

FIG. 21. Vickers M41 Photoplan microscope with 35 mm autowind camera. The photomultiplier is inserted at the top left corner of the limb and links with the J35 exposure unit for fully automatic photomicrography. Note lamp housings for both epi- and trans illumination. See also Figs. 22 and 23. (Courtesy Vickers Instruments Limited, York.)

system is the Nikon F camera with photomic T head, linked to the microscope by a simple microscope adaptor tube which is a metal tube, with a single field lens, capable of being coupled to the microscope and to the camera.

A simple system of this sort has the advantage that there is only a single glass element between the photo-eyepiece and the film plane so that there are fewer surfaces to collect dirt, reflect light or introduce errors. The disadvantages are the possibility that camera shutter and mirror movements may be translated to the microscope, image viewing must be via the camera viewing screen and light measurement for exposure determination may not be possible. All these disadvantages can, however, be overcome. If the camera is attached to a suitable stand it may be lowered over the eyepiece so that there is no physical contact, and in the case of the Nikon F

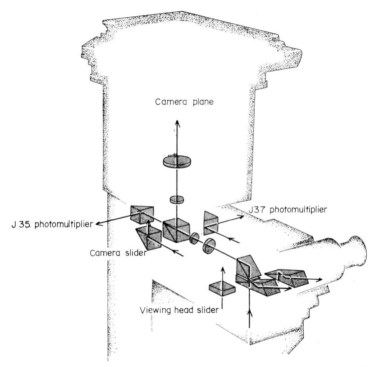

Camera plane

J37 photomultiplier

J 35 photomultiplier

Camera slider

Viewing head slider

Fig. 22. Diagram of the optical arrangement for photography with the Vickers M41 Photoplan microscope. A large number of combinations is possible between the viewing head, photomultipliers, and the camera plane, employing the camera prism slider and the viewing head slider. (Courtesy Vickers Instruments Limited, York.)

system the "through-the-lens" light metering capability of the Photomic T head can be used for exposure determination. Of course, to focus the image, the camera must be of reflex type and a ground glass screen with central clear glass area (Nikon Type C screen) is needed.

One of the great advantages of this system is the fact that the camera can be used for all normal laboratory photographic needs, including macrophotography (see Section II.B above), and the only additional requirement for photomicrographic use is the simple adaptor tube available

at moderate cost, and a type C screen. The arrangement for photomicrography is shown in Figs. 24 and 45, and photomicrographs of bacteria
agar on layers, obtained with it, are shown in Fig. 25.

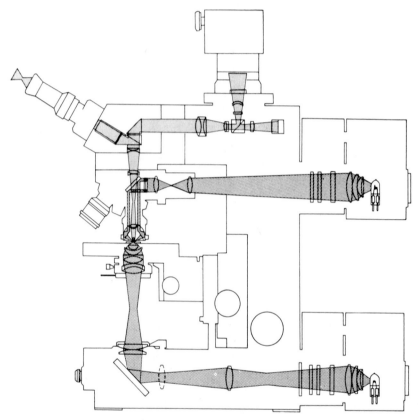

FIG. 23. Diagram of the Vickers M41 Photoplan showing the optical arrangement
for incident and transmitted illumination, and incident dark ground. (Courtesy
Vickers Instruments Limited, York.)

IV. PHOTOGRAPHIC EMULSIONS

The photographic "image" is the result of a physico-chemical change
which has been brought about in a sensitive layer on a suitable substratum,
and the process of change terminated by appropriate chemical treatment.
The physical agent is electro-magnetic radiation, the energy of which causes
interactions between chemical moieties in the emulsion, so that there is a
relationship (though not a simple one) between the quantity of radiation
falling on a given area of emulsion, and the amount of chemical change

FIG. 24. Simple microscope adapter tube for Nikon F. (Courtesy Nippon Kogaku, Tokyo.)

produced. Many naturally occurring substances are affected by radiation, but a substance suitable for photography must react rapidly to small total quantities of radiation and behave in a predictable way. The chemicals which most adequately meet these requirements are the halides of silver to which may be "coupled" other chemicals required for particular properties.

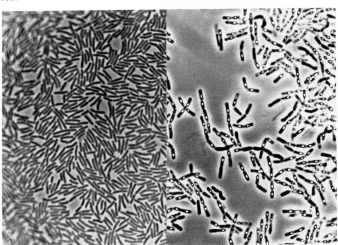

FIG. 25. Left hand half-frame: *Escherichia coli* growing on nutrient agar, photographed by phase contrast using a "poor" lens combination, achromatic objective and Abbe condenser. Note the image curvature and poor edge definition. Right hand half-frame: *Bacillus cereus* PmD2, phase contrast photomicrograph using a fluorite objective and achromatic condenser to give improved image quality and flatter field.

A. Black and white emulsions

The photographic emulsion is made by precipitation of silver bromide from a mixture of potassium bromide and a silver salt (usually nitrate) in liquid gelatin. Some silver iodide is also usually incorporated. The "emulsion" is then "cured" with heat and chemicals until the right crystal (grain) size is attained, as well as the desired sensitivity and contrast characteristics. The former increases, the latter decreases, with heat treatment. The prepared emulsion is then coated on a backing of acetate, glass or paper, depending upon the type of stock being made. There is a protective layer of gelatin laid over the "grain" layer and there is a coating of dyed gelatin (anti-halation backing) on the "underside" of the film to absorb transmitted light and prevent image diffusion by reflected radiation. This layer also prevents curling of the film which would result from the expansion and contraction of gelatin were it applied to one surface only. Both the top and bottom "coats" may be applied on a thin substratum of hardened gelatin to enable proper adhesion of the coating to the base (acetate, glass, etc.).

1. *Colour sensitivity—types of film*

Silver halides are sensitive to all the higher energy radiations of the electromagnetic spectrum and it is therefore possible to make photographic images using, for example, beams of electrons, X-rays, or ultraviolet rays— all of which are invisible to the eye; the eye being visually insensitive to these radiations (though it is physiologically very sensitive and is damaged by them). On the other hand the eye is visually sensitive, to a greater or lesser degree, to all the wavelengths in the range from about 4000 Å to 7000 Å with a maximum at about 5500 Å (see Fig. 26). The sensitivity ranges of "unsensitized" silver halide emulsions do not cover the spectral range of the eye, a silver chloride emulsion reacting only as far as the violet of the visible spectrum, and silver bromide/silver iodide emulsions only as far as the blue-green, 5000 Å. In order to increase the sensitivity of the film to cover the spectral range utilized by the eye, sensitizing dyes are added to the emulsion, and it is now possible to produce emulsions which are sensitive even into the infra-red region of the spectrum (Fig. 27). It will be noted from Fig. 26, however, that a sensitized (panchromatic) emulsion does not have the same relative colour sensitivity as the eye, being more sensitive to blue and red and less sensitive to green, than is the eye. This means that photographs of blue and red objects will appear relatively too dark, while green objects are relatively too light. To obtain a better photographic tonal range, more nearly that recorded by the eye, a green filter must be used. On the basis of the different spectral sensitivities of film emulsions it is possible to group them into a number of broad classes; the following are useful within the context of this Chapter.

(a) *Process films*. When tone rendition is not required, for example, in the copying of graphs, documents, line-drawings, etc., an unsensitized emulsion may be used. These emulsions are slow, but of very high contrast and fine grain, and since they are not sensitive to radiation in the longer wavelength half of the visible spectrum, they can be handled in orange or red light. Orthochromatic (partially sensitized emulsions, Fig. 27) are also available in this category.

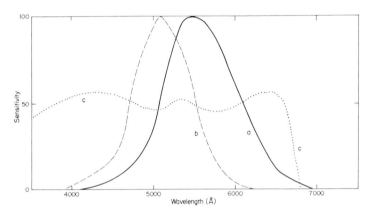

FIG. 26. The spectral sensitivity of the eye and of a film emulsion. The sensitivity range of the eye at high illumination levels is given by the "photopic curve", a; and at low illumination levels by the "scotopic curve", b; the sensitivity distribution of a typical panchromatic emulsion is given by c.

(b) *Normal Panchromatic*. These have a "balanced" sensitivity, the manufacturer's aim being to achieve maximum green sensitivity with lesser but equal blue and red sensitivity; but the relative sensitivities for the different colours never quite achieve that of the eye (Fig. 27(c)). These emulsions are comparatively slow, of medium to high contrast, and fine grain and are the best type for photomicrography, except in especially difficult circumstances.

(c) *Fast Panchromatic or Hypersensitive*. These emulsions have an extended red sensitivity, high speed reactions and reduced contrast properties. They are also "grainier" than the slower variety, increased grain-size being necessary for fast response. They are not as well balanced as the former class, being rather over-sensitive in the red, but are consequently more useful for Tungsten light photography which is rich in red radiation. For

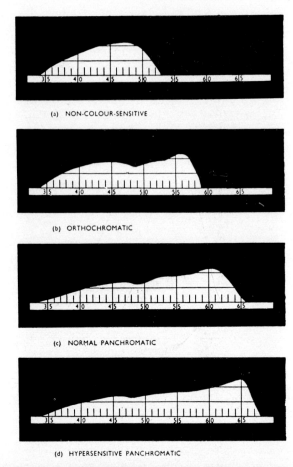

(a) NON-COLOUR-SENSITIVE

(b) ORTHOCHROMATIC

(c) NORMAL PANCHROMATIC

(d) HYPERSENSITIVE PANCHROMATIC

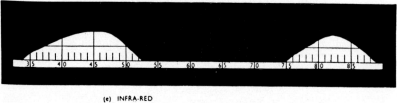

(e) INFRA-RED

FIG. 27. Wedge spectrograms of typical emulsions of each of the principal classes of colour sensitivity (to tungsten light at 2850°K).

"normal" tone rendering they must be used in conjunction with filters of appropriate spectral transmission.

(d) *Polaroid-Land film.* This film may be produced with extremely high-speed emulsions, even up to 30,000 ASA. Basically it has a fast panchromatic emulsion which is developed *in situ* by squeezing the contents of a satchet of developer over the sensitive layer after exposure. The chemicals then diffuse through to the "positive" layer where the unused silver halide, unconverted to silver in the original exposure, is reduced to silver in the "positive" layer to yield a "translated" positive image, which is then "fixed".

(e) *Infra-red emulsions.* These films which are especially sensitive to long wavelengths (Fig. 27(e)) are finding increased use in biological work because of the high penetration of infra-red rays. Infra-red photomicrography may reveal structures within opaque objects, e.g. insect eggs, not normally visible to the eye. Such emulsions must be cool-stored as they may become heat-fogged in time.

2. Latent image formation and development

When a silver bromide "grain" is exposed to light some of the silver bromide is reduced to metallic silver, the amount depending on the energy and duration of the radiation. If the exposure is extremely short, no reduction may take place, but at slightly longer exposures a few atoms of silver may be produced. These silver atoms form the *latent image* of the emulsion. This latent image is then developed in a chemical solution (the developer) whose function is to utilize the silver atoms as a catalyst for the reduction of further "molecules" of silver bromide to silver. The greater the original exposure (quantity of light incident, see below) of a particular area of film the greater the number of catalyst atoms in the latent image and, therefore, the faster the rate of reduction to silver in that area during development; so the greater the degree of blackening during the development time. Conversely, when there are very few "centres", and maybe not even on all grains, the slower the reduction and the greater the "spaces" between reduced points. Such areas are consequently less "blackened" and more "translucent" as a result of development. The unreduced silver bromide would continue to undergo reduction if the film was exposed to light, or left in the developing solution, so to prevent the reduction process continuing beyond the desired point, the unconverted silver halide must be washed out of the emulsion. The fixing solution (sodium or ammonium thiosulphate) serves this purpose and when all the unconverted halide has been dissolved away by the fixer, the film may be safely exposed to light.

It is not possible to deal here with the various types of developer which are obtainable, but it should be noted that they vary in their ability to reduce the halide within reasonable development times. This means that a particular emulsion given a particular exposure, and therefore having particular latent image characteristics, may require vastly different periods of development depending on the developer used. Some developers may be unable to produce a satisfactory developed image from the particular latent image formed. In other words the sensitivity of the emulsion, or the effective film speed, will depend on the reducing capacity of the developer. Very vigorous developers with high reduction rates can also lead to increase in the developed grain size. It is essential that the photographer should select a film emulsion and developer formula that are compatible for the type of work in hand, and the only way to find out what the particular effect will be is to experiment with films and developers under recorded conditions. The task is made easier, however, if certain characteristics are known in advance.

3. *Sensitometry and the characteristic curves*

As mentioned above, when light falls on a photographic emulsion it elicits a response depending on the sensitivity of the emulsion, and the study of the relationship between the response and the light is called sensitometry.

(a) *Brightness and exposure.* Whatever the subject presented to a camera it may be defined solely in terms of the brightness of the various parts as "viewed" by the camera. The image depends for its formation on these variations in brightness. Similarly, the positive print is simply a pattern formed from areas of differing brightness. Brightness is measured in "candles per square centimetre" or "lamberts". The incident light falling on an opaque surface is measured as "illumination", the reflected light as brightness or luminance.

The international standard unit of light emission now used is the candle. The candle is the luminous intensity of a source which emits light energy at $1/60$ the intensity emitted by 1 cm^2 of a perfectly radiating body at the freezing point of platinum (2042°K). Light flow is measured in lumens and this is the flux of light on to each unit square of the inside surface of a hollow sphere of unit radius with a point source of luminous intensity one candle at its centre. A uniform diffuser is a surface with a reflection or transmission factor which is independent of the angle of illumination; the brightness of such a diffuser emitting one lumen per square foot is termed a foot-candle. One foot-lambert $= 1/\pi$ candles/ft^2; one lambert $= 1/\pi$ candles/cm^2. In the case of a perfect diffuser, illumination and brightness (luminance) are equal, but no perfect diffusers in fact exist.

The variations in brightness of a subject depend on the variations in illumination on it and upon the different reflection characteristics of different areas and the angle of view. The ratio of the maximum to the minimum brightness is called the brightness range of the subject. The purpose of sensitometry is to relate the range of brightness of the subject to the range of brightness of the print and this will, of course, depend upon the response of the negative as well as of the printing paper.

When a negative is exposed, the response of the film is proportional to the product of the illumination (I) and the exposure time (t), so that the exposure, $E = I.t$. This is the reciprocity law and implies that the same response of the negative may be obtained if either the illumination is doubled and the exposure time halved, or vice versa. That is, there is a reciprocal relationship between I and t. Unfortunately, under certain conditions this relationship is not obeyed and the term reciprocity failure is applied to these situations. Since reciprocity failure must often be considered in photomicrography more will be said about it later.

It should be noted that, although the exposure time t is controllable (by the shutter mechanism) the value of I for any area of the film depends upon the brightness of the corresponding area of the subject, so that, under normal conditions any given negative will have received an almost infinite number of different exposures (E).

Since t is measured in seconds and I in lumens per square foot or per square metre, E is expressed as lumen-seconds per square foot or per square metre. The latter is also called metre-candle-seconds. Since in common parlance t is usually called exposure, to avoid its confusion with E, t should be referred to as "camera exposure".

(b) *Measuring the response of the negative.* After development the negative will show different degrees of blackening depending upon the variations in E. These degrees may be measured as transmission T, where

$$T = \frac{I}{I_0}$$

I_0 is the light incident upon any area and I the light transmitted through it. T is always less than 1 and normally expressed as a percentage.

The *opacity* (O) is the reciprocal of T, and the density (D) is the logarithm of the opacity so that

$$D = \log O = \log \frac{1}{T} = \log \frac{I_0}{I}$$

Density is the unit of "blackening" usually used and its value increases with the degree of blackening, i.e. increase in E. The density is proportional to the quantity of deposited silver, so that a doubling of the value of D represents a doubling in the amount of deposited silver. Density units are also appropriate because the subjective evaluation by eye of equal gradations of "blackness increase" corresponds with equal values of density increase as measured instrumentally.

(c) *The characteristic curve.* If the measured density D is plotted against the exposure E, the response for any particular film under specified conditions of development will be represented by a curve of the type shown in Fig. 28. If, instead of this semi-log plot, a log/log plot is constructed, a

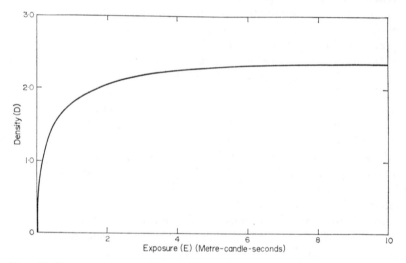

Fig. 28. Response curve of an emulsion obtained by plotting density against exposure. (Courtesy Ilford Limited, London.)

much more useful curve of the relationship between blackening and exposure is derived (Fig. 29). This plot of D against log $I.t$ is the characteristic curve for that emulsion under the specified conditions. One of the great benefits of this type of curve is that it indicates clearly the response of the emulsion at very low and very high exposure values (E, not to be confused with camera exposure t, although this does affect E). In other words the characteristic curves give information on the behaviour of the film when under- and over-exposed.

It will be seen that the minimum recorded density is always above zero since some grains, although not producing a latent image can nevertheless be reduced on development. This minimum value of density is the fog level of the emulsion for the development conditions used. Beyond the threshold value of exposure (A, Fig. 29) increase in E, and so log E, gives slowly increasing values of D until a maximum value for the rate of increase in D with log E is reached at B. The region of the curve below B is called the toe. From B to C is the straight line portion of the curve when the relationship between D and E is constant. The shoulder from C to M, the maximum possible density, represents a range of exposure when the rate of increase in density decreases (until it stops at M). Beyond M, at

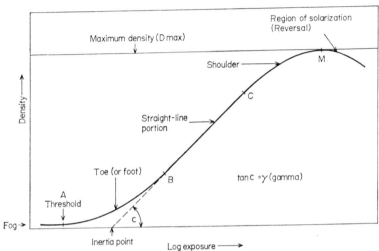

FIG. 29. The characteristic curve for a negative material. (Courtesy Ilford Limited, London.)

very high values of E the density may actually decrease, a phenomenon known as solarization. The slope of the straight-line section BC (tan c) is called the gamma value of the curve and is obviously related to the contrast range of the film, but the γ-value relates only to the region BC and does not give information on the response of the film to under- and over-exposure. Since the log E scale includes negative log values it is usually found that log E is plotted in relative log units; the shape of the curve remains the same.

(d) *Effect of development.* There is no absolute way of defining film speed (see below (g)) and the effective film speed, as well as other characteristics, depend on the development conditions. The way development affects the characteristic curve is shown in Fig. 30. In this figure the centre curve represents response of a film when "normally" developed. If the same film is used for identical camera exposures of the same subject so that the range of log E values is the same, and each latent image is the same, nevertheless over-development or under-development will result in greatly different characteristic curves as indicated in Fig. 30. It is obvious that the over-developed emulsion has the highest gamma and will, therefore, produce the most contrasting film, while under-development produces a low gamma curve or soft negative. In the former case the shadow areas (toe region) will give increased density to some extent and a great increase in density in the highlights and the overall effect is a great increase in the density range. In the latter case the density range can be seen from the

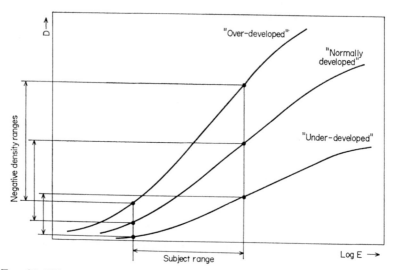

FIG. 30. Effect of development on gamma value and on tonal (density) range. (Courtesy Ilford Limited, London.)

ordinate to be very restricted and the densities of the highlights now approach the densities in the shadows, and the result is a thin negative of poor contrast. Normal development is intermediate between these two in its effects on response.

(e) *Effect of exposure variation.* For any particular emulsion under a fixed set of development conditions the shape and position of the response curve is constant but the portion of the curve which corresponds to any particular photograph area will depend on the values of E. These values will depend on the brightness range of the subject and upon the camera exposure given. For a subject of low contrast the range of log E will be small and, therefore, great variations in exposure are possible without the Density values falling too low (on the toe) or too high (on the shoulder). In other words it is more difficult to under- or over-expose, or alternatively for such a subject there is a great exposure latitude (Fig. 31(a)).

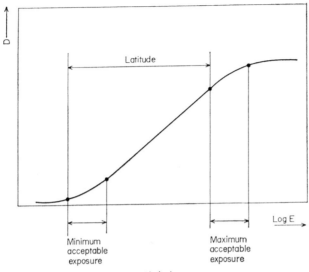

(a) Subject of low contrast-great latitude

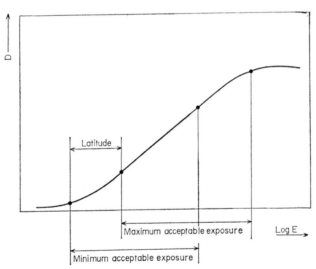

(b) Subject of high contrast-little latitude

FIG. 31. Effect of subject contrast (brightness range) on film latitude. (Courtesy Ilford Limited, London.)

On the other hand when the brightness range of the subject is great then even slight variations in E (which can be due to differences in camera exposure) may mean that a portion of the range falls on the toe or the shoulder of the curve. If the exposure range includes the toe of the graph then shadow detail will be lost; if it includes the shoulder of the graph then highlight detail will be lost, i.e. there will be little or no tone gradations in the darkest and brightest subject areas. In such circumstances the film, for such a subject, has little exposure latitude (Fig. 31(b)).

In practice it is better to aim for the brightness range of the subject to elicit film response in the lower (less dense) part of the characteristic curve, so long as the lower end of the "toe" is not included. The reason is that the resulting "thin" negative will still preserve the tone gradation of the subject, but will be sharper than an exposure range at the higher densities. In the latter case deeper layers of grains are reduced and light scattering by grains leads to loss of image definition and "crispness".

(f) *Reciprocity failure.* As mentioned above the law $E = I.t$ does not always hold good since the response of the film also depends on the rate at which radiant energy is supplied and not simply on the total amount of energy. All films suffer from reciprocity failure but it is a serious consideration only at very high or very low levels of illumination. At extremely high or extremely low levels of illumination the response of the film is less than the law forecasts and the film behaves as if it were slower. The practical consequence is that the exposure time must be increased to allow the film to exhibit its "proper" sensitivity. This is a phenomen well known in photomicrography where the amount of light available, e.g. in fluorescence work, may be of a very low level necessitating an exposure of many minutes. Manufacturers of equipment designed for exposure (camera exposure) determination from these low levels take this into account either in the design of the instrument itself, or else provide tables enabling the user to apply an exposure increase factor manually (see Section VII below).

Also, at very high intensities of illumination, requiring very short exposures there is a net decrease in contrast, while at low intensities requiring very long exposures there is a net increase in contrast, that is, a relative increase of γ-value. In determining response under extreme conditions of this sort the exposure scale may be obtained by keeping t constant, and varying I, in which case it is an *intensity scale*; alternatively, if a low I is kept constant and the values of E obtained by varying the camera exposure t, a *time scale* is obtained. The results will not be the same and the appropriate scale must be used for practical applications. For example, the exposure factor required for the use of a filter will depend on whether the increase in E is achieved by opening the aperture (f stop) to increase I, or by increasing the camera exposure t. In this latter case, the filter factor depends largely on the exposure time required for the unfiltered negative. Quotations of filter factor are qualified as to whether they are intensity scale based or time scale based. In photomicrography, E variations would be achieved by increase in lamp intensity, which also alters the colour temperature of the source; or by increase in the exposure time. The latter is more likely under difficult conditions since the light intensity which may be used can be limited by other factors such as the rating of the source, heating, injury to specimen, etc.

From the foregoing considerations it should be clear that the ideal negative will be produced on a film with appropriate development such that the minimum density is about 0·2 to 0·3 above the fog level, while the maximum density is around 1·1 to 1·2.

(g) *Film speed.* There is no absolute value of film speed for any film and the effective speed will depend upon a number of factors. However, all manufacturers give a film speed rating for their products as a guide to the user. Various systems of rating have been suggested at one time or another and the basis of rating has been changed from time to time. A number of different systems have been used, all based on a reference point on or derived from a particular characteristic curve. Perhaps the most commonly

TABLE IV

Comparison of film speed ratings

BS ASA	DIN	Scheiner	Weston I & II	APEX
20	14°	22°	8	2·5
40	17°	25°	16	3·5
50	18°	26°	20	4
64	19°	27°	25	4·5
80	20°	28°	32	4·5
100	21°	29°	40	5
125	22°	30°	50	5·5
160	23°	31°	64	5·5
200	24°	32°	80	6
250	25°	33°	100	6·5
320	26°	34°	125	6·5
400	27°	35°	160	7
500	28°	36°	200	7·5
640	29°	37°	250	7·5
800	30°	38°	320	8
1000	31°	39°	400	8·5
2000	34°	42°	800	9·5
4000	37°	45°	1600	10·5

used are the ASA (now identical with the BS arithmetic) and the DIN systems. The latter is a logarithmic scale and there is no simple conversion factor from one to the other which covers the whole range. The relative values of some of the speed ratings on various scales are given in Table IV.

In the arithmetic systems ASA and Weston in Table IV, a doubling of the number rating represents a doubling of the speed, and it will be noticed that three units in the log scales DIN and Scheiner are equivalent to a doubling of speed. The Apex is a new logarithmic system not yet generally accepted, in which a doubling in film speed is represented by one unit of log increase.

Tungsten light is relatively richer in red and poorer in blue radiation than is daylight. This implies that the relative response of photosensitive materials, both film and the sensory cells of exposure meters will react differently to the two types of light. The effective speed of an emulsion is slower in tungsten light by a factor depending on the type of emulsion, the main difference being shown by the "blue sensitive" emulsions (Table V).

TABLE V

Relative response of film emulsions in daylight and artificial light

Sensitive material	Speed ratio Daylight : Tungsten	Difference in log rating
Non-colour-sensitive	4 : 1	6°
Orthochromatic	2 : 1	3°
Normal panchromatic	1·5 : 1	2°
Fast panchromatic	1·25 : 1	1°

(h) *Response of photographic papers.* Sensitometric methods can also be applied to photographic papers and characteristic curves derived. In this case, however, the density values plotted are the reflection density values defined by

$$D = \log \frac{1}{R}$$

where R is the reflection factor of the material; that is the ratio of the light reflected by the image to the light reflected by the base (paper). The general shapes of the characteristic curves are similar to those obtained with film, the main differences being that under normal development conditions (and good storage of stock), fog is absent; there is a longer toe and shorter straight-line portion, and the γ-value is relatively higher than for the same emulsion used as film. The physical character of the surface

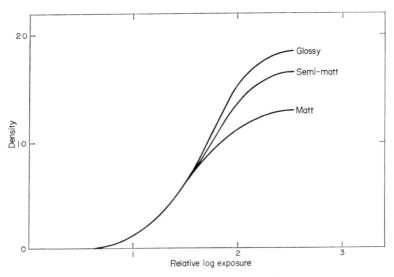

FIG. 32. Effect of surface finish on the characteristic curves for papers.

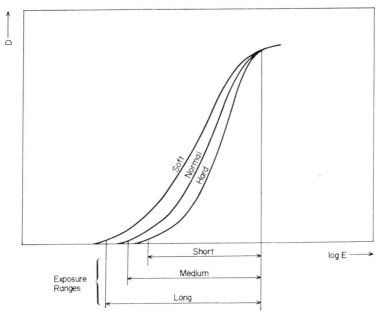

FIG. 33. Characteristic curves for papers of different contrast grades.
(Courtesy Ilford Limited, London.)

affects the reflection, and so the recorded density, giving different curves for different "finishes" as shown in Fig. 32.

In addition, as with film, different contrast grades are possible, the characteristic curves for hard papers being steeper (more contrast) and slower (longer exposure time required) than for the softer papers, Fig. 33. Similarly, for any grade, the contrast diminishes with decreasing development time, equivalent to a lower γ-value, and full development, or development to completion, does not occur. For good prints it is necessary to have a minimum development time (about $1\frac{1}{2}$ min) and in practice one should

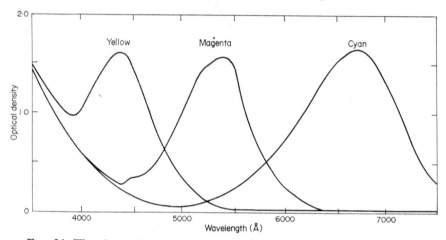

FIG. 34. The absorption curves of a typical set of dyes used in a colour film. Each dye transmits freely in the long wavelength region to the right of the absorption peak. There is a positive and increasing absorption beyond a minimum on the short wavelength side. The cyan and magenta dyes have significant secondary absorptions in the shorter wavelength visible which results in imperfect colour reproduction.

fix on a time and maintain this as routine so that print quality is easier to compare between exposures (enlarger). A time of 2 min is a good standard and easy to time on a dark-room clock.

In general a negative developed to a low gamma with exposure in the low density range (toward the toe) should be printed on a hard paper to increase the density range, and vice versa.

Papers also suffer from reciprocity failure, probably most often noticed in the extremely long exposures required when making "giant" enlargements, especially from fairly dense negatives, another reason for keeping the exposure range of the negative towards the "thin" end of the characteristic curve. While a certain amount of "correction" can be achieved in printing there is no substitute for a really good negative.

B. Colour materials

Modern colour film design is based on a system of subtractive primary colours. Experiment will show that any colour can be produced by appropriate admixtures of the additive primaries, red, green and blue light. White light is, of course, a mixture of all the colours and it is easy to see that any particular colour may be obtained from it by filtering out, i.e. subtracting, the unwanted wavelengths such that the mixture which remains is the desired colour. Any filter required to subtract a particular colour from a mixture will be the complementary of that colour and the complementaries of the additive primaries red, green and blue, are cyan, magenta and yellow. Since cyan dye subtracts red, and yellow subtracts blue, then a combination of cyan and yellow will transmit green, and so on. If the third, magenta, which subtracts green, is combined with the other two then all the additive primaries are subtracted and the result is therefore "black". The actual wavelengths absorbed by the three subtractive dyes are shown in Fig. 34, from which it will be seen that the system cannot be perfect since there is some absorption at short wavelengths for all three which leads to "secondary absorption" in undesirable portions of the spectrum.

In the manufacture of colour film many emulsion layers are needed. The top colour sensitive layer is sensitive to blue light only; below this is another coloured layer which acts as a yellow filter subtracting blue light and preventing it from reaching the lower layers. The middle layer is sensitive to blue and green, while the bottom layer is sensitive to blue and red. Since the yellow layer subtracts blue the latter two react to green and red respectively. Each emulsion layer also has a colourless (to avoid filtration) chemical "coupler" which is developed during processing into the coloured complementary of the colour to which that layer is sensitive. If a colour negative is being produced then this image is formed in the complementaries of the true colours, the reversal to the subject colours being brought about in the printing. On the other hand, if reversal colour material is being processed the reversion from complementaries to subject colours is brought about in a single processing procedure. The changes which occur are illustrated in Figs. 35 and 36, between pp. 334–335.

1. Negative film

In Fig. 35 it can be seen that a black subject area will affect none of the layers of film, since no light energy is emitted. The top blue sensitive layer will respond to the white and blue area, the green sensitive layer to the white and green area, and the red sensitive layer to the white and red area. On development exposed silver halide in the various layers will be reduced and the oxidized colour developer will react with the colour couplers to form, in the appropriate places, the dyes which are complementary in colour to the exposure radiation. The bleach process then removes the yellow filter layer and the silver image leaving only a dye image in complementary colours to the subject colours.

The printing paper, similarly, has three colour sensitive layers, but the yellow filter layer is not needed as the red and green sensitive layers are almost insensitive to blue. When light shines through the negative on to the paper the cyan area acts as a subtracting filter, therefore allowing the non-complementary green and blue to pass and form latent images in the green sensitive and blue sensitive layers. Similarly, where the negative image is magenta, the red sensitive and blue sensitive layers are exposed, and so on. Negative areas corresponding to white and black have all primaries and no primaries developed in the negative, respectively, so the reaction for the paper will be no primaries and all primaries developed. When the developed paper is viewed in white light by the eye, the colour interpreted will depend on the wavelengths reflected to it, therefore an area with developed magenta and yellow dyes will subtract their complementaries green and blue so that only the red wavelengths reach the eye and the red area corresponds to the red area of the subject. Similarly for the other areas.

2. *Reversal colour film*

The first stage of the reversal colour film process is the same as for the negative, except that the first stage developer does not convert the couplers to their coloured form, so that there is a silver image in the same layers as before (cf. Figs. 35(a) and 36(a)) but the dyes are still in leuco form. The film is now exposed to a bright light to expose all the silver not exposed in the camera (Fig. 36(b)). In the second or colour developer, the first exposed areas still do not develop colour, but the newly exposed areas are acted upon by the second developer to produce dye in those layers which were exposed on the second occasion only. All the silver images and the yellow filter are now dissolved away as before and the material is now a collection of subtractive filters. For example, a blue subject area would give a latent silver image in the blue sensitive layer on exposure in the camera, but not in the red and green sensitive layers which are screened by the yellow filter layer. After the first development silver image is formed in this layer but no colour. On second exposure silver images are now formed in the green sensitive and red sensitive layers. The second development produces silver images and complementary colour dyes in the green sensitive (magenta dye) and red sensitive (cyan dye) layer. When the silver and other interfering materials are dissolved away the film in that area is a combination

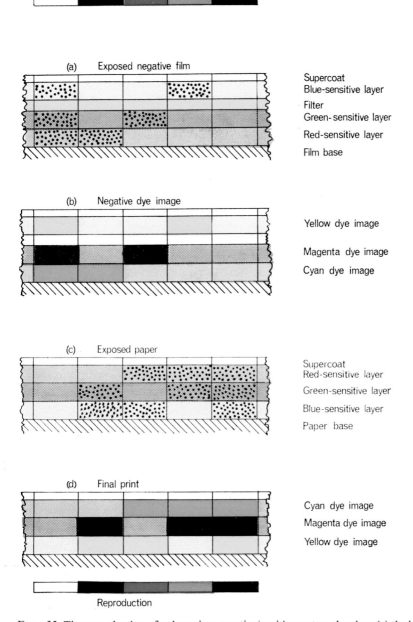

Fig. 35. The reproduction of colours in a negative/positive system showing: (a) the latent image in the exposed negative film; (b) the dye image in the processed negative film; (c) the latent image in the exposed print material; (d) the final print. (From "Photography for the Scientist", Ed. C. E. Engel, Academic Press.)

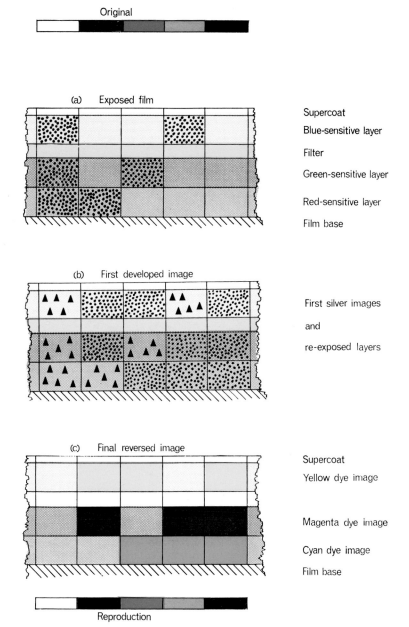

FIG. 36. The reproduction of colours by a reversal film showing: (a) the latent image in the exposed film; (b) the negative silver image after development, and the latent image produced by re-exposure; (c) the dye image after the colour development. The uncoupled colour forms are shown lightly shaded here and in Fig. 35, although in fact they are normally colourless. (From "Photography for the Scientist", Ed. C. E. Engel, Academic Press.)

of magenta and cyan filters which subtract green and red from the transmitted projection light so giving a blue image, the subject colour, in that
area.

3. Colour temperature

The relative distribution of spectral energy in a continuous spectrum
(as is emitted by most incandescent solids and liquids) is described by its
"colour temperature". The colour temperature of an emitting source is
the absolute temperature in degrees Kelvin (°K) to which liquid platinum
must be heated in order to emit light with the same energy (and so, wavelength) distribution. At low temperatures the long wavelengths predominate, and as the temperature is raised the balance of the distribution
veers progressively towards the high energy, blue, end of the spectrum.
Each source may, therefore, be typified by a particular colour temperature
value in degrees Kelvin, the higher the colour temperature the more the
quality of the emitted light approaches that of daylight. The colour
temperatures of various sources are given in Table VI.

TABLE VI

Colour temperatures of various light sources

Source	Colour temperature	
	°K	(Mired)
Domestic lamp, 60 W	2800	(357)
Low-voltage microscope lamp, 6 V, 15 W at 6 V	2850	(351)
Projection lamps	2900–3200	(345–313)
Photoflood (normal use, life 100 h)	3200	(313)
Quartz-iodine lamp (12 V, 100 W)	3300	(303)
Photoflood (overheated, life 2 h)	3400	(294)
Carbon arc (depending on carbon)	3800–5500	(263–182)
Tungsten photo-flash bulb, clear	3800	(263)
Sunlight	ca. 5500	(182)
Xenon burner, XBO 162	6050	(165)
Xenon flash tube	ca. 6000	(167)
Daylight, cloudy sky	6500	(154)
Daylight, cloudless sky (blue)	ca. 20000	(50)

The mired value of a source is obtained by dividing one million by the
colour temperature of the source, so that the higher the latter the lower the
former. Mired is the abbreviation for micro reciprocal degrees.

(a) *Colour temperature modification by filters.* Since the distribution of
wavelengths transmitted by filters can be determined it is possible to give
mired values to filters which are related to the degree to which they change
the colour balance of the light emitted by the source.

To raise the colour temperature a decrease in overall mired value is needed so that the blue correction filters are rated in negative values, while conversely, to lower the colour temperature a positive mired change is required, and this may be achieved by the orange range of filters. Some colour correction filters are listed in Table VII.

TABLE VII

Colour correction filters and the mired shifts they produce

Kodak Wratten			
82C	−45 mireds		
82B	−32	,,	Blue range
82A	−21	,,	
82	−10	,,	
81	+ 9	,,	
81A	+18	,,	
81B	+27	,,	Orange range
81C	+35	,,	
81D	+42	,,	
81E	+52	,,	
Ilford			
829	−73	,,	
830	−93	,,	Blue
831	−50	,,	
Agfa/Gevaert			
CT B1-16		Blue	
CT O1-20		Orange	

Filter mired values are additive so that, for example, a 53 mired change can be produced by combining filters 82A and 82B and if these were used with a low-voltage lamp of 2850°K, the mired value would be reduced from 351 to 298, equivalent to increasing the colour temperature to

$$\frac{10^6}{298} = 3356°K$$

or the colour temperature of a quartz-iodine lamp. Filtration necessarily means absorption of radiation, so that the brightness of the source is diminished and for photomicrography this means the exposure must be increased. The lamp loading must not be changed after the calculation of the required filter since a change in output means a change in colour temperature. A filter combination producing a change of more than 100 mireds should be avoided and rather a film rated for a colour temperature response nearer that of the source should be used instead. This means that for tungsten lamp colour photomicrography the "artificial light emulsions" should be used, and daylight emulsions should only be used with the high colour temperature sources such as the xenon burners. In modifying colour temperatures of illumination for photography one should aim for mired values of film and source to differ by not more than 5–10 mired.

TABLE VIII

Types of reversal colour film

Make	Film	Sizes	Speed ASA, DIN	Light	Processing
Agfa	CT18	135, 120	50, 18	Daylight	Lab.
	CK20	135, 120, sheet	80, 20	3200°K	User
GAF	GAF 64	135, 120, 126	64, 19	Daylight	User
	GAF T/100	135	100, 21	3200°K	User
	GAF 200	185	200, 24	Daylight	User
	GAF 500	135, 120	500, 28	Daylight	User
3M	3M Color Slide	135, 126	50, 18	Daylight	Lab.
Gevaert	Gevacolor R5	135, 120	50, 18	Daylight	Lab.
Kodak	Ektachrome E-3 ⎫	120	50, 18	Daylight	User
	Ektachrome B E-3 ⎭	sheet	32, 16	3200°K	User
	Ektachrome X E-2	All	64, 19	Daylight	User
	High-speed Ekta- chrome E-2	135	⎧ 160, 23 ⎩ 125, 22	Daylight 3200°K	User User
	Kodachrome II	135	25, 15	Daylight	Lab.
	Kodachrome IIA	135	40, 17	3400°K	Lab.
	Kodachrome X	135	64, 19	Daylight	Lab.
Orwo	Orwocolor UT16	Various	32, 16	Daylight	User
	Orwocolor UK14		20, 14	3200°K	User
	Orwochrome UT21	Various	100, 21	Daylight	User
Perutz	Perutzcolor C18	135, 120	50, 18	Daylight	Lab.
	Perucolor C-PAK	126	64, 19	Daylight	Lab.
Telko	Telcolor	Cut film	50, 18	Daylight	User

(b) *Colour film stocks.* A selection of colour film material and their main features are listed in Tables VIII and IX. Films that require "Lab" processing must be returned to the manufacturers or other authorized laboratory. In many cases the user may perform processing and kits of chemicals may be available, or the solutions may be made up from published formulae. Where a film may be used both in daylight and tungsten the appropriate colour balance must be achieved by filtration for one or other usage and a loss of film speed will be involved unless correction is made. Re-ratings are not given in the table in all cases and the manufacturer's

TABLE IX

Types of colour negative film

Make	Film	Sizes	Speed ASA, DIN	Light Balance	Processing
GAF	GAF Color Print	135, 126	80, 20	Day	Lab.
Agfa	CN17 and ⎫ CN17 M ⎭	135, 127, 120, sheet	40, 17	Day to 3200°K	User
	CN14	135	20, 14	Day to 3200°K	User
3M	3M Color Print	135, 120, 126, ⎫ 127 ⎭	80, 20	Day	User
Gevaert	Gevaecolor N5	135, 127, 120 ⎫ sheet ⎭	32, 16	Day	Lab.
	Gevaecolor N3	Various	20, 14	3200°K	Lab.
	Scientia-color	35, sheet	2, 5	3200°K	Lab.
Kodak	Ektacolor L	Sheet	10–25, 11–15*	3200°K	User
	Ektacolor S	120, sheet	80, 20	Day	User
	Kodacolor	135, 828, 127 ⎫ 120 ⎭	32, 16 16, 13	Day 3200°K	User User
	Eastmancolor 5251	35	50, 18 32, 16	3200°K Day + Wratten No. 85	User User
Orwo	Orwocolor NC16 NC17	Various	32, 16	Day to 3200°K	User
	Orwocolor NK16	Various	32, 16 25, 15	Day 3200°K	User
Perutz	Perucolor	135, 126	80, 20	Day	User
Telko	Telcolor		32, 16	Day to 3200°K	User

* Rating depends on camera exposure.

literature should be consulted. Also, film types and speed ratings are varied at frequent intervals and current data should be consulted to ensure that no recent changes have been made.

V. ILLUMINATION FOR PHOTOMICROGRAPHY

While it is possible to use natural light—either direct sunlight or diffuse daylight—for photomicrography, both have disadvantages as to quality and quantity, sunlight being particularly variable, and artificial sources which can be strictly controlled are preferable. The luminosity of the light source should not be below about 1000 candles/cm², otherwise the scale of reproduction may be too restricted or the required exposure too long. The intensity should if possible be variable. For metal filament lamps this is possible within limits by means of a voltage regulator, but changes in power will, of course, give changes in colour temperature. Where a gas burner is used and the intensity is not electrically controllable, neutral density filters must be used to limit intensity as required.

A. Types of illuminator

Since colour film is manufactured in two main ranges of sensitivity, for 3200°K (tungsten type) and around 6000°K (daylight type), it is convenient to choose lamps which give colour temperatures in these ranges. It must be remembered that the colour temperature of tungsten lamps changes considerably with the applied voltage, so that for any operating voltage the colour temperature of the light emitted should be measured, and, if necessary, corrected by the use of balancing filters (see Table VII). An instrument for measuring and correcting for colour temperature is described below (Section VII.B, 7). If such an instrument is not available the lamp should be run at a voltage which produces the appropriate temperature balance as specified by the lamp manufacturer and adjustments for intensity are then made by adding neutral density filters (which do not alter the colour balance). Even when using at the specified voltages, it is important that the voltage in the supply should not vary, which may be possible if other equipment in the laboratory or building draws heavily on the same supply. Variations in the current have a much lesser effect.

It is not true that a point source is the most suitable for photomicrography and the emitter should have a sufficiently large area of homogeneous intensity so as to fill the entry pupil of the microscope. In fact no filament lamp is perfect, as there will be inhomogeneities due to the spaces between the coils of the incandescent wire. Also, the geometrical shape is usually a square obtained by flattening the coil, rather than a circle which would match the aperture of the optical system. Nevertheless, modern light sources are adequate for their purpose and poor results are due usually to faults of alignment or adjustment rather than to the lamps themselves.

A number of different types of lamp have already been considered in Chapter I, Volume 5A of this Series, and there is no need to reiterate them here. Also, when living material is being observed a number of other factors, such as the heating of the specimen, must be considered and these have been referred to in Chapter X, Volume 1 of this Series.

The life of a tungsten lamp will depend to some extent upon its method of use, in particular upon the voltage at which it is operated, and although it is sometimes necessary to over-rate a low-voltage lamp in particular photomicrographic situations, this should be done to the minimum extent and for the minimum time. A decrease in the life of the lamp must be expected under such conditions. Also, as the lamp deteriorates its brightness and colour temperature will decrease for any applied voltage and the condition of the lamp should be considered, especially for colour photomicrography.

The tungsten iodine lamps which are now readily available overcome many of the difficulties associated with the older type of filament lamp and are especially useful for photomicrography. The changes in operating characteristics with applied voltage for a 12 V 100 W tungsten iodine lamp are given in the graph in Fig. 37.

As mentioned before (this Series, Volume 5A, Chapter I) sources other than filament lamps may be used, and in some cases must be used. While the carbon arc is unsurpassed for its homogeneity at high luminous intensity it suffers from serious drawbacks. It has an enormous heat dissipation, is expensive to use and must be carefully controlled so that the electrode distance is maintained as the carbon is consumed. It has the great advantage, on the other hand, of supplying a spectrum which may be used for visible, ultraviolet, infrared and fluorescence microscopy.

Xenon high-pressure lamps can be run on d.c. or a.c. current, preferably the former, and achieve very high luminosities and colour temperature. The XBO type lamp has a colour temperature independent of the supply current and a continuous spectrum similar in quality to average daylight. It is a poor u.v. emitter and not suitable for u.v. fluorescence microscopy. Like the xenon flash tube its quick reaction enables it to be operated in pulses, e.g. for time lapse sequences. The xenon flash tube is the ideal illuminator for the photography of fast moving objects as it has an extremely high luminosity, up to 10^6 cd/cm^2, operating for times as short as 1/5000 sec, which nevertheless enables exposure times as short as 1/500 sec.

The mercury high-pressure burner has advantages not present in the xenon burner, but suffers from many disadvantages which the latter does not have. Its main advantage apart from its luminous intensity is the strong emission in the u.v. region. It emits strongly at 2540 Å, and at 3650 Å and 3660 Å in the longer u.v. wavelengths with another emission line in the blue region at 4050 Å. It is the ideal source, therefore, for fluorescence microscopy. It does not give a continuous spectrum, but a few isolated lines in the visible region and none in the red, so is the worst form of light for conventional photomicrography. Also, it cannot and should not be pulsed, as it requires an appreciable warm-up time.

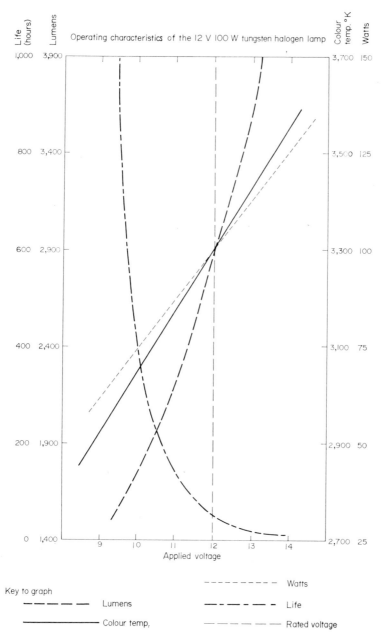

FIG. 37. Operating characteristics of a 12 V 100 W tungsten halogen lamp. (Courtesy Vickers Instruments Limited, York.)

B. System of illumination

For best results in photomicrography the Kohler system of illumination must be used. This system and the way in which it is set up have been dealt with earlier (this Series, Volume 5A, Chapter 1) and it will be assumed that the reader has mastered it in detail. However, even when Kohler illumination has been arranged and the microscope focused for optimum optical conditions, there are two items to which insufficient attention is often paid. The first is the condenser. For good photomicrographic results the condenser should be of the aplanatic-achromatic type as described in Volume 5A, Chapter 1 of this Series. Simpler (and considerably cheaper) condensers which are not corrected for spherical and chromatic aberration reduce image contrast and cause colour casts in colour photomicrography. For the best results, too, the numerical aperture of the condenser must be matched to that of the objective and this means that for high N.A. objectives, the condenser must also be under homogeneous immersion.

The second neglected point concerns the use of the iris diaphragms. The field iris diaphragm should be opened only just enough to illuminate the field of view. Especially where stained tissue sections are involved the light refracted or diffracted from areas of the specimen which are outside the field of view cause deterioration in image quality. Similarly, the condenser iris diaphragm must be adjusted so that the objective system is just filled with light, that is, the aperture of the illuminating cone is matched to the aperture of the objective in use. This is best done by removing the eyepiece and closing the condenser diaphragm until it is just visible within the field when looking down the tube at the back of the objective. These adjustments are to be made when all the optical elements are optimally focused.

VI. FILTERS AND THEIR USE

This subject has already been dealt with in relation to microscopy in Section IX of Chapter 1, Volume 5A of this Series. From what has also been said in this Chapter the importance of the use of filters for light balancing or colour correction as well as for the adjustment of intensity, will readily be appreciated. Heat absorbing filters should always be used with intense sources, especially where living tissue is involved. The use of a green filter with a transmission band around 5500 Å is recommended for photomicrography on black and white (panchromatic emulsions) because it helps to balance the spectrum reaching the film, giving it a range similar to that of the eye; and there is the additional value that this is the wavelength at which the various optical elements have their maximum correction. For further information on filters the reader should consult this Series, Volume 5A.

VII. EXPOSURE TIME DETERMINATION

By this is meant the camera exposure referred to earlier and not the determination of the response of the film. Until fairly recently the determination of exposure times was a somewhat haphazard affair requiring a number of trial exposures before a suitably exposed negative could be selected. There is now a large variety of electrical and electronic gadgets which facilitate the process and give very acceptable results, without film wastage, except under the most difficult of circumstances. Nevertheless, instruments may break down and it is still necessary to know the best way of making an "exposure test strip".

A. The test strip method

The aim in exposure time determination is to choose a camera exposure such that the film exposures (log $I.t$) fall on the linear part of the gamma curve, or the linear part plus the upper part of the toe. Since the gamma curve is a log curve, the range of trial exposure times should also be a log sequence, assuming always that I remains constant. In the case of 35 mm black and white film a separate frame may be used for each exposure trial, but with cut film the best comparisons and least waste are obtained if a single sheet of film contains the complete exposure trial range.

In practice the slider of the film pack is pulled back as far as possible to reveal the whole area of the film and an exposure is given. The slider is then pushed in 1/6 of the way and a second exposure time allowed and so on. The exposure time steps are calculated so that the total exposure time per "strip" of film is part of an overall logarithmic time series. The system is made clear by the example in Table X.

After development the film is examined and the strip which shows the best tone gradation is selected for the final exposure under the same

TABLE X

**Exposure time determination by the test strip method
for photographic plates**

Strip No.	Slide position	Exposure time each position	Cumulative exposure time per strip
1	Fully open	$\frac{1}{4}$ sec	$\frac{1}{4}$ sec
2	1/6 closed	$\frac{1}{4}$ sec	$\frac{1}{2}$ sec
3	2/6 closed	$\frac{1}{2}$ sec	1 sec
4	3/6 closed	1 sec	2 sec
5	4/6 closed	2 sec	4 sec
6	5/6 closed	4 sec	8 sec
	Fully closed		

conditions of lamp intensity, development, etc. Strips which are over-exposed will show loss of detail in the highlights, i.e. excessive density without sufficient tone separation especially in the darkest areas of the film. Under-exposed strips will be thin in overall density and will show no tone separation in the "shadow" areas, i.e. larger areas of zero or near zero density (i.e. fog level).

B. Exposure meters

Within the last ten or 15 years a wide range of exposure meters designed especially for photomicrography has become available. The list given below is not comprehensive but has been selected to give the microscopist some idea of the various types of instrument available and to enable him to determine the type most suited to his individual needs. No attempt is made to give fully detailed working instructions as these are always obtainable from the manufacturer and, in any case, all the instructions in the world are no substitute for experience gained in use. It is always a good idea to request the supplier to loan the instrument for a few days so that the micro-scopist can judge for himself its ease of application and precision.

1. The Microsix-L exposure meter

This instrument by Ernst Leitz (Wetzlar) most nearly resembles the conventional exposure meter used in everyday photography, but has a greatly enhanced sensitivity (Fig. 38). The "photosensor" is a photo-resistor energized by a small battery; when exposed to light there is a change in the conductivity of the resistor. The measuring "eye" accepts an angle of 30° and the meter can therefore be used as a conventional exposure meter for photography as well.

The needle zero setting can be checked and reset, if necessary, and the condition of the battery can also be checked. Once these have been done the basic procedure is as follows. The film speed is selected for black-and white or colour, on the film speed dial (3b, Fig 38) and the photo-resistor head placed on one of the measuring sites. The possible sites for measuring are: eyepiece tube, with or without eyepiece, focusing telescope, special site on micro-attachment, or directly on the ground glass screen of a large-format camera. Any of these positions will serve so long as the meter dials are calibrated for use in such positions. A switch on the "head" enables scale selection for "bright" or "dim" specimens. The pointer settles on a particular part of the indicator scale (1, Fig. 38) and this value is then set on the yellow rotating scale (2c). The exposure time required is then read from the exposure time scale (2a) opposite the calibration value (3a). The calibration values are the "stop" numbers which would be

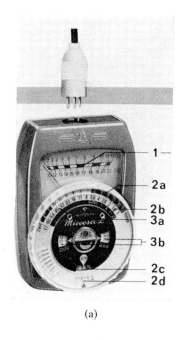

(a)

(b)   (c)

Fig. 38. Leitz Microsix-L. (a) Measuring instrument: 1, scale; 2a, exposure times; 2b, film speeds; 2c, number scale 1–21; 2d, light value scale (photomicrography); 3a, calibration value; 3b, film speed dial. (b) Measuring head with measuring eye. (c) Reverse of (b) showing selector switch. (Courtesy E. Leitz, Wetzlar.)

used in ordinary photography. Full instructions for calibrating are supplied with the meter.

2. *Remiphot exposure meter* (Reichert, Wein)

This is an exposure meter primarily designed for use in photomicrography and is fitted to the eyepiece tube in use (Fig. 39).

FIG. 39. The Reichert Remiphot exposure meter, which is attached directly to the ocular tube. The sensing is performed by semi-conductor photo-resistors. (Courtesy C. Reichert, Vienna.)

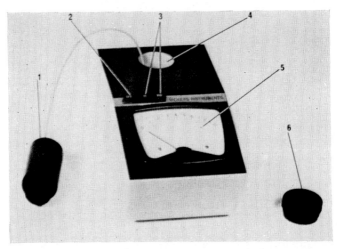

FIG. 40. Vickers cadmium sulphide photometer. 1, photometer head; 2, "off" and battery check switch; 3, light value selector switches; 4, stowage well for photometer head; 5, battery check mark; 6, protective cap. (Courtesy Vickers Instruments Limited, York.)

3. *Cadmium sulphide cell photometer* (Vickers, York)

This photometer possesses a micro-ammeter which measures the current that flows in a circuit when light falls on the surface of a cadmium sulphide cell (Fig. 40). The photometer head fits standard R.M.S. eyepiece tubes. Power is supplied by a 15 V dry cell and a switch on the top of the case enables the state of the battery to be checked.

The photometer must be calibrated for use with any particular film and development conditions. To calibrate the instrument the microscope and camera are set up under a fixed set of conditions, illumination, film, etc., and a $\log_2$ series of exposure times used to take a number of photographs or make a test strip negative as described under A above. The test negative which is adjudged to give the correct density range on development is then used as a standard against which to compare the instrument reading. In other words the photomicrographic system is set up as before and the photometer head inserted in the side tube of the Vickers camera reflector body (see Fig. 16). The instrument is then switched on and the deflection of the ammeter needle is used to derive a corresponding light value from a table which is supplied. The exposure time of the test negative is then entered in a table against that light value. The table is then completed for the other light values, doubling or halving the exposure time for each unit light value. A separate calibration table must be constructed for each type of film and development.

Once calibrated the photometer can be used to derive exposure times by setting up as before, obtaining the value of the ammeter needle deflection and using this to select a light value from the table supplied. This value is then entered in the calibration table (light value/exposure time) and the exposure time read off.

The instrument may also be used to determine exposure by insertion in the eyepiece tube of any microscope and compensating factors to account for eyepiece magnification or camera bellows extension can be read from a table or calculated from formulae.

4. *Ikophot M exposure meter* (Zeiss, Oberkochen)

In the Ikophot M system the detector for the exposure meter is a CdS photoconductor cell built into a slide which slots into the focusing eyepiece tube of the Zeiss CS camera attachment. In use the meter may be calibrated by making a series of test exposures and using the calibration number adjudged to give the best result, or, the manufacturer's advice on use of calibration number may be followed. The film speed is set on the dial in the usual way (see Fig. 41(a)), the red index tag is set to the calibration number for the film format in use (it is at 2·5 in Fig. (41a)), the detector is then slotted into the focusing eyepiece light path and the centre button is

(a)   (b)

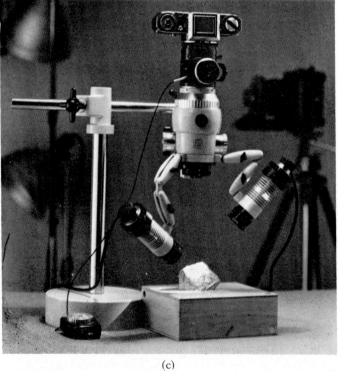

(c)

FIG. 41. (a) Ikophot M exposure meter (without CdS detector); (b) with attached flash exposure ring; (c) Ikophot M in use on the Zeiss Tessovar photomacrographic zoom equipment. The meter is attached to the detector slide which is in the "read" position in the focusing eyepiece tube. (Courtesy Carl Zeiss, Oberkochen.)

depressed and then released when the needle remains stationary. The milled outer ring is then turned until the "eye" in the pointer covers the needle (Fig. 41) and the exposure required is read from the dial opposite the red index tag. The second (black) index tag (upper tag, Fig. 41) indicates the exposure to be used for subjects with dark background, e.g. in dark field, polarization or fluorescence photomicrography.

With the flash exposure ring fitted (Fig. 41(b)) the Ikophot M may be used to determine the conditions required for exposure with the Zeiss microflash unit. A single flash at full 60 W sec power is equivalent to a 1 sec exposure with the 6 V, 15 W lamp when operated at 5·5 to 6 V and the flash exposure ring is, therefore, fixed with the zero of the full power scale (outer figures, Fig. 41(b)) opposite the 1 sec exposure time. The meter is then set as for use with the viewing light of the flash unit at 5·5–6 V and the neutral density filters required to reduce the light appropriately for use of the flash exposure is read against the red index tag, i.e. 0·12 at full power or 0·25 at half power in Fig. 41(b). Details of the actual filter combination to be used are obtained by reading from a table which is supplied.

The detector head takes a partial field reading in the centre of the format and this area is marked by concentric etched circles in the eyepiece field screen, where the formats are also indicated. The head may also serve as the sensor for automatic exposure setting when used with the CS-matic shutter and Prontor CS-matic control unit (not shown).

The Ikophot M system may also be used with the new Zeiss Tessovar photomacrographic zoom system, which is illustrated in Fig. 41(c). The Tessovar permits magnifications within the range 0·8× to 12·6× in continuously variable gradations permitted by the zoom lens, and the exposure at any magnification is easily obtained since the detector is situated in the "exit radiation". A variety of camera backs may be fitted to the shutter unit by means of coupling rings.

5. J35 *Automatic exposure unit* (Vickers, York)

A photo-multiplier head is affixed in the side tube of the Vickers camera reflector body or in the appropriate position on the M41 Photoplan microscope, and the current from this feeds a resistance capacitance charging circuit. The equipment is illustrated in Fig. 16(c). The shutter is opened and the exposure timing sequence is started when the exposure button is pressed. A capacitor controlling the electro-magnetic shutter is charged in proportion to the amount of light delivered to the photo-multiplier. When the capacitor charge reaches a pre-determined value, influenced by the film speed selector, the shutter closes and, with the autowind camera, the film is automatically wound on.

In use the two main requirements are that the film speed is selected (the range is from 5 to 3200 ASA), and the micro-ammeter needle deflection must fall within the operational range for the automatic use of the instrument. A 3-range photometer scale knob ($1\times$, $10\times$, $100\times$) enables very wide differences in specimen brightness to be accommodated. Exposures are automatic in the range from 1/50 sec to 15 min. Longer exposure times can be made manually.

Compensation for reciprocity failure must be made when colour films are used under fairly dim lighting conditions and a table is provided giving the decrease in ASA (and hence setting) for a number of different colour films under increasingly poor lighting conditions.

6. *J37 Photometer timer* (Vickers, York)

This is an extremely sensitive photometer which permits two separate operations to be carried out. In the first stage the brightness of a very small part of a field, down to 1/500 of the format area, may be recorded and an exposure time determined with the aid of a slide-rule. The appropriate exposure time for the film speed and lighting conditions is then set on the instrument manually and the exposure button pressed to make the exposure. After the pre-set time has elapsed the shutter will close and the film automatically wind on, if the autowind camera is used. The equipment is illustrated in Fig. 42.

Fig. 42. Vickers J37 high sensitivity photometer timer. The photomultiplier head is not shown. (Courtesy Vickers Instruments Limited, York.)

The great advantage of this instrument is its ability to make accurate partial field determinations, so that, say, an exposure may be made specifically for a small area of a cell which is fluorescing dimly against a dark ground. On the Vickers M41 microscope special provision is made for its use. A prism slider enables all the light to be directed to the photomultiplier for measurement purposes, and subsequently all the light may be directed to the camera to make the photograph. All the light may, alternatively, be passed to the viewing eyepiece with its graticule so that centration of the specimen for partial field integration of light from a small central area may be accurately carried out. In addition there is the possibility of integrating light over the whole field and the concentric circles enable integration over $1/10$, $1/100$ and $1/500$ of the field area.

An ammeter denotes the range in which the instrument will operate, and the sensitivity ranges can be selected in 10-fold steps from 1 to $10^5 \times$. Automatic exposures based on the times derived by using the slide-rule are automatic up to 14 min; longer times may be given manually by moving the time selection switch to a "T" position.

7. *Photomicrographic exposure meter EMM-V* (Olympus, Tokyo)

This is the only instrument known to the author which is both a colour temperature and an exposure time determiner. The instrument is illustrated in Fig. 43. To perform the colour temperature determination and balance it for the type of colour film being used the procedure is as follows. The microscope is set up for photomicrography and the specimen focused, etc. The specimen is then removed, however, leaving the arrangement otherwise as before, and the colour temperature sensing head is inserted in the appropriate side-arm of the Olympus camera body. The camera body slider is set to the C.T. position, the changeover dial knob (Fig. 43) is then set to position C.T.R. (Colour Temperature Regulator) and the colour temperature dial is set at the colour temperature value of the film emulsion to be used. Six temperatures from 3000°K to 6000°K may be selected. The needle deflection on the colour temperature scale of the ammeter is then noted and the brightness of the lamp adjusted until the needle returns to zero, at which point the colour temperature of the lamp will be equivalent to that for the film. If this is not possible within the permitted voltage limits of the lamp, the colour temperature may be modified by inserting one or more of three different blue light balancing filters (-45, -100, -200 Mired) and then adjusting the lamp voltage until the meter needle returns to zero. Once the filter combination and lamp voltage have been set these adjustments must not be changed. The specimen is then replaced.

After the specimen is re-focused, the film speed dial is set to the ASA rating (16–200, BW; 32–400, colour) and the changeover knob (Fig. 43) positioned depending upon the Olympus camera in use. The changeover dial is set to High, the exposure meter sensing head is inserted in place of the C.T. head and the camera slider set to the exposure meter position. The meter needle deflection is then read. The Low position may be selected on the changeover dial to get a reading if a longer than "one sec" exposure is indicated. Direct read-out of exposures from 1/50 sec–32 sec

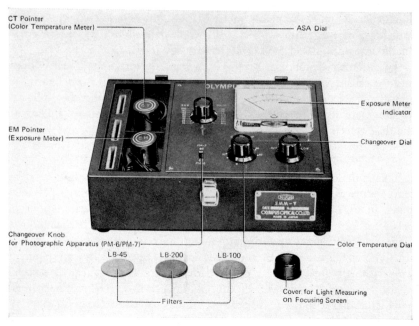

Fig. 43. Olympus EMM-V, colour temperature and exposure determination apparatus. The three blue correction filters of −45, −100 and −200 Mired value are supplied as standard. (Courtesy Olympus Optical Company Limited, Tokyo.)

for Olympus PM-6 and PM-7 cameras may be obtained. The exposure time denoted is set on the camera shutter and exposure made in the usual way by cable release. The instrument, therefore, determines exposure time, it does not control it.

It is possible, however, to use the instrument with other types of equipment since the colour temperature and exposure meter heads may be inserted into the microscope eyepiece tube. Olympus provides charts from which to determine values for use with their other cameras (including exposure times for Polaroid 3000 ASA film), when the EMM-V has to be used in this way. It therefore seems that the device may be used with any micro-scope/camera arrangement so long as the appropriate calibrations are made and charts drawn up for the user's own equipment and film.

8. *"Kinoaufsatz"* (Wild, Heerbrugg)

This is a focusing and light sensing device designed originally for use with cine and TV cameras. It clamps on to the eyepiece sleeve over the photo-eyepiece in the usual way. The graticule in the focusing telescope is ruled with a double-line cross-hair and two rectangles corresponding respectively with the fields photographed when used with 50 mm. and 70 mm focal length cine camera lenses. The attachment was designed specifically for use with the Paillard–Bolex 16 mm camera but may be used with other similar cameras. For TV camera use a deflecting prism acces-sory can be fitted into the depression on the top of the "Kinoaufsatz" to deflect light horizontally into the TV camera lens. Axial centration should be checked on the monitor by comparison with the telescope field of view. The various parts of the attachment are illustrated in Fig. 44.

The attachment includes two partly transparent swing-out prisms. One allows 50% of the light to reach the camera, the rest being deflected into the focusing telescope. The second prism passes 95% to the camera and only 5% to the telescope. Below the prisms is a photocell which may be swung into the light path in order to measure the light intensity and derive an exposure time. The lever controlling the photocell movement is coupled to the lever controlling the prisms in such a way that the meter reading can only be taken when the 95% prism is in the optical path. On release of the photocell lever the cell is automatically retracted by a spring leaving the 95% prism in the path ready for use. The 50% prism is used only to obtain increased illumination in the telescope for focusing under "low light" conditions or for film exposure when data superimposition on a particular frame is required. For this purpose a projection tube (Fig. 44(a), 7) can be clamped to the attachment opposite the telescope and an image on 35 mm film (or similar) projected on to the film plane.

When the 50% prism is used the exposure time must be doubled.

To determine the exposure time required the "Kinoaufsatz" is used in conjunction with a transistor amplifier and micro-ammeter (or with the Wild Exposure Meter S). The output from the photocell is fed to the amplifier by means of a cable attachment and the amplifier itself has a 4-range selector giving a gain of up to $400 \times$. The range is selected so as to obtain deflection on the needle of the micro-ammeter within the scale. From the ammeter reading and range selected a value is obtained, which can be entered in a calibration table supplied, covering an ASA range from 3–2500. The exposure time is read from the appropriate column.

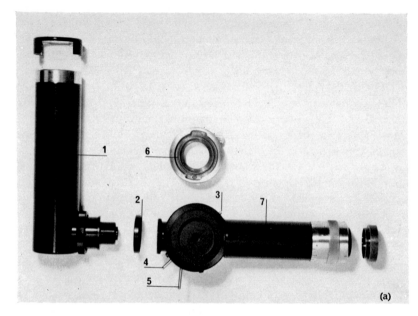

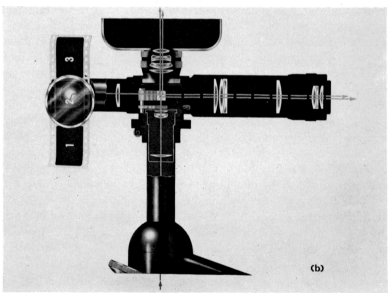

FIG. 44. The Wild 'Kinoaufsatz' (cine/TV focusing attachment). (a) View from "top"; note the central depression into which the camera lens mount fits. 1, projection tube; 2, protective cover; 3, connection for photocell; 4, photocell lever; 5, prism lever; 6, clamping ring; 7, focusing telescope. (b) Cross section of "Kinoaufsatz" in use. (Courtesy Wild, Heerbrugg, Switzerland.)

The author has found this table to be very accurate and the values printed have always yielded good results.

While the "Kinoaufsatz" is primarily designed for use with cine and TV cameras it may also be used for photomicrography with other types of camera. The essential requirement is that the lens of the camera should be of a diameter and focal length such that it may be brought sufficiently close to the top of the attachment so that vignetting is avoided. The type of lens mount may be crucial since a "deep-set" lens cannot be brought close enough. If the external diameter of the mount is greater than 34 mm it cannot be inserted in the depression designed to house it. The attachment can readily be used with a camera fitted with the Nikon simple attachment tube described above (III.B, 5). For this use the chrome clamping ring of the tube must be unscrewed.

It should be noted that in using the "Kinoaufsatz" it is not necessary that the camera lens or attachment tube should be in physical contact with the "Kinoaufsatz"; in fact it should not. This avoids the transfer of vibration to the microscope during shutter operation.

VIII. TIME LAPSE SYSTEMS

All the major manufacturers of microscopes now supply time lapse cinemicrographic control equipment of varying degrees of complexity. The camera required is usually a 16 mm camera, either with an electric motor drive or with the possibility of having a drive-motor coupled with the winding handle, as with the Bolex. The control equipment is usually electronically based although synchronous-clock systems are still manufactured, for example, by W. Vinten Limited of London, who also specialize in 16 mm cameras for this purpose. As a rule an electromagnetic shutter is fitted, either as an external item or built into the camera itself. Most of the equipment available allows a very wide selection of intervals between photographs from a few seconds to many minutes or even hours. There is usually also provision for flash synchronization so that intense illumination need not be maintained between exposures. The exposure times are themselves variable from fractions of a second to many minutes.

Although individual workers have designed and constructed apparatus for time lapse photomicrography on 35 mm stock, the cameras used are often "cinema" cameras using the cinema frame format of 24×18 mm and not the 36×24 mm format. It is possible, however, to make a "hybrid system" for 35 mm format time lapse from commercially available equipment, and such a system is described here in the belief that it may be of interest to researchers who have found the 16 mm format too small for work in bacteriology.

A system which will permit a reasonable time span for the experiment consists of the Nikon F camera with 250 exposure motor-drive back. The camera can be coupled to the microscope by means of one or other of the microflex attachments described earlier, or the simple attachment tube (Fig. 24) may be used. If the latter, then exposure determination is performed using the built-in meter of the Photomic T prism head mentioned earlier. The motor is set for single frame operation and, after the shutter speed has been set manually, the whole sequence of exposure and film wind-on may be triggered automatically by an intervalometer. The Nikon camera motor is powered from a 12 V d.c. supply and the shutter may be operated

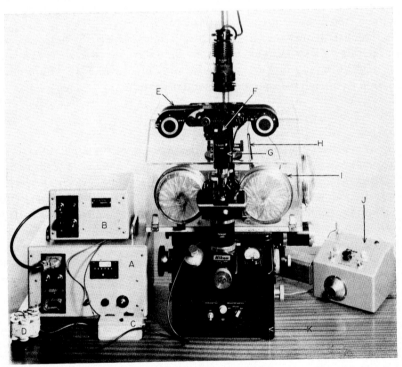

FIG. 45. Time lapse photomicrographic apparatus for 35 mm format, and Nikon Model M inverted microscope. The Nikon F camera is fitted with the 250-exposure motor drive back which may be controlled via a switching relay box by the Vinten (or other) intervalometer and mains unit which are shown. A, Vinten mains unit with frame counter; B, Vinten intervalometer; C, Nikon relay box; D, 12 V d.c. source, E Nikon 250-exposure motor drive back; F, Nikon F camera with photomic T prism and exposure meter; G, Simple microscope attachment tube; H, Thermometer and thermistor sensing probe; I, Incubator box; J, Hot air blower; K, Nikon inverted microscope model M.

by closing the circuit via a relay box. The circuit may be closed manually by depressing a button on the relay box or electrically through a relay operated by an intervalometer.

The timing mechanism may be electronic or electro-mechanical based on a synchronous electric motor. In the Vinten intervalometer (W. Vinten Ltd., London, England), for example, a synchronous electric motor drives two spindles with replaceable cams, which operate micro-switches as they rotate; the intervals being determined by the number of notches in the periphery of the cams. The Nikon motor-drive F-250 camera may be operated by the Vinten intervalometer via the Nikon relay box and such an arrangement is shown in Fig. 45.

Since only 250 frames (36×24 mm) are available, the film will have to be changed when they are consumed. However, if one frame is "shot" every 2 min, then over 8 h of experimental observation may be accommodated on a single film. A 2 min interval is adequate for the recording of bacterial divisions during colony growth as it is extremely difficult to estimate fission times to an accuracy greater than ± 1 min. It is possible to process film strips of 250-frame length in the laboratory using a Kodak Model 100 "Spiral film processing outfit".

If time lapse photography is to be used then physiological conditions must be appropriate for the material being photographed and one or other form of microculture technique is necessary. This subject has been discussed in Chapter 10 of Volume 1 of this Series to which the reader is referred in the hope that it may assist him in avoiding some of the problems associated with these techniques.

REFERENCES

Cowen, B. C. (1969). *Proc. R. microsc. Soc*, **4**, 71–73.
Heunert, H. H. (1968). *In* "Photography for the Scientist" (Ed. C. E. Engel), p. 421. Academic Press, London.
Martinsen, W. L. M. (1968). *In* "Photography for the Scientist" (Ed. C. E. Engel), p. 451. Academic Press, London.
Ruthmann, A. (1970). "Methods in Cell Research". G. Bell & Sons, London.

BIBLIOGRAPHY

(Many references to papers on specific aspects of the subject matter of this Chapter will be found in the books listed.)

Allen, R. M. (1958). "Photomicrography". Van Nostrand, New Jersey.
Barabás, J. and Vadász, J. (1966). "Mikroszkópos fényképezés". Muszaki Könyvkiadó, Budapest.
Boy, M., Luhmann, H., and Schweinitz, J. (1961). "Foto- und Filmtechnik in der Medizin". VEB Fotokinoverlag, Halle.

Clark, G. L. (1961). "Encyclopedia of Microscopy". Chapman and Hall, London

Bergner, J., Gelbke, E., and Mehliss, W. (1966). "Practical Photomicrography". Focal Press, London and New York.

Eastman Kodak Co. (1962). "Photography through the Microscope". Rochester, New York.

Engel, C. E. (Ed.) (1968). "Photography for the Scientist". Academic Press, Inc., London.

Faragó, M. (1954). "Mikroszkóp és Mikrophotografálás". Könnyuipari Kiadó, Budapest.

Heunert, H. H. (1959). "Praxis der Mikrophotographie". Springer Verlag, Berlin.

Ilford Ltd. (1958). "Manual of Photography". Ilford, Essex.

Lawson, D. F. (1960). "The Technique of Photomicrography". Newnes, London.

Linssen, E. F. (1952). "Stereo Photography in Practice". Fountain Press, London.

Longmore, T. A. (1955). "Medical Photography". Focal Press, London and New York.

Michel, K. (1957). "Die Mikrophotographie". Springer Verlag, Wien.

Needham, G. (1958). "The Practical Use of the Microscope, including Photomicrography". Blackwell, Oxford.

Rose, G. G. (Ed.) (1963). "Cinemicrography in Cell Biology". Academic Press, London and New York.

Ruthmann, A. (1970). "Methods in Cell Research". G. Bell and Sons, London.

Schenk, R., and Kistler, G. (1960). "Mikrophotographie". S. Karger, Basel.

Stade, G., and Staude, H. (1958). "Mikrophotographie". Akad. Verlagsgesellschaft Geese and Portig, Leipzig.

Stevens, G. W. W. (1957). "Microphotography: Photography at Extreme Resolution". Chapman and Hall, London.

Szabó, D. (1967). "Medical Colour Photomicrography". Akadémiai Kiadó, Budapest.

Tupholme, C. H. S. (1961). "Colour Photomicrography with a 35 mm Camera". Faber and Faber, London.

Note: A list of microscope manufacturers will be found at the end of Chapter 1 in Volume 5A of this Series.

rtest sure ible auto- c sure trol	Type of cell	Warning light for			Trial exposure switch	Measurement of small area detail	Setting to compensate for dark-ground
		Shutter open	End of film	Too much light			
ec	Vacuum photocell	Yes	No	No	Yes	No	No
ec	Vacuum photocell	Yes	No	No	No	Yes 5%	No
) sec	Photo-multiplier	Yes	Yes	By meter reading	No	Yes 5%	No
) sec	Photo-multiplier	Yes	Yes	Yes	Yes	Yes 1%	Yes
) sec	Cadmium sulphide	Yes	No	No	No	No	No
ec	Cadmium sulphide	Yes	No	By meter reading	No	No	No
sec	Cadmium sulphide	Yes	Yes	Not relevant	Not relevant	Yes 1·5%	No
sec	Photo-multiplier	Yes	No	By meter reading	No	No	No
sec	Photo-multiplier	Yes	No	By meter reading	No	No	No
sec	Cadmium sulphide	Yes	Yes	By meter reading	Yes	No	Yes
sec	Gas-filled photocell	Yes	No	No	No	No	No

Position of cell	Type of cell	Warning light for		Setting to compensate for dark-ground	Macro system available
		Shutter open	Too much light		
At one side of plate	Vacuum Photocell	Yes	No	No	Yes
At one side of plate measures area 1% of 9 × 12 cm format	Photo-multiplier	Yes	Yes	Yes	Yes
Whole frame integrated	Photo-multiplier	Yes	By meter reading	No	Yes
Whole frame integrated	Gas filled Photocell	Yes	No	No	Yes

Author Index

Subject Index

A

Acetone, solvent for bacteriostat assessment, 216

Acridines, bacteriostatic activity, 213

Actinomycetes, *see also under specific names*, genetic analysis, 116-125

Adox colour film, characteristics, 337, 338

Aeration, bacteriostatic activity and, 230

Aerosols,
    evaluation of bactericidal activity of, 250-252
    room sterilization with, 176

Agar cup method, of bacteriostat assessment, 216-217

Agar imprint technique, for replica plating, 244-246

Agaricus bisporus, spores as inocula, 198

Agars, phage-typing of *Staph. aureus* and, 8-9

Agfa colour film, characteristics of, 337, 338

Agricultural disinfectants, activity assessment of, 240

Air disinfection, 176, 250-252

Airborne organisms, assessment of bactericides against, 250-252

Alfalfa mosaic virus, 146

Algae, inoculation of, 175

Alkaloid production, inoculation with *Claviceps* for, 201

am mutations, *see under* Amber.

Amber (am) mutations, 130, 136

p-Aminobenzoic acid, 229

Ampicillin, assessment of activity of, 226

α-Amylase production, inoculation with *Aspergillus oryzae* for, 170, 201

Anabaena, inoculum for, 175

Anacystis, inoculum for, 175

Anacystis nidulans, genetic analysis in, 127

Animal viruses, genetic studies of, 141-145

Animals (*see also under specific names*), for *in vivo* assessment of antibacterial activity, 259-260, 261-262

Ansco colour film, characteristics of, 337-338

Antagonism, between antibacterial substances, 220, 221

Anthrax, disinfectants against, 240

Antibacterial activity, (*see also under* Bactericidal *and* Bacteriostatic), assessment of, 211-276
    bactericidal activity testing *in vitro*, 233-252
    bactericides and bacteriostats, concepts, 213-215
    bacteriostatic activity testing *in vitro*, 215-233
    preservation and preservatives, 253-259
    testing *in vivo*, 259-273

Antibacterial agents,
    assessment methods, 211-276
    evaluation principles, 212-215
    formulation of, 212
    types of, 211-212

Antibiotic Assay Discs, 218

Antibiotic diffusion tests, 196, 216-219

Antibiotics, *see also under specific names*, assessment of effect on growth, 226

Antiseptic, definition of, 212

A.O.A.C., *see* United States Association of Official Agricultural Chemists.

Arithmetic series, of dilution, 228

Arthrospores, as inoculum, 162, 164, 171

Ascobolus, tetrad genetic analysis, 57, 69

A. immersus, genetic analysis in, 69

Ascomycetes, *see also under specific names*, 61-69

double-stranded, 134–140

single stranded, 140

DNA viruses, *see also under specific names,*

animal, 143–144

bacterial, 133, 134–140

plant, 145

Diascopic illumination, in macrophotography, 283–285

Dictyostelium discoideum, genetic analysis in, 80, 81

Diffusion methods, for assessment of antibacterial activity, 216–219

Dilution tests, for bacteriostatic activity assessment, 222–224

Dinitrophenol, as fungal spore germinant, 198

Disinfectants,

assessment of activity, 234, 235, 236–244

definition of, 211

Disinfection, continuous flow, 252

Dispersing agents for bacteriostats, 216, 230

"Ditch" plate method, for bacteriostat assessment on solid media, 216–218

DNA, *see* Deoxyribonucleic acid.

Dogs, phage-typing of staphylococci from, 23

Domestic lamp, colour temperature of, 335

Dominance tests in genetic analysis,

Aspergillus nidulans, with, 47, 65

basis of, 47–48

Chlamydomonas reinhardii, in, 78

parasexual cycle in, 56–57

Saccharomyces cerevisiae, in, 72

Salmonella typhimurium, in, 109

Streptomyces coelicolor, in, 124

transduction systems in prokaryotes and, 90

Ustilago maydis, in, 76

Dose-response curves, in antibacterial activity assessment, 226–227

Double-stranded DNA phages, *see also under specific names,*

genetic studies, of, 134–140

Drinks, preservation against microbial spoilage, 253, 254, 257

E

Emericellopsis, genetic analysis in, 66

Emulsifying agents, for bacteriostatic agents, 216

Enteric bacteria, *see also under specific names,* genetic analysis in, 94–111

Enterotoxic food-poisoning, 2

Epi-illumination techniques, 285

Epistasis, genetic analysis test, 50

Escherichia coli,

antibiotics effective against, assessment of, 226

conjugation system of, 82

disinfectants active against, early tests of, 233

epistasis and radiation-sensitive mutants of, 50

genetic analysis in, 94–106

infection with, experimental, 263, 270

pH changes during growth and bacteriostasis, 230

phages of, 134–138

photomicrograph of, 317

spoilage microbe, 254

sulphonamide mechanism in, 229

textile disinfection test organism, 249

transduction by TI phage, 91

Ethanol, bacteriostatic activity and, 216

Ethylene glycol,

antibacterial activity assessment, 251

sterile room bactericide, 176

Euglena, inoculum for, 175

Eukaryotic microbes, *see also under specific names,* genetic analysis, 50–81

general considerations, 50–57

genetic distance and recombination frequency, 34

particular organisms, examples of, 61–81

tetrad analysis, 34, 57–61

Exponential growth phase, inoculation techniques ensuring, 159–209

Exposure meters, for photomicrography, 344–355

Exposure time determination in photomicrography, 343–355

light-meters for, 344–355

Exposure time—*cont.*
 test strip method for, 343–344
Eye lens, in photomicrography, 285–288

F

F⁻ mating strain, of *E. coli*, 96
Female and male strains of bacteria, 83
Ferrania colour film, characteristics, 337–338
Fertility types, and genetic analysis of bacteria, 83–84, 112
Filamentous fungi, *see also under specific names,*
 genetic analysis in, 73–75
 inoculation techniques for, 163–165, 169–174
Film speed,
 black and white film, 325, 329–330
 colour film, 337–338
Filter paper discs, for antibiotic assessment, 218
Filters, in photomicrography, 342
"Fine-structure mapping", genetic, 32, 42–45, 56, 65, 71–72, 124
Fixed length cameras, for photomicrography, 304–310
p-Fluorophenylalanine, haploidization and, 52, 73
Focal length, of macrophotographic lenses, 279–282
Focusing telescopes, in photomicrography, 288–290
Fog level, of photographic emulsion, 324
Fomes, mycelial inoculum for, 172
Food-poisoning, enterotoxic, phage-typing for, 2
Foods, preservation against microbial spoilage, 253, 254, 257
Foot-and-mouth disease, disinfectants against, 240, 241
Formaldehyde,
 blanket disinfectant, as, 249
 room sterilization by, 176
Formulation, of antibacterial substances, 212
Fowl pest, disinfectants against, 240, 241

Functional genetic tests, *see* Complementation *and* Dominance.
Fungi, *see also under specific names,*
 genetic analysis in, 56–59
 ideal inocula for, 163–165
 inoculum size effects on, 197–205
 mycelia as inocula for, 171–174
 selection and standardization of inocula for, 169–174
 spores as inocula for, 164, 169–171, 174
Fusarium, spores as inocula for, 171
F. oxysporum, genetic analysis in, 66

G

Gastro-intestinal tract infection, assessment of antibacterial activity against, 271–272
Genetic analysis in micro-organisms, *see also under specific names*, 30–158
 conjugational mapping in prokaryotes, 83–84
 eukaryotic microbes, 50–81
 genetic mapping, the basis of, 34–47
 non-selective analysis in prokaryotes, 94
 parasexual cycle in eukaryotes, 51–57
 prokaryotes, 81–128
 segregation and, 32–34
 tetrad analysis in eukaryotes, 34, 51, 57–61
 transduction and, 84–91
 transformation and, 91–94
 viruses, 128–147
Genetic approach to biology, 32–33
Genetic distance, recombination frequency and, 34–39
Genetic fine-structure, the basis of genetic mapping, 42–45
Genetic mapping, *see also* Genetic analysis *and* Gross mapping,
 recombination independent, 127–128
 Aspergillus nidulans, in, 61–63
 Bacillus subtilis, in, 113–116
 Chlamydomonas reinhardii, in, 77
 coliphages, in, 134–138
 Neurospora crassa, in, 66–68

Lambda phage, *see under* Phage.

Laminar flow equipment, 175, 177

Leitz Aristophot bellows camera, 297, 304–305

Leitz macro-ring illuminator, 298–299

Leitz Microsix-L exposure meter, 344–345

Leitz Mikas attachment, 289–290

Leitz Milar lens, 281–282

Leitz Summar lens, 281–282

Lidwell phage-typing apparatus, 14–15, 16

Lindra thalassiae, ascospores as inocula for, 174

Linkage estimations, four-factor crossing for, 117–121 tetrad analysis for, 61

Linkage groups, in genetic mapping, 42, 53–54

Linkage maps, in genetic analysis, 32, 33, 42, 119–121

Liquid media tests, bacteractidal activity assessment by, 233–234

A.O.A.C. tests, 237–239

agricultural disinfectant tests, 240–241

Chick-Martin test, 236–237

Kelsey-Sykes test, 241–243

phenol coefficients, 233–240

Rideal-Walker test, 235, 236

sporicidal tests, 243

tuberculocidal tests, 243–244

bacteriostatic activity assessment by, 224–233

combined action tests, 232–233

culture conditions affecting, 229–232

dose-response curves and, 226–227

growth assessment for, 232

methods, 224–227

microculture techniques for, 224

punch card records of, 224, 225

serial dilutions for, 224–226, 228–229

test preliminaries, 228–229

Lisboa test tube, dairy disinfectant assessment test, 248

Listeria monocytogenes, experimental infection with, 263

Localized infections, for *in vivo* assessment of antibacterial activity, 267–272

corneal ulcers, 269

gastro-intestinal tract, 271–272

intravenous injection for, 270

muscle, 268

urinary tract, 270–271

wounds and in the skin, 267–268

Loci sequencing, *see also* Genetic mapping, multi-factor crosses and, 39–41

Lysis, phage-typing of *Staphilococcus aureus* and, 2–28

Lytic enzymes, clostridial spore liberation with, 169

M

Macro-bellows extensions, 296, 302

Macro-dia apparatus, for macrophotography, 297–298

Macrophotography, cameras for, 279–285, 294–301

colour films for, 333–338

diascopic illuminator for, 285

exposure meters for, 344–355

filters for, 342

image formation in, 291

macro-dia apparatus and, 297–298

macro-ring illuminator for, 298–299

microscopes for, 281–283, 299–302

photographic emulsions for, 318–338

supplementary lenses for, 278–279

Macro-ring illuminator, in macrophotography, 298–299

Magenta dye, of colour film, 332–335

Male and female strains of bacteria, 83, 96

Manganese, carry-over in fungal conidia, 193

Mapping, *see* Genetic mapping.

Marker frequency analysis, in *B. subtilis*, 114, 115, 127

Marker sequencing, in genetic analysis of *Streptomyces coelicolor*, 117–122

Markers, for animal viruses and phages, 129–130